Transglutaminase

Developments in molecular and cellular biochemistry

Najjar, Victor A., ed.: Biological Effects of Glutamic Acid and its Derivatives, 1981. ISBN 90-6193-841-4

Najjar, Victor A., ed.: Immunologically Active Peptides, 1981. ISBN 90-6193-842-2

Najjar, Victor A., ed.: Enzyme Induction and Modulation, 1983. ISBN 0-89838-583-0

Transglutaminase

edited by

VICTOR A. NAJJAR
Division of Protein Chemistry
Tufts University School of Medicine
Boston, Massachusetts, U.S.A.

and

LASZLO LORAND
Department of Biochemistry, Molecular and Cell Biology
Northwestern University College of Arts and Sciences
Evanston, Illinois, U.S.A.

Reprinted from *Molecular and Cellular Biochemistry* Volume 58, 1984

1984 **MARTINUS NIJHOFF PUBLISHERS**
a member of the KLUWER ACADEMIC PUBLISHERS GROUP
BOSTON / THE HAGUE / DORDRECHT / LANCASTER

Distributors

for the United States and Canada: Kluwer Boston, Inc., 190 Old Derby Street, Hingham, MA 02043, USA
for all other countries: Kluwer Academic Publishers Group, Distribution Center, P.O. Box 322, 3300 AH Dordrecht, The Netherlands

Library of Congress Cataloging in Publication Data

Main entry under title:

Transglutaminase

(Developments in molecular and cellular biochemistry)
'Reprinted from Molecular and cellular biochemistry, volume 58, 1984.'
1. Transglutaminases--Addresses, essays, lectures. i. Najjar, V. A. (Victor A.), 1944- . ii. Lorand, Laszlo, 1923- . iii. Series.
QP606.T73T73 1984 574.19'25 83-23835
ISBN 0-89838-593-8

PRINTED IN THE NETHERLANDS

Contents

Preface

Even a casual search of recent literature would show that the involvement of transglutaminases in a great variety of biological reactions has captured the imagination of many investigators. The present volume of *Molecular and Cellular Biochemistry* is a reflection of this growing interest.

Transglutaminases catalyze the post-translational modification of proteinbound glutamine residues which either causes protein cross-linking by γ-glutamyl-ϵ-lysine peptide bridges or covalent incorporation of small molecular weight amines. These reactions could have far-reaching effects on altering the physical states and biochemical behaviors of the extra-cellular or intra-cellular systems. Examples could be drawn from clotting in blood or in seminal fluid, from keratinization, from cell aging, or from cell activation.

Within the framework of the definition, the transglutaminase name denotes not just a single enzyme but a large group of enzymes with distinct characteristics. Multiple forms are often found in a given tissue or cell type, but, as yet, we do not know the reasons for this diversity.

Editors: Laszlo Lorand, Ph.D.
Victor A. Najjar, M.D.

Molecular and Cellular Biochemistry 58, 9–35 (1984).

Transglutaminases

Laszlo Lorand and Sylvia M. Conrad
Department of Biochemistry, Molecular and Cell Biology, Northwestern University, Evanston, IL 60201, U.S.A.

Summary

This paper is intended as a background to the topic of transglutaminases, while focusing on current ideas regarding the biological roles of these enzymes. Specifically, the following topics are discussed: geometry of forming γ-glutamyl-ϵ-lysine cross-linked structures; energetic considerations; the γ-glutamyl-ϵ-lysine cross-link; amine incorporation assays; artefactual incorporation of amines in cells and tissue homogenates; synthetic substrate systems; regulation of transglutaminase activities; strategies for probing transglutaminase-mediated events in biological systems; the blood clotting paradigm; transglutaminase and cell aging: the Ca^{2+}-enriched human erythrocyte; transglutaminase and cell activation: the thrombin-stimulated human platelet and the fertilized sea urchin egg.

Introduction

This paper is intended as a background to the topic of transglutaminases, while focusing on current ideas regarding the biological roles of these enzymes (see also Refs. 1 and 2).

The 'transglutaminase' designation was first used by Heinrich Waelsch and co-workers about thirty years ago (3) to describe a Ca^{2+}-dependent transamidating activity present in the liver and in some other tissues of the guinea pig, which could be measured by the covalent incorporation of amines (e.g. histamine, putrescine) into proteins (e.g. casein), and also by a reaction between $^{\alpha}N$ and $^{\alpha}C$-blocked glutamine peptides (such as benzyloxycarbonylglutaminylvaline ethylester) and hydroxylamine (4):

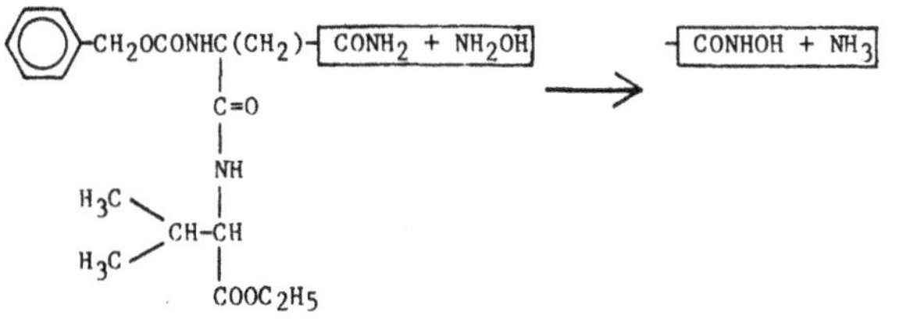

The transglutaminase name is an obvious misnomer, since it is *not* free glutamine but only compounds with γ-glutaminyl residues occupying endo-positions which serve as substrates[1].

In the absence of small molecular weight amines, a cross-linking between endo-γ-glutaminyl and endo-ϵ-lysyl residues of proteins may take place. This results in the formation of γ-glutamyl-ϵ-lysine or ϵ(γ-glutamyl)lysine isopeptide side-chain bridges, suggesting the more descriptive 'endo-γ-glutamine:ϵ-lysine transferase' name. The Enzyme Commission recommends the 'R-glutaminyl-peptide:-amine-γ-glutamyl-transferase' designation (EC 2.3.2.13).

With protein substrates, transglutaminases can thus catalyze the two nucleophilic displacement reactions shown in Fig. 1, leading either to protein cross-linking (as in reaction A) or to the incorporation of small primary amines into the protein substrate (as in reaction B)[2].

The biological significance of the cross-linking of

[1] Lysine also shows very little substrate activity.

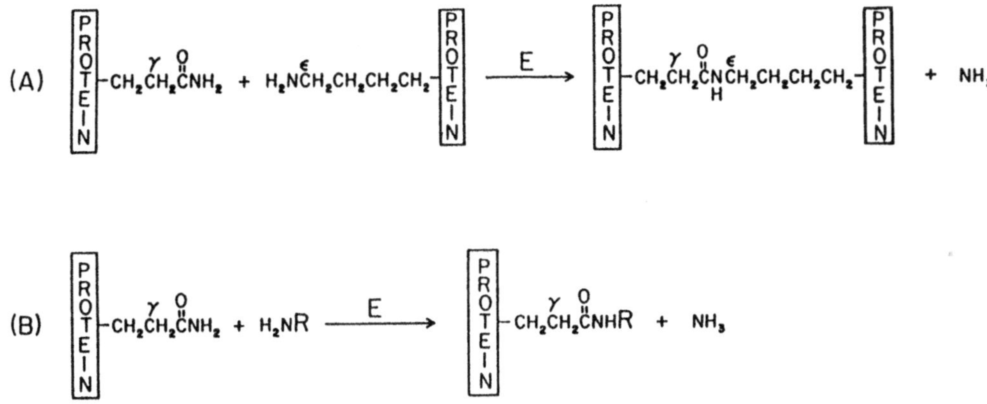

Fig. 1. Two known biological reactions are catalyzed by transglutaminases: (A) cross-linking of proteins, creating γ-glutamyl-ε-lysine side chain peptides, and (B) incorporation of small molecular weight amines into γ-glutamine sites of proteins. Reactions (A) and (B) compete against each other.

protein units by even a few γ-glutamyl-ε-lysine bridges is nowhere more obvious than in human blood clotting where a disorder of fibrin stabilization may give rise to life-threatening hemorrhage (for a current review, see Ref. 7). As shown in Fig. 2, the low elastic modulus of plasma clots, inadequate to stem the loss of blood in patients with hereditary deficiency of fibrin stabilizing factor (Factor XIII), can be augmented as much as five-fold by supplementation with the purified zymogen of the cross-linking enzyme (8). Conversely, the elastic modulus of the clot formed in normal human plasma can be reduced five-fold when competing small molecular weight amines are added to prevent the development of ε-lysyl bridges.

Dramatic physical changes accompany the transglutaminase-mediated cross-linking of protein units in other biological systems as well. Clotting of *Homarus* plasma (9) or the generation of copulation plug derived from the seminal fluid of rodents (10) can be cited as examples relating to *extra-cellular* phenomena, whereas the irreversible rigidification of cell membrane in CA^{2+}-enriched human erythrocytes (11) and in keratinocytes (12) are the results of the cross-linking of proteins by *intra-cellular* transglutaminases.

In sharp contrast with reaction A, the biological consequences of reaction B in Fig. 1 are far less obvious. Nevertheless, the expectation may well be justified that the post-translational modification of proteins by small molecular weight primary amines, be it a drug or a natural polyamine, is indeed an important function for transglutaminases.

As far as drugs are concerned, isoniazid, used in antituberculosis therapy, and hydralazine, used as an anti-hypertensive agent, are the best studied examples (13):

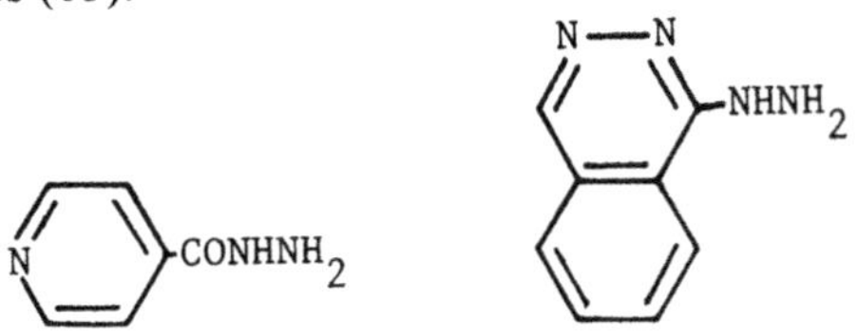

Current textbooks of pharmacology offer no enlightenment, yet the possibility of transglutaminase-dependent modification of proteins by such drugs may carry farreaching consequences: a given enzyme might become more or less active following the covalent incorporation of a drug into its structure; turnovers of proteins could be affected; antigenic properties might be altered which could conceivably lead to a loss of tolerance against one's own proteins and precipitate an autoimmune response. That the latter could be an important consideration is suggested by the fact that in several hemorrhagic disorders following the sudden appearance of some inhibitor of fibrin stabilization (7), there was a history of treatment with isoniazid, and the circulating inhibitor was an IgG of great specificity directed either against Factor XIII itself

[2] There are some reports suggesting the possibility that small molecular weight acyl compounds such as benzyloxycarbonylglutaminylvaline ethylester (5) and t-cinnamoylthiocholine (6) might also react with proteins in a transglutaminase-dependent manner, presumably by amidating ε-lysine sites.

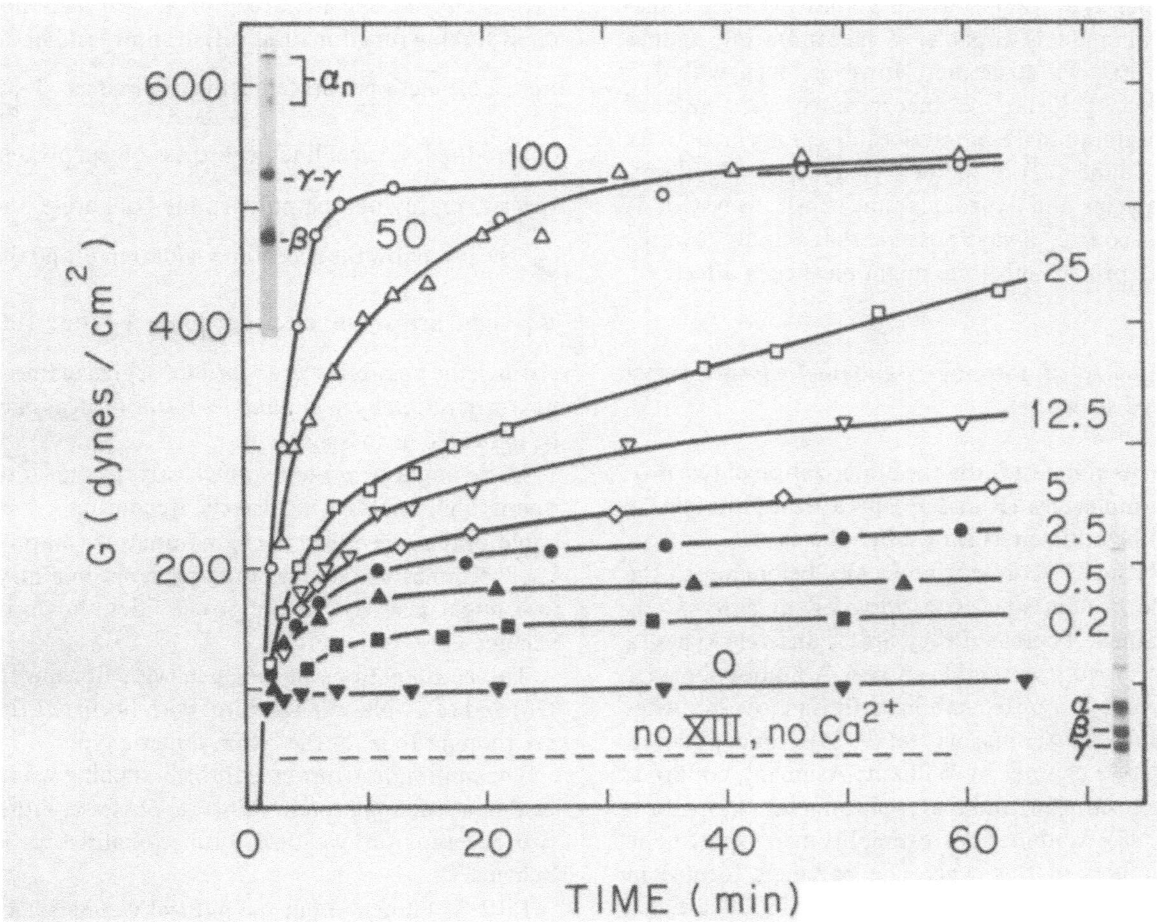

Fig. 2. Increase in the elastic modulus (G) of clot in a Factor XIII-deficient plasma, as a function of supplementing with measured doses of the purified Factor XIII zymogen (8). Added concentrations of the latter, based on active site titrations with iodoacetamide, ranged from 0.2×10^{-9} M to 100×10^{-9} M and are indicated on the graph ($\times 10^{9}$) for each experiment. The line marked 0 (▼) at about 100 dynes/cm² represents the strength of the clot formed in Factor XIII-deficient plasma, with thrombin and Ca^{2+} added but without supplementation with purified Factor XIII. The broken line at about 40 dynes/cm² is the elasticity of the clot obtained in Factor XIII-deficient plasma just with the addition of thrombin. Fibrinogen content was 1.88 mg/ml. The inserts show the electrophoretic profiles for reduced clots in sodium dodecylsulfate for the 65 min sample from the deficient plasma (▼; lower right) and for the 60 min sample from the 10^{-7} M Factor XIII supplemented material (○; upper left hand).

(14, 15) or against the fibrinogen/fibrin substrate (16). Considering the structures of other drugs such as phenelzine, C_6H_5-$CH_2CH_2NHNH_2$ and a variety of monoamine oxidase inhibitors (Lorand, unpublished results) and what we know about the amine specificity of transglutaminase (17), the iatrogenic issue is a very serious one, indeed[3].

[3] Recently some more compounds such as actinomycin D and adriamycin were added to this list (18).

The post-translational modification of protein structures by naturally occurring polyamines (histamine, putrescine, spermidine, spermine) or by their monoacetylated derivatives is an equally intriguing possibility. Alterations in enzyme activities could ensue (19), and, in fact, it has recently been reported that ornithine decarboxylase was 'down-regulated' by the covalent incorporation of putrescine (20). However, this claim by Russell regarding the inactivation of ornithine decarboxylase has been seriously questioned (21). As shown by Lockridge (22) and confirmed later by Birkbichler et al. (23), purified guinea pig liver transglutaminase is capable of incorporating amines into its own structure. However, even with 2–3 moles of histamine incorporated per mole of transglutaminase, no noticeable change in activity was found (22) in the benzyloxycarbonylglutaminylglycine and hydroxylamine substrate system. It is, of course, always possible that activity towards some protein substrate might have been affected.

Geometry of forming γ-glutamyl-ε-lysine cross-linked structures

In its simplest form, the dimerization of two protein molecules (P and P′) by a transglutaminase may be outlined as shown in Scheme A.

If P and P′ represent molecules belonging to the same protein species, a *homologous ensemble* is obtained, whereas if they are of different types, a *heterologous ensemble* ensues. A homopolymeric example is a pure, stabilized fibrin clot (24); whereas the lobster plasma clot (9, 25)[4], rat copulation plug (see chapter by Williams-Ashman), collagen-fibronectin (26) and the α_2-plasmin inhibitor-fibrin (27, 28) combinations exemplify heteropolymeric structures of this type. The polymers formed in Ca^{2+}-enriched human erythrocytes (29, 30) are also fused heterologous structures.

A given protein is considered bifunctional if it carries enzyme-specific acceptor as well as donor cross-linking functionalities. If, in shorthand notation, a bifunctional protein is symbolized as 

(where the | vertical line, regardless of the protein species, represents the polypeptide backbone, the  heavy arrow the acceptor γ-glutamine, and the  light arrow the electron donor ε-lysine side chains), the possibility of an infinite degree of linear polymerization by γ-glutamyl-ε-lysine bridges may be indicated as in Scheme B.

There might be proteins which carry acceptor or donor functionalities exclusively. Accordingly, they could only serve either as chain initiators (marked I) or terminators (T), and their presence in a mixture might give rise to the possibilities shown in Scheme C.

The combinations of collagen with fibronectin (26) and of α_2-plasmin inhibitor with fibrin (27, 28) are thought to be of the latter, dimeric type.

The minimum requirement for branching would be that a reacting protein substrate possesses either two acceptor or two donor functionalities as in Scheme D.

Little is known about the natural geometric arrangement of constituent protein units in any of the biologically occurring polymeric ensembles. Enzyme-dependent titration and labelling of γ-glutamyl sites (see reaction B in Fig. 1) with dansylcadaverine or with ^{14}C-glycine ethylester led to the finding that the cross-linking sites in fibrin were situated near the C-terminus of γ-chains and the mid-portion of the wandering α-chain domain of

4 We have recently shown (Myhrman, Bruner-Lorand and Lorand, unpublished results) that actually more than one protein (i.e., not just fibrinogen) in the plasma of the *Homarus americanus* is involved in clotting.

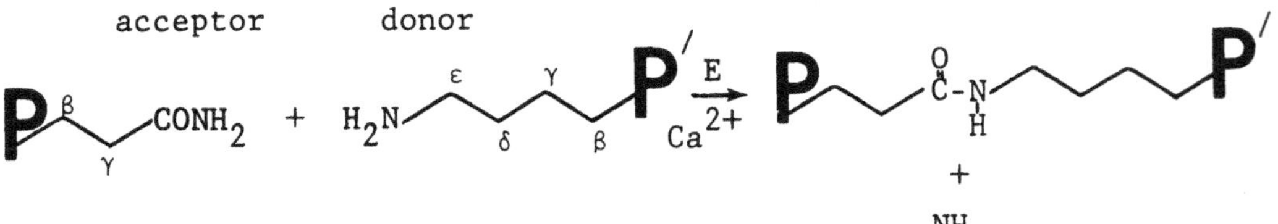

Scheme A

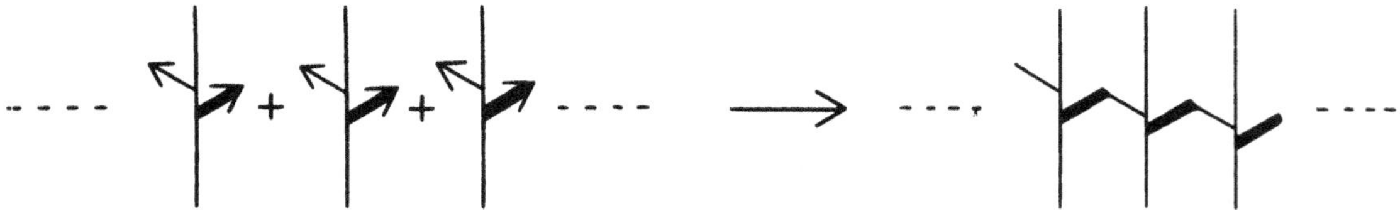

Scheme B

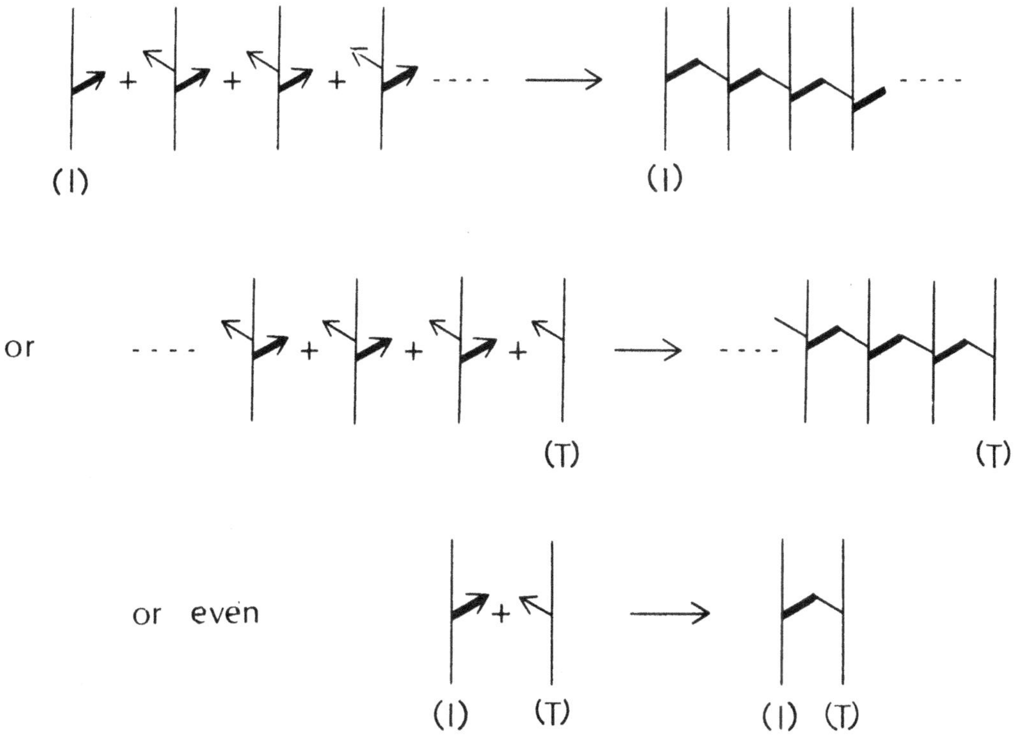

Scheme C

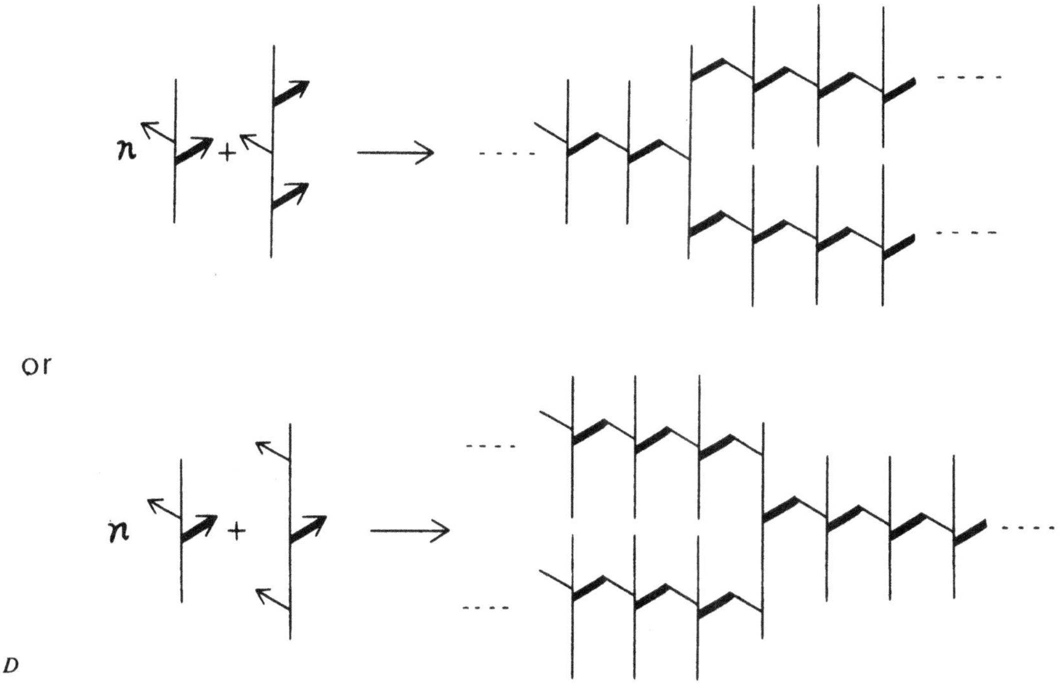

Scheme D

the protein (31–34) (Fig. 3). Reciprocal cross-linking of γ-chains generates an end-to-end γ-γ' dimer in the direction of the fiber axis, whereas formation of peptide cross-links between α-chains gives rise to polymeric α-chain units, promoting side-by-side polymerization.

The site-specific, transglutaminase-directed labelling of proteins in γ-glutamine residues by dan-

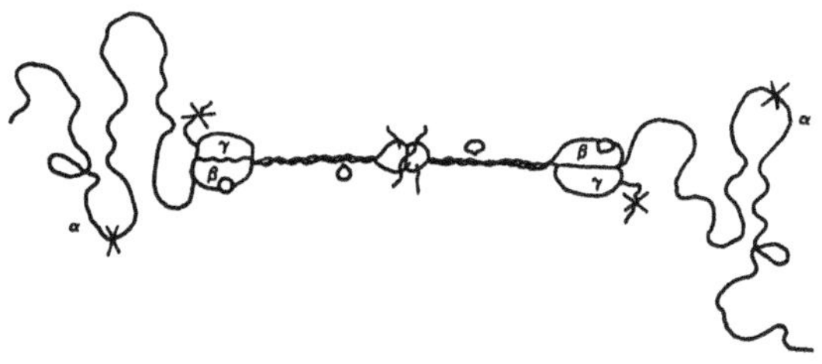

Fig. 3. Cross-linking sites in human fibrin. Fibrinogen has an $(\alpha_2\beta_2\gamma_2)A_2B_2$ composition, where A and B represent the fibrinopeptides at the N-termini of the α and β chains on the central nodule of the molecule (called E domain). These are the peptides cleaved by thrombin during clotting [i.e. $(\alpha_2\beta_2\gamma_2)A_2B_2 \rightarrow \alpha_2\beta_2\gamma_2 + 2A + 2B$] which, however, leaves the linear arrangement of nodules essentially unchanged. The Factor XIII_a-reactive cross-linking sites of the protein (marked by X) are located near the C-terminal ends of the short γ-chains and in the midsection of the C-terminal, wandering portion of the long α-chain, both protruding from the two distal nodules (called D domains). Amine substitution, as mediated by Factor XIII_a, occurs much faster with fibrin than with fibrinogen, indicating that removal of fibrinopeptides serves to unmask the cross-linking sites of the protein. Furthermore, the reaction takes place in an ordered manner with substitution of γ-sites proceeding faster than that of α-sites.

sylcadaverine (35) or by other labelled amines has become a general strategy for probing the cross-linking domains of proteins. Human plasma fibronectin, for example, is known to contain two regions reactive towards fibrinoligase[5]; one near the C-terminal, inter-chain disulfide bridge in the molecule and another involving the third residue from the N-terminal end (36–40).

No methods are available for disassembling γ-ϵ cross-linked polymeric structures without destroying the backbones of the constituent units themselves. In other words, we have no means of preferentially cleaving the γ-ϵ peptide bond without breaking α peptides. In complex hybrid systems, such as found in the membranes of Ca^{2+}-enriched human erythrocytes and in keratinocytes, even the identification of constituent proteins (not to mention an evaluation of the stoichiometry of combination) proved to be a very laborious task. However, immunochemical methods were found to be the most helpful (41). The approach is based on the idea that γ-ϵ linked polymers still retain some of their antigenic determinants of their individual protein constituents and that the probing of shared epitopes by crossed immunoelectrophoresis or by Western blotting can reveal the composition of the ensemble. By methods of this type, it could be shown that the cross-linked polymer of Ca^{2+}-enriched human red cells comprised band 3, spectrin, ankyrin, band 4.1 protein, and possibly also hemoglobin and catalase (Bjerrum, Hawkins, Griffin and Lorand, unpublished results).

[5] Synonymous with coagulation Factor XIII_a or plasma transglutaminase.

Energetic considerations

The reactions, as written in Fig. 1, would proceed with very little change in free energy ($\Delta G° \sim O$) and should be considered reversible. In both instances a γ-amide bond is broken: $-\overset{\displaystyle O}{\overset{\|}{C}}-NH_2$ and a γ-amide bond is made: $-\overset{\displaystyle O}{\overset{\|}{C}}-\underset{\underset{\displaystyle H}{|}}{N}-$. In the physiological milieu, of course, the NH_3 leaving group would undergo immediate protonation: $NH_3 + H^+ \rightarrow NH_4^+$, which would promote reaction to the right. However, even by adding large amounts of extraneous amines, reversibility for protein cross-linking reactions, as shown in Fig. 1A, is often difficult to demonstrate because of clotting or phase transitions. Actual reversal has been reported only for the Factor XIII_a-catalyzed dimerization of fibrin with α_2-plasmin inhibitor (27).

However, a back exchange of protein-bound amides can be more easily shown. In the example given in Fig. 4 (42), first succinylated β-lactoglobulin was covalently attached to arylamine glass beads, then reacted with ^{14}C-histamine in the presence of guinea pig liver transglutaminase and Ca^{2+} which modified available γ-glutamine sites on β-lactoglobulin (R). After unbound radioactivity was removed by washing, the beads were mixed with transglutaminase in the presence of cold histamine and Ca^{2+}, and the release of radioactivity into the supernatant was measured. It can be seen that the delabelling was transglutaminase-dependent and that it was augmented four-fold when histamine was added (i.e., hydrolytic delabelling was about 25% of the total achievable with an amine present). These reactions are illustrated in Scheme E.

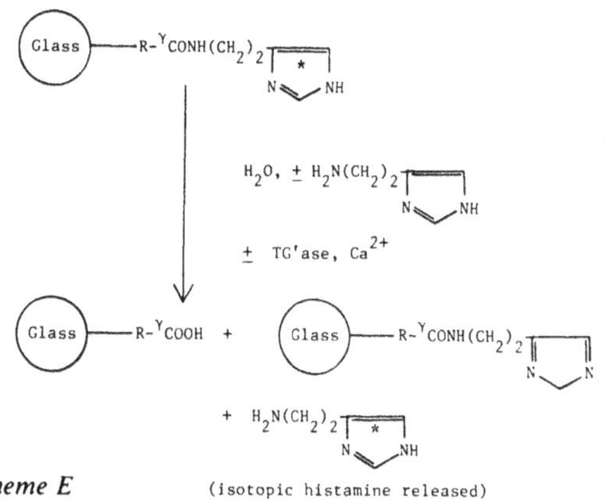

Scheme E

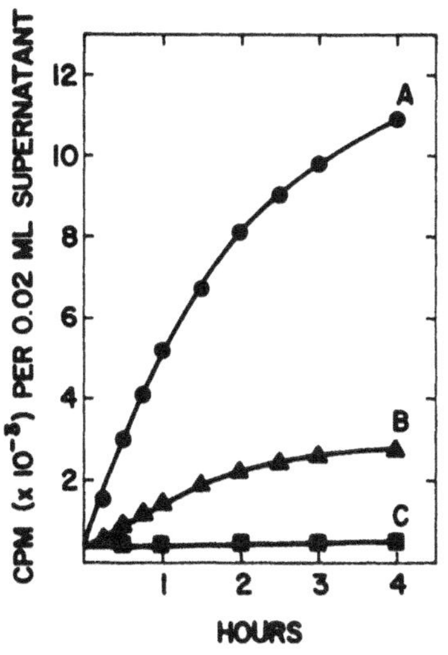

Fig. 4. Progression curve for transglutaminase-catalyzed removal of label from ^{14}C-histamine-succinylated-β-lactoglobulin-arylamine glass (42). The reaction mixture (0.15 ml) contained 53 mM Tris-HCl (pH 7.5), 2.4 units transglutaminase/ ml, and 20 mg ^{14}C-histamine-succinylated-β-lactoglobulin-arylamine glass (5 500 cpm/ mg). Incubation was at room temperature (23 °C). The radioactivity in 0.02 ml samples of supernatant was determined. *A:* Complete system with 50 mM cold histamine. *B:* Complete system without added cold histamine. *C:* Same as *A*, but without enzyme. The 4 hr value of 11×10^3 cpm represents the release of 75% of the total bound radioactivity.

The γ-glutamyl-ϵ-lysine cross-link

As indicated in relation to Fig. 2, even the few γ-ϵ side chain bridges in a blood clot (ca. 1 mole γ-ϵ peptide/ 80 000 g protein) (35, 43–45) have a profound effect on physical properties. The frequency of such cross-links varies greatly from one system to the next, which is probably a reflection on biological requirements. The guinea pig copulation plug (10) and the cornifying envelope in keratinocytes (12) represent the highest order of cross-linking of this type, involving as much as 30% and 18% of lysyl residues, respectively.

Characterization and quantitation of the γ-glutamyl-ϵ-lysine dipeptide is still a laborious task since it can only be identified following the total digestion of the polymer by a series of proteolytic enzymes. More significantly, a single analytical system alone does not yield unambiguous results. The choice is between employing two or more independent systems for isolating the dipeptide on the basis of different principles of separation, or further analyzing the purified product by hydrolysis or derivatization. In our laboratory, we developed three different chromatographic systems (35, 43, 46) and, presently, we rely on the parallel examination of proteolytic digests by a direct ion exchange chromatographic procedure and by an HPLC technique following pre-column derivatization with o-phthalaldehyde (46). Each system must have the capacity of resolving trace quantities of the γ-glutamyl-ϵ-lysine dipeptide against at least a thousand times higher concentration of all other products of protein digestion.

It must be remembered, however, that not all γ-glutamyl-ϵ-lysine producing reactions are due to transglutaminase. In addition to the reaction catalyzed by this enzyme (I), there is at least one other well-known possibility (II) for forming such side-chain bridges between protein units (47, 48) (see Scheme F).

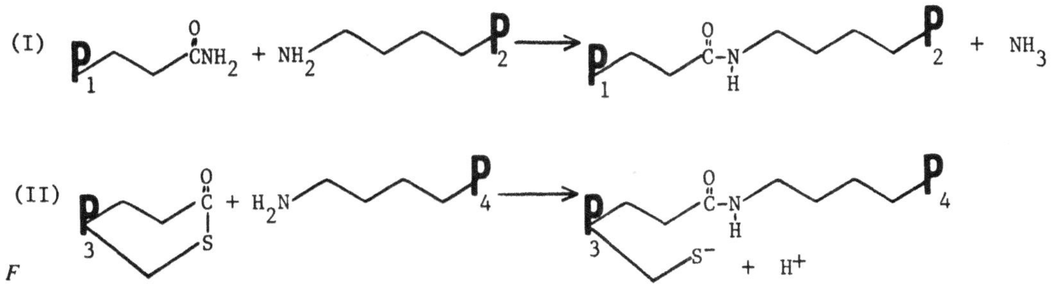

Scheme F

Though both I and II would give rise to γ-ϵ peptides, there are a number of differences between the two reactions: (i) Reaction I proceeds only in the presence of transglutaminase, also requiring Ca^{2+}, whereas no enzyme is needed in reaction II. (ii) In reaction I, formation of the γ-ϵ peptide is accompanied by the liberation of NH_3, whereas in reaction II the leaving group is a cysteine thiol which remains protein bound on P_3. (iii) Different Glx residues are involved in the two reactions; a glutamine in I and an internal glutamyl thioester in II. As exemplified by α_2-macroglobulin (49), both types of Glx functionalities could be present in the same protein. (iv) Small molecular weight amines (H_2NR) could replace the ϵ-NH_2 lysyl residues (in P_2 and P_4) in both I and II and could become incorporated into P_1 and P_3, while inhibiting cross-linking to other proteins. However, there may be considerable differences in H_2NR specificities. From what we know about the pathway of transglutaminase (E-SH)-catalyzed reactions, deduced from studies with synthetic substrates to be discussed later, we can postulate the existence of an acylenzyme intermediate for reaction I as in Scheme G, and the amine specificity depends on the affinity of the small molecular weight nucleophile for the acylenzyme intermediate, as shown in Scheme H.

Thus, amine incorporation by transglutaminase always follows a Michaelis-Menten type of saturation kinetics (17, 35). By contrast, we are unaware of any report indicating a prior complexation step for the incorporation of amines by α_2-macroglobulin, the C3 or the C4 components of complement, all of which contain similar γ-glutamyl thioester bonds (47). In these situations, the velocity of amine incorporation may depend only on a bi-molecular encounter between these proteins and the H_2NR. Due to the different nature of the two reactions, even amines of similar pK_a's (e.g., methylamine and dansylcadaverine) can be expected to differ as reactants in I and II.

Amine incorporation assays

The history of research on transglutaminases is intimately tied to assays based on the incorporation of isotopically labelled amines (e.g., putrescine, histamine, spermine, cadaverine) into trichloracetic acid-precipitable proteins (3). In spite of its many pitfalls, the Ca^{2+}-dependent incorporation of amines remains the cornerstone for detecting and measuring transglutaminase activities in tissue homogenates. Various modifications of the original methodology are in use. A significant improvement was the blocking of the ϵ-lysine residues of test proteins (e.g., casein) so that no self cross-linking of the protein substrate could occur; acetylation, succinylation (50), dimethylation (51) or guanidylation (52) were found to be useful in this regard. Filter paper (53) and chromatographic techniques (54) are employed for the efficient and simultaneous handling of a large number of test samples and for the rapid separation of protein-bound radioactivity from free amine.

The synthesis of dansylcadaverine in this laboratory (35, 55) opened up new avenues for studying transglutaminases by fluorescence methodologies. Dansylcadaverine has excellent affinity for all acylenzymes of this kind ($K_{M,app}$, ~10^{-4} – 10^{-5} M at pH 7.5); its incorporation into casein and into many other proteins is accompanied by a large blue-shift and by an increase in the quantum yield of dansyl group emission. These circumstances made it possible to devise continuous rate assays (50) for measuring transglutaminase activities and to apply his-

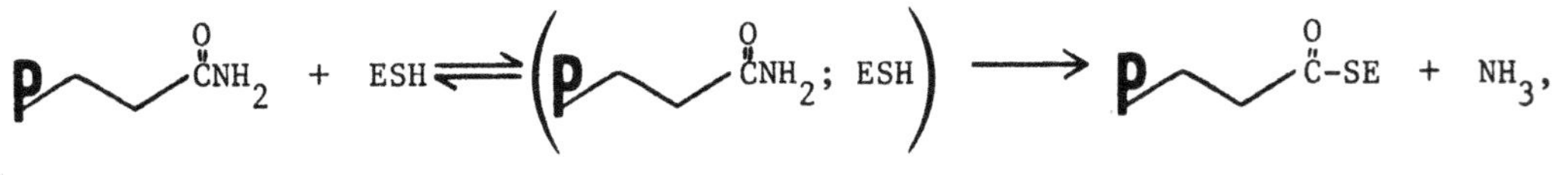

Scheme G

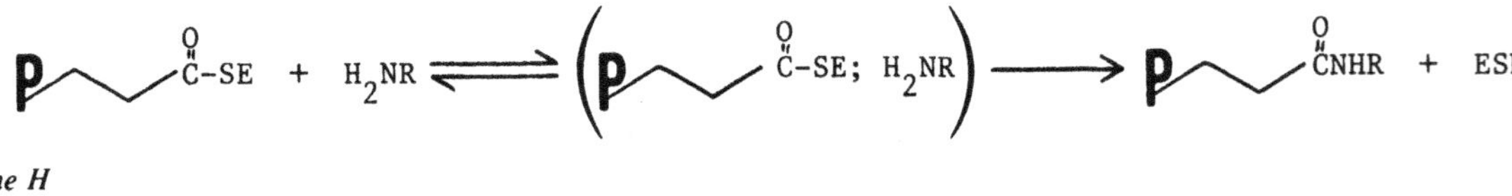

Scheme H

tological manipulations for the staining of enzymes in tissue slices (56)[6]. The dansylcadaverine reaction also permitted activity staining of transglutaminases on electrophoretic gels (58, 59). As such, purified guinea pig liver transglutaminase, which had previously been represented to comprise only a single enzyme, could be resolved by non-denaturing electrophoresis and dansylcadaverine staining into two differently charged components[7] (Fig. 5). In fact, *multiple forms* of transglutaminases can be detected in many tissues. In guinea pig prostate alone, four different transglutaminases could be identified (Fig. 6) and three activities could be discerned in homogenates of the sea urchin egg (61).

Artefactual incorporation of amines in cells and tissue homogenates

Following incubation of human peripheral lymphocytes with isotopically labelled putrescine, activation by phytohemagglutinin caused incorporation of label from putrescine into cell proteins (preferentially into one of 18 000 MW) (62, 63). However, as it was later shown, this incorporation was not due to transglutaminase action, but to the formation of protein-bound hypusine by an unrelated sequence of events (64).

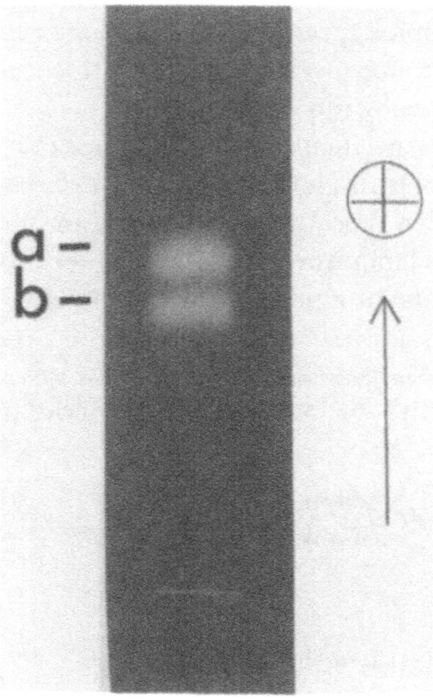

Fig. 5. Multiple forms of transglutaminase in fresh guinea pig liver homogenate as revealed by dansylcadaverine-specific fluorescent activity staining following non-denaturing electrophoresis (59).

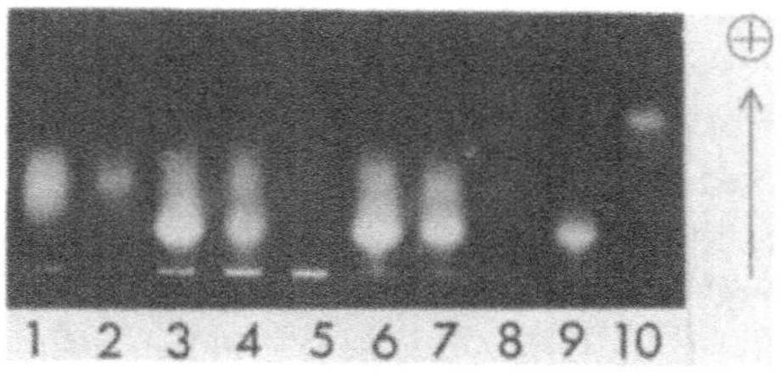

Lobe	Lanes		
	Fluid	Homogenate	High Speed Supernatant
Anterior	1	3	6
Dorsal	2	4	7
Lateral		5	8
Seminal Vesicle			10

Lane 9. Purified transglutaminase I from anterior lobe

Fig. 6. Multiple forms of transglutaminases in the guinea pig prostate, as shown by activity staining with dansylcadaverine (139).

Spurious incorporation of a label originating from amines may also be encountered if the amine (e.g., putrescine) is oxidized in the tissue to an aldehyde. Among other products, a Shiff base could ensue. In situations where interference by amine oxidases is suspected (e.g., in pathological serum samples[8] or when serum is collected during pregnancy) (65, 66), specific inhibitors of the oxidase must be included in the test system.

Another possibility for the artefactual, transglutaminase-independent incorporation of amines (putrescine, histamine, etc.) into proteins is contained in a recent report regarding the action of myeloperoxidase which converts RNH_2 to $RNCl_2$ in tissues. The latter products can react with proteins very rapidly (67).

[6] A recent report deals with other fluorescent and colored derivatives of cadaverine for such purposes (57).

[7] Such preparations of guinea pig liver transglutaminase are also known to contain a proteolytic contamination (60).

[8] In connection with assaying Factor XIII in a patient's plasma, we have recently encountered (Lorand and Velasco, unpublished data) an anomalously high incorporation of label from ^{14}C-putrescine, which could be abolished by the addition of methylglyoxalbis(guanylhydrazone), an inhibitor of diamine oxidase.

Synthetic substrate systems

Efforts to synthesize a large number of $^{\alpha}$N-blocked glutaminyl peptides (68) followed the path set by the early work of Waelsch and collaborators (4, 69). The water soluble compound, benzyloxycarbonylglutaminylglycine (though its affinity even for guinea pig liver transglutaminase is quite poor; $K_{M,app} = 0.05$ M), proved to be useful for detecting transglutaminases. Sensitivity can be greatly improved if, instead of employing it with hydroxylamine, it is used in combination with dansylcadaverine (54):

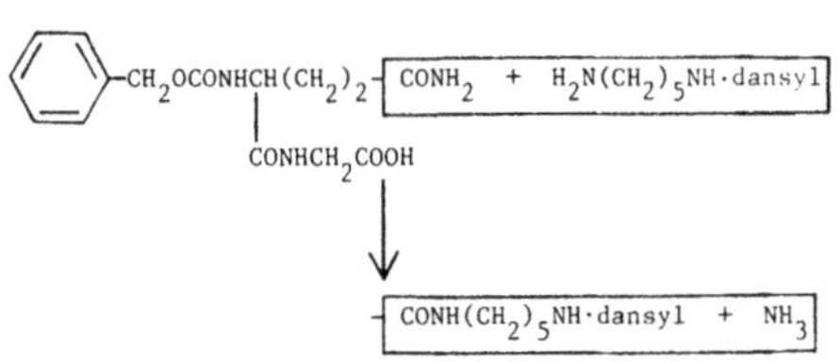

Synthetic substrates have been most helpful from the point of view of elucidating the pathway of catalysis of transglutaminases. Significant progress was made in recent years with regard to two such enzymes: guinea pig liver transglutaminase and human Factor XIII_a. Folk and co-workers used nitrophenylesters such as p-nitrophenylacetate or trimethylacetate (60, 70–73), whereas Lorand and collaborators synthesized a variety of thioesters such as t-cinnamoyl- or β-phenylpropionyl thiocholine (58, 74–77) for their studies. The thiocholine derivatives are not only more water soluble and very much more stable than the p-nitrophenylesters[9], but the thioesters are also considerably more specific as substrates. In addition, they offer a wide range of analytical possibilities, including fluorescence. This is illustrated in Scheme I for the β-phenylpropionylthiocholine-dansylcadaverine substrate pair where the formation of thiocholine can be measured by on-line reaction with 5,5′-dithiobis(2-nitrobenzoic acid), production of the acid can be followed in a pH′ stat and the water-insoluble coupling product can be quantitated by fluorescence, following continuous extraction into a heptane phase layered above the aqueous reaction mixture. Chromatographic procedures are also available for measuring residual dansylcadaverine as well as the formation of the amide product (54). Hydrolysis of transcinnamoylthiocholine can be studied by spectral changes in the UV (74).

Some kinetic data are given as illustration in Tables 1 and 2. The former relates to the steady-state hydrolysis of a number of thioesters by human Factor XIII_a, whereas Table 2 pertains to the question of amine specificity, based on the acceleration of thiocholine production in the presence of different amines by this enzyme.

Amine specificity is one of the most characteristic aspects of transglutaminase-catalyzed reactions, as it has been repeatedly emphasized in publications from this laboratory since 1968 (35). Affinity for the side chains of amines can be demonstrated (i) by

[9] p-Nitrophenylesters are akin to anhydrides, with a free energy of hydrolysis of ΔF° at pH 7 = 13 010 cal mole^{-1} (78).

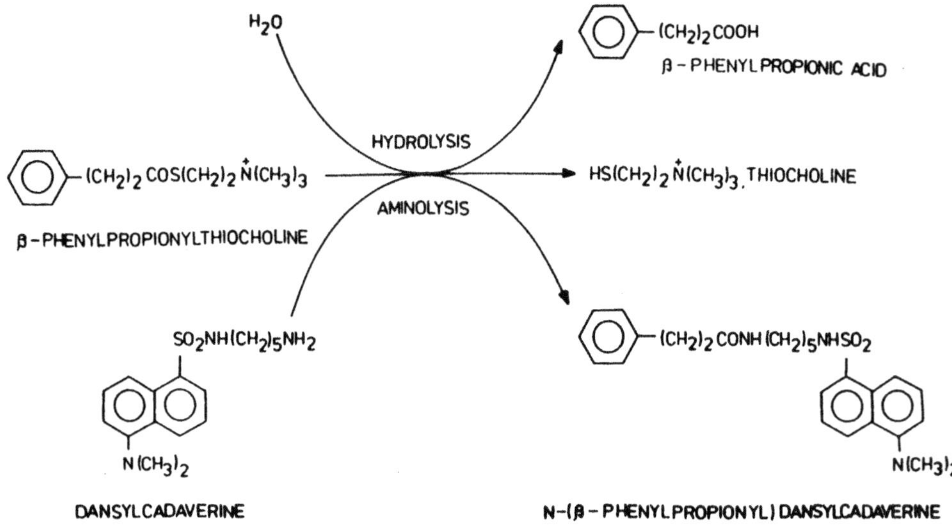

Scheme I

Table 1. Kinetic constants for the non-enzymatic and for the fibrinoligase-catalyzed steady-state hydrolysis of thiol esters at pH 7.5 and 25 °C (75).

	Non-enzymatic	Enzymatic		
		$K^{hyd}_{m,app}$	k^{hyd}_{cat}	$k^{hyd}_{cat}/K^{hyd}_{m,app}$
	$sec^{-1} \times 10^6$	$M \times 10^3$	sec^{-1}	$M^{-1}\ sec^{-1} \times 10^{-3}$
$C_6H_5(CH_2)_2COS(CH_2)_2N(CH_3)_3$ + I- β-Phenylpropionylthiocholine iodide	1.3	0.29	0.21	0.74
$C_6H_5(CH_2)_2COS(CH_2)_2N(CH_3)_2 \cdot HCl$ 2-Dimethylaminoethanethiol β-phenylpropionate hydrochloride	58.5	0.18	0.13	0.74
$C_6H_5CH = CHCOS(CH_2)_2N(CH_3)_{3\ 4}$ I- *trans*-Cinnamoylthiocholine iodide	1.3	0.10	0.08	0.80
$C_6H_5CH = CHCOS(CH_2)_2N(CH_2C_6H_5)(C_2H_5)_2$ + Br- 2-Benzyldiethylaminoethanethiol *trans*-cinnamate bromide	8.0	0.25	0.08	0.32

Table 2. Enhancement of the steady-state formation of thiocholine in the reactions of five different amines with 1.5 mM β-phenylpropionyl-thiocholine iodide, at 50 mM calcium chloride, E_0 = 0.12 μM, pH 7.5, and 25 °C[a] (75).

Amine substrates	v^{am}_{lim}/E_0	$K^{am}_{m,app}$
	sec^{-1}	$M \times 10^4$
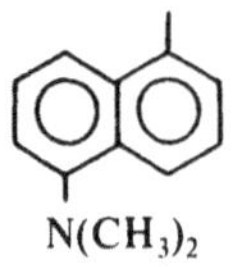 $SO_2NH(CH_2)_6NH_2$ ... $N(CH_3)_2$ *N*-(5-Aminopentyl)-5-dimethyl-amino-1-naphthalenesulfonamide; dansylcadaverine	1.7	8
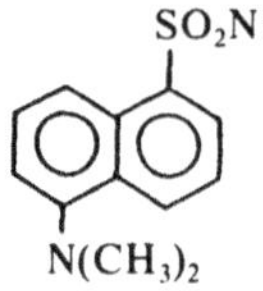 $SO_2NH(CH_2)_2S(CH_2)_2NH_2$ ... $N(CH_3)_2$ *N*-(5-Amino-3-thiapentyl)-5-di-methylamino-1-naphthalenesul-fonamide; dansylthiacadaverine	1.7	4
 $SO_2NHCH(CH_2)_4NH_2$, C = O, O, CH_3, CH_3 *N*-α-*p*-Tosyl-L-lysine methyl ester	1.4	11
$NH_2(CH_2)_4NH_2$ Putrescine	2.1	77
$CH_3(CH_2)_3NH_2$ *n*-Butylamine	1.2	430

[a] Limiting aminolytic enhancement rates (v^{am}_{lim}) were calculated from Lineweaver-Burk transforms of data after subtracting the purely hydrolytic velocity. Amine concentrations required to bring about half of the limiting velocities are denoted as $K^{am}_{m,app}$.

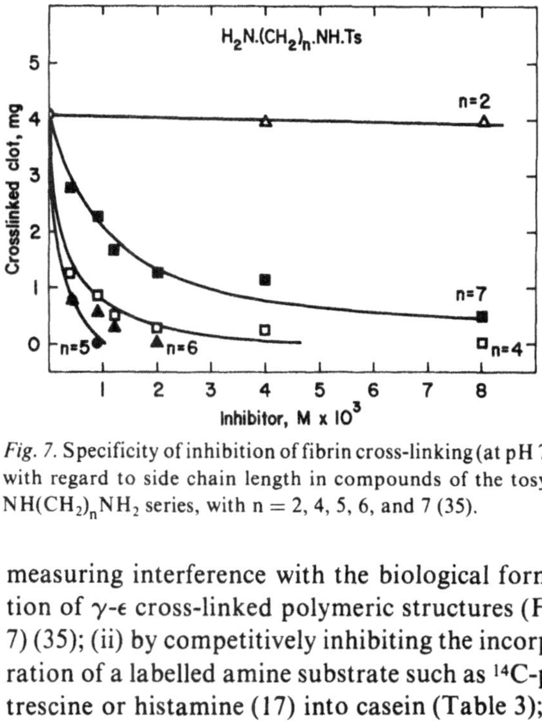

Fig. 7. Specificity of inhibition of fibrin cross-linking (at pH 7.4) with regard to side chain length in compounds of the tosyl · $NH(CH_2)_nNH_2$ series, with n = 2, 4, 5, 6, and 7 (35).

measuring interference with the biological formation of γ-ϵ cross-linked polymeric structures (Fig. 7) (35); (ii) by competitively inhibiting the incorporation of a labelled amine substrate such as ^{14}C-putrescine or histamine (17) into casein (Table 3); or (iii) as shown in Table 2 (75), by examining reactions in fully synthetic substrate systems.

Parameswaran and Lorand (77) recently synthesized the acceptor substrate, dansyl-(N-methyl-β-alanyl)thiocholine:

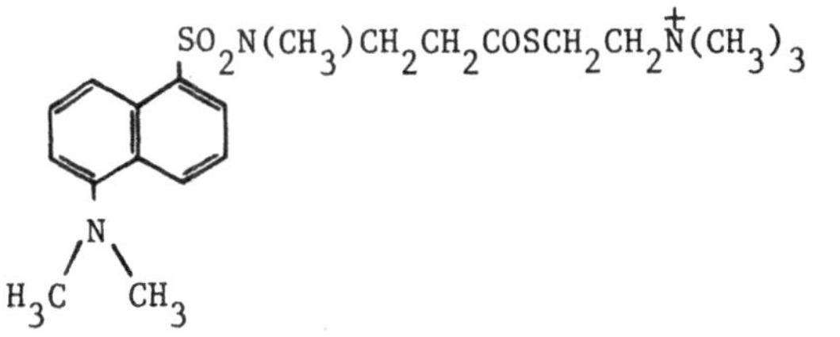

Table 3. Apparent relative affinities of amines for guinea pig liver transglutaminase, as determined by competitive inhibition kinetics for the incorporation of labelled putrescine into N,N′-dimethylcasein at pH 7.5, 37 °C (17, 53). Affinities are given in relation to dansylcadaverine (III) for which the reciprocal apparent inhibition constant ($K^{-1}_{i,app}$) is referred to as 100.

Compound	Structure		Relative affinity measured	Relative affinity corrected[a]
A.	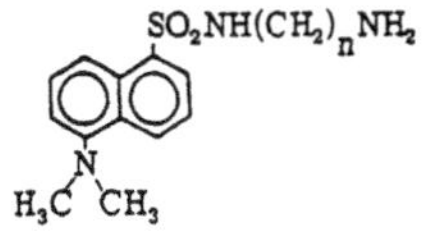			
	n			
I	3		2	2
II	4		60	60
III	5		100	100
IV	6		44	44
B.	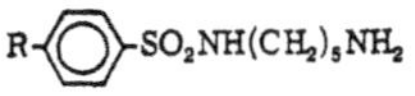			
	R			
V	-H		28	28
VI	$-CH_3$		25	25
VII	$-OCH_3$		23	23
VIII	-I		35	35
IX	$-NO_2$		34	34

[a] Corrected for difference of the pK_a of the primary ammonium ion of the compound relative to that of dansylcadaverine.

Compound	Structure		measured	corrected
C.	$Ar-Y-(CH_2)_5NH_2$			
	Ar	Y		
VI	$p-H_3CC_6H_4-$	SO_2NH	25	25
X	$p-H_3CC_6H_4-$	NHCO	6	6
XI	$2,4-(NO_2)_2C_6H_3-$	SNH	47	47
XII	$2,4-(NO_2)_2C_6H_3-$	SO_2NH	105	105
XIII	$2,4-(NO_2)_2C_6H_3$	NH	20	20
IX	$p-NO_2C_6H_4-$	SO_2NH	34	34
XIV	$p-NO_2C_6H_4-$	NHCO	24	24
D.	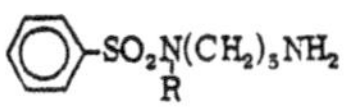			
	R			
V	-H		28	28
XV	$-CH_3$		24	24
XVI	$-C_2H_5$		29	29
XVII	$-CH(CH_3)_2$		17	17
XVIII	$-CH_2C_6H_5$		31	31

Compound	Structure		measured	corrected
E.	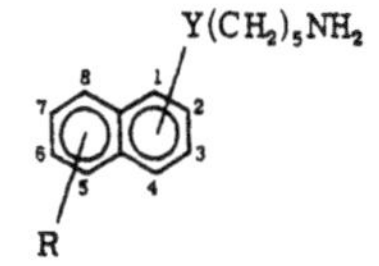			
	R	Y ($-SO_2NH-$)		
XIX	-H	1-	37	37
III	$5-N(CH_3)_2$	1-	100	100
XX	$5-(CH_3)_2NCH_2-$	1-	41	41
XXI	$5-(CH_3)_2N(CH_2)_3-$	1-	54	54
XXII	-H	2-	40	40
XXIII	$6-CH_3-N-$ 	2-	57	57
F.	$R-(CH_2)_5NH_2$			
	R			
XXIV	$(C_6H_5CH_2)_2\overset{\oplus}{N}H-$		156	16
XXV	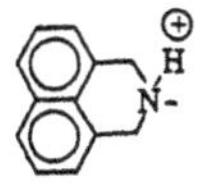		422	42
XXVI	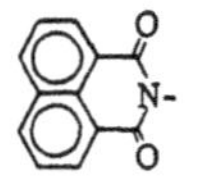		33	33

Compound	Structure		measured	corrected
G.	$R-SO_2NHCH_2CH_2-Y-CH_2CH_2NH_2$			
	R	Y		
VI	$p-H_3CC_6H_4-$	CH_2	25	25
XXVII	$p-H_3CC_6H_4-$	S	181	18
XXVIII	$p-H_3CC_6H_4-$	$C(CH_3)_2$	2	2
VII	$p-H_3COC_6H_4-$	CH_2	23	23
XXIX	$p-H_3COC_6H_4-$	S	103	10
XXX	$2,4,6-(CH_3)_3C_6H_2-$	CH_2	63	63
XXXI	$2,4,6-(i-C_3H_7)_3C_6H_2-$	CH_2	95	95
XXXII	$2,3,4-Cl_3C_6H_2-$	S	198	20
XXXIII	$p-C_6H_4-N{=}NC_6H_4-$	S	223	22
XIX	naphthalene-1-	CH_2	37	37
XXXIV	naphthalene-1-	S	214	21
XX	$5-(CH_3)_2NCH_2$-naphthalene-1-	CH_2	41	41
XXXV	$5-(CH_3)_2NCH_2$-naphthalene-1-	S	183	18
III	$5-(CH_3)_2N$-naphthalene-1-	CH_2	100	100
XXXVI	$5-(CH_3)_2N$-naphthalene-1-	S	767	77
XXXVII	$5-(CH_3)_2N$-naphthalene-1-	$C(CH_3)_2$	11	11
XXXVIII	$5-(CH_3)_2N$-naphthalene-1-	SO	21	2
XXXIX	$5-(CH_3)_2N$-naphthalene-1-	SO_2	20	2

which allowed yet another approach for studying nucleophile specificities in reactions with amines as well as with alcohols (Table 4). The latter reaction also made it possible to obtain a meaningful pH profile for the first time for human Factor XIII_a (Fig. 8). The steady-state kinetic analysis led to the conclusion that the reaction proceeded along an acylenzyme intermediate pathway. Since the active center is known to be a cysteine thiol (79, 80)[10], the reaction with the enzyme (ESH) can be illustrated by equations 1–4:

$$(1)\quad RCOSR' + ESH \underset{k_{-1}}{\overset{k_1}{\rightleftharpoons}} (RCOSR'\cdot ESH) \xrightarrow{k_2} RCOSE + R'SH$$

$$(2)\quad RCOSE + H_2O \xrightarrow{k_3} RCOOH + ESH$$

$$(3)\quad RCOSE + H_2NR'' \underset{k_{-4}}{\overset{k_4}{\rightleftharpoons}} (RCOSE\cdot H_2NR'') \xrightarrow{k_5} RCONHR'' + ESH$$

$$(4)\quad RCOSE + HOR''' \underset{k_{-6}}{\overset{k_6}{\rightleftharpoons}} (RCOSE\cdot HOR''') \xrightarrow{k_7} RCOOR''' + ESH$$

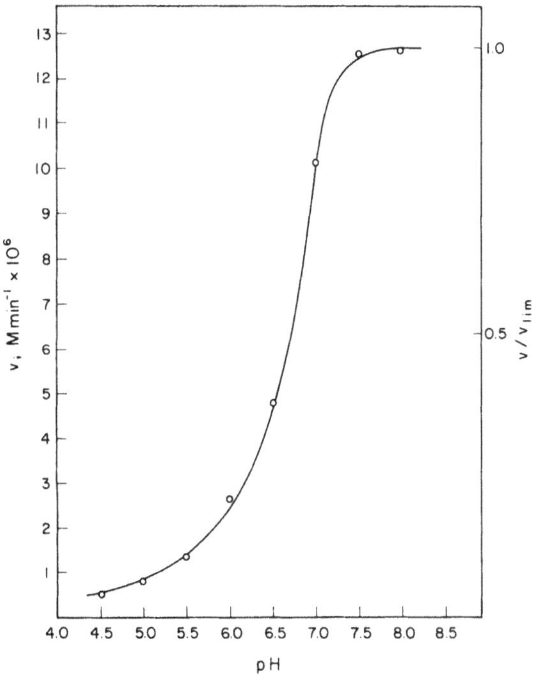

Fig. 8. Effect of pH on the steady-state velocity of methanolysis of [dansyl-(N-methyl-β-alanyl)]thiocholine iodide (S_I) by fibrinoligase in 0.1 M sodium acetate, 27 °C, and 30 mM calcium chloride (77). The reaction was carried out in 0.1 ml of aqueous phase containing 0.1 M sodium acetate, 1.86 mM S_I, 1.24 M methanol, 0.5% N,N-dimethylformamide, thrombin-activated plasma Factor XIII ($e_0 = 0.36\ \mu M$), and finally 30 mM calcium chloride.

[10] The claim of 'half of the sites' type of reactivity for Factor XIII_a by Chung et al. (81) cannot be supported by investigations from this laboratory (80).

The acylenzyme intermediate[11] (RCOSE) formed in eq. 1 could either react with water (as in eq. 2), with an amine (as in eq. 3) or with an alcohol (as in eq. 4), giving rise to three different products. The kinetic specificity of amines is governed by their affinities for the acylenzyme intermediate (i.e., by $K_b = k_{-4}/k_4$) as well as by their reactivities (i.e., k_5), and the k_5/K_b expression is a suitable index of amine specificity.

If one subscribes to the view that kinetic analysis of specificity corresponds to a 'template mapping' of the environment of active center domains of an enzyme, the conclusion is justified that the acceptor and the donor substrates must fit into rather tight crevices on Factor XIII_a and on transglutaminases. It is as if the carbonyl group of the first substrate (i.e., γ-position in glutamine) had to span a fixed distance from some surface anchoring toward the catalytic sulfhydryl group, and as if the amine functionality of the second substrate (i.e., ε-position in lysine) also had to reach a certain length to be able to react with the acylenzyme intermediate. An illustration of these concepts is given in Fig. 9. For example, dansylcadaverine would be bound through its aryl moiety to some site, defined as a 'secondary binding site' (59), on the surface of the acylenzyme and with the pentylamine side chain accommodated in a crevice to orient it for reaction with the activated carbonyl group of the first substrate.

Regulation of transglutaminase activities

In spite of the fact that some of the enzymes discussed show a great deal of specificity for preferred protein substrates (e.g., Factor XIII_a reacts with fibrin nearly ten times faster than with fibrinogen) (33, 82), transglutaminases could not have evolved without some regulatory safeguards. Activation of an enzyme of this type at the wrong time and wrong place could have disastrous consequences. Thus, in nature they are found as inactive *zymogens* or as *latent enzymes.* Elucidating the mechanisms of activation of zymogens and latent forms is essential for the understanding of biological controls.

[11] A recent publication by Folk (60) describes the isolation of the trimethylacetyl intermediate of guinea pig liver transglutaminase in small yield.

Table 4. Steady-state kinetic constants for the fibrinoligase (human plasma Factor XIII_a; $E_0 = 0.3$ to 0.72×10^{-6} M) catalyzed reactions of dansyl-N-methyl-β-alanylthiocholine iodide (S_1) with amines or alcohols as second substrates (S_2) in the presence of 30 mM $CaCl_2$ (27 °C; 0.5% dimethyl formamide) (77).

S_2	K_M ($M \times 10^3$) for S_1	K_M ($M \times 10^3$) for S_2	k_{cat} (s^{-1})	k_{cat}/K_M (for S_2) (s^{-1} M^{-1})
amines				
CH_3NH_2	1.1[a]	9.1[a]	0.45[a]	50[a]
	1.1[b]	12.5[b]	0.5[b]	40[b]
$CH_3(CH_2)_3NH_2$	1.1[c]	13.3[c]	0.18[c]	14[c]
$(CH_3)_3COC(=O)NH(CH_2)_5NH_2$	1.1[a]	7.7[a]	0.8[a]	104[c]
t-Boc-cadaverine	1.1[a]	0.95[a]	0.64[a]	674[a]
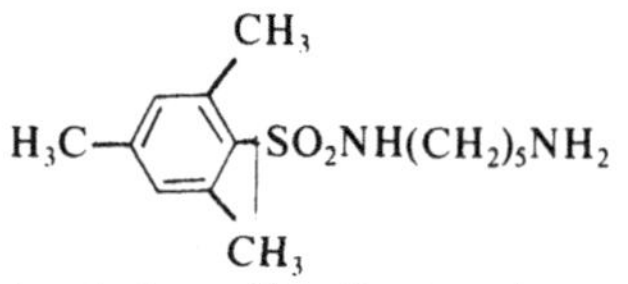 (mesitylenesulfonyl)cadaverine	0.9[a]	0.05[a]	0.21[a]	4200[a]
	K_M ($M \times 10^3$)	K_M (M)		
alcohols				
CH_3OH	1.5[d]	0.4[d]	0.72[d]	1.8[d]
	1.3[e]	0.4[e]	1.14[e]	2.85[e]
CH_3CH_2OH	0.6[d]	0.8[d]	0.2[d]	0.25[d]
	1.3[e]	1.1[e]	0.48[e]	0.44[e]
	0.7[f]	0.8[f]	0.35[f]	0.45[f]
	1.4[g]	1.1[g]	0.37[g]	0.34[g]
$CH_3(CH_2)_3OH$	0.9[d]	0.65[d]	0.15[d]	0.23[d]

[a] 0.05 M Tris-HCl, pH 7.5 at $\mu = 0.15$.
[b] 0.1 M sodium acetate, pH 7.5, at $\mu = 0.21$.
[c] 0.05 M *N*-methylmorpholine hydrochloride, pH 7.5, at $\mu = 0.14$.
[d] 0.1 M sodium acetate, pH 7.0, at $\mu = 0.19$.
[e] 0.05 M *N*-methylmorpholine hydrochloride, pH 7.5, at $\mu = 0.12$.
[f] 0.05 M Pipes-Na^+, pH 7.5, at $\mu = 0.2$.
[g] 0.05 M Pipes-Na^+, pH 7.0, at $\mu = 0.2$.

There are examples for *sequestering* the enzyme from the substrate, with contact occurring between them only as the need arises. An interesting case in point is the clotting of the seminal fluid of rodents, where enzyme and substrates are secreted by two different lobes of the prostate and the two become mixed only in the ejaculate (see chapter by Williams-Ashman). Another example of compartmentalization is found in *Homarus* blood clotting where the transglutaminase, which is responsible for catalyzing the polymerization of several plasma proteins, is contained within the amebocytes and is released only as the cells disintegrate (Fig. 10). Interestingly, human platelets, unlike the *Homarus* cells, do not release transglutaminase by contact activation.

Whether examples can be found for the sequestration of transglutaminase from its substrate within the same cell, and whether redistribution of the enzyme from one sub-cellular compartment to another might take place as a result of stimulation (e.g. from the membrane bound to soluble phase), is an interesting question.

The conversion of Factor XIII (fibrin stabilizing factor) in vertebrates to Factor XIII_a (fibrinoligase) is the best studied model for *zymogen activation* thus far. In as much as both the cross-linking enzyme and the main target substrate (fibrin, derived from fibrinogen) are generated in the fluid phase (plasma) of the same tissue (blood), the zymogen form must have represented the best evolutionary solution poised for rapid activation. Details of the regulation of Factor XIII conversion are so interesting that we shall have to discuss them later under a separate heading. Here, it should suffice to state that the zymogen (Z) first undergoes a *limited proteolysis* with thrombin, but that the hydrolytically

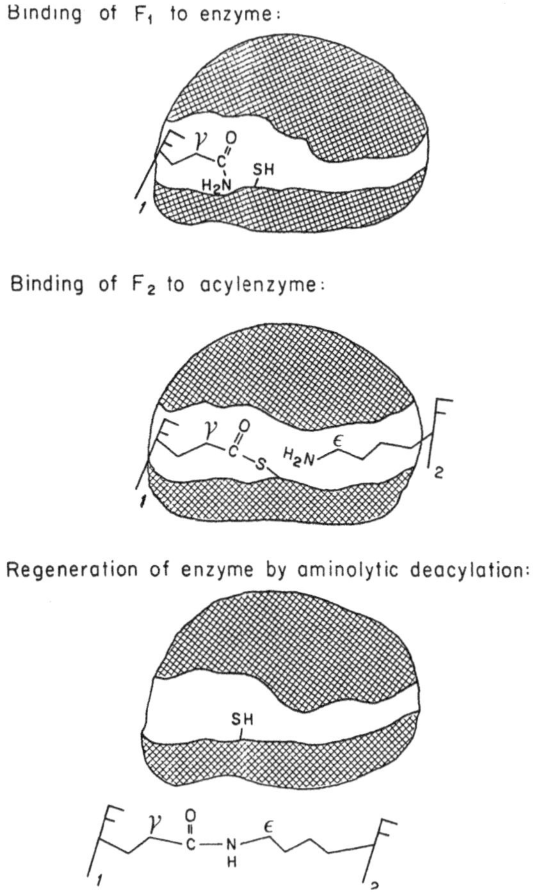

Fig. 9. Intermediate steps for the fibrinoligase (Factor XIII_a)-catalyzed cross-linking of two fibrin molecules (F_1 and F_2) by a γ-glutamyl-ε-lysine peptide bond.

modified zymogen (Z′) is still inactive. The active center-cysteine of the enzyme (E) is unmasked only in the presence of Ca^{2+}. As such, this type of activation model is represented by a sequence of: $Z \xrightarrow{\text{protease}} Z' \xrightarrow{Ca^{2+}} E$.

Ca^{2+} seems to be a *universal activator* for all latent transglutaminases studied thus far (Latent enzyme $\xrightarrow{Ca^{2+}}$ Active enzyme), and an example for three such enzymes is given in Fig. 11. The Ca^{2+}-requirement of plasma Factor XIII is a special case and will be discussed separately. The intracellular transglutaminases show an apparent requirement for half-maximal activation in the 10^{-4} M to 5×10^{-4} M Ca^{2+} range, and many different cells are known to accumulate this much Ca^{2+} during their life cycles. As a rule, however, this is

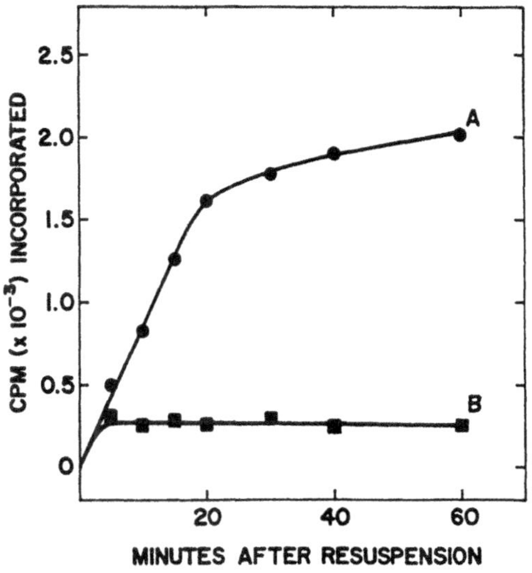

Fig. 10. Time-dependent release of lobster hemocyte transglutaminase under clotting conditions (42). Plasma (10 ml) was extracted from a chilled lobster into a chilled syringe. It was divided into two equal portions and centrifuged at 12 000 × **g** for 10 min at 0 °C in plastic centrifuge tubes. The pellets were resuspended with 2 ml of 0.4 M NaCl–20 mM $CaCl_2$. One tube (A) was maintained at 23 °C (i.e. at clotting condition); the other (B) was kept at 0 °C. Release of the enzyme was followed by taking 20 μl samples of supernatant for assaying by the filter paper method of Lorand et al. (53), using β-lactoglobulin as test protein for the incorporation of ^{14}C-putrescine.

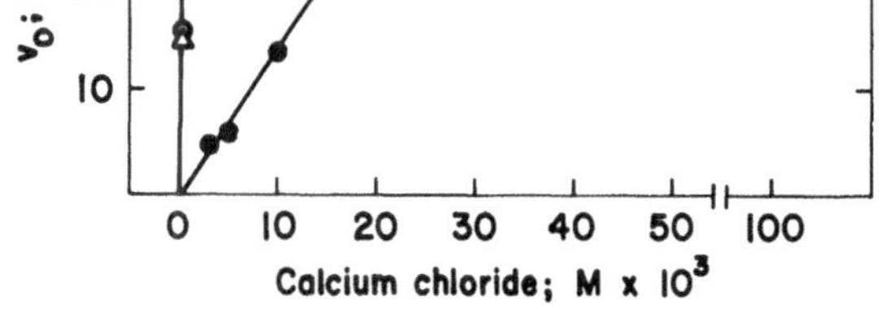

Fig. 11. Apparent Ca^{2+} ion requirement for three transamidating enzymes (76). The aqueous reaction mixtures contained 8.2 mM β-phenylpropionyl-thiocholine iodide, 1.6 mM dansylcadaverine, 50 μg of platelet transglutaminase/ml (△), 0.50 μM guinea pig liver transglutaminase (○), or 0.32 μM thrombin-activated human Factor XIII (●), and $CaCl_2$ at the concentrations shown on the abscissa, added last to initiate amide formation (ordinate), which was measured from the steady-state increase of fluorescence in n-heptane, layered above the aqueous phase.

not the Ca^{2+}-concentration where calmodulin is most effective in regulating enzyme activities (83); nevertheless, there is a recent abstract by Puszkin et al. (84) suggesting a relationship between transglutaminase and calmodulin.

The Ca^{2+} needed for transglutaminase activation may enter the cell from the outside, or it may arise by virtue of redistribution from intra-cellular stores of Ca^{2+}, e.g. by destruction of mitochondria. Augmentation of free Ca^{2+} may also be caused by a drop in concentration of a metabolite (e.g., ATP) which would otherwise bind Ca^{2+}. In addition, *Zn^{2+} ions* are known to compete against Ca^{2+} and can inhibit transglutaminase with a K_i of about 10^{-7} M (Figs. 12 and 13) (85); thus, alteration of the Ca^{2+}/Zn^{2+} ratio in some cells could be a critical factor.

Once transglutaminase is generated, various *activators* and *inhibitors* could exercise a regulatory role. So far, we have no idea as to the nature of such compounds in the biological milieu, but *in vitro* experiments might serve to highlight some possibilities. Since transglutaminases operate by a cysteine thiol active center mechanism (ESH), disulfide-containing metabolites (RSSR; e.g., cystamine),

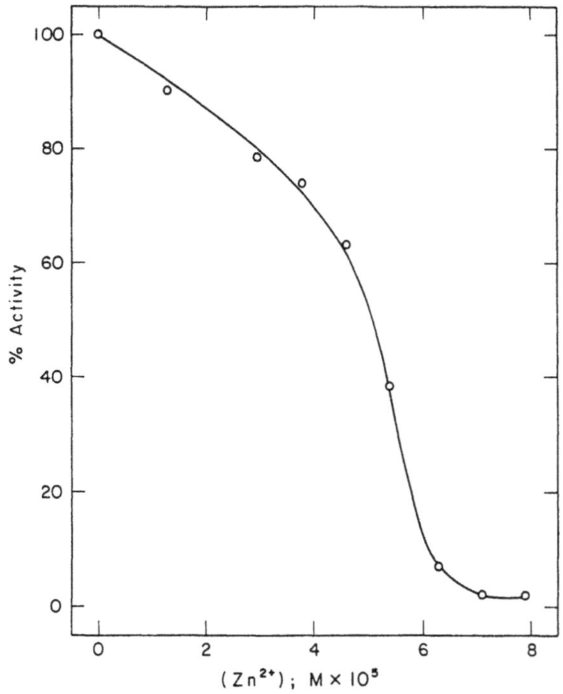

Fig. 13. Inhibition of guinea pig liver transglutaminase by Zn^{2+}-ions (85). Amide formation between β-phenylpropionylthiocholine and dansylcadaverine was measured at pH 7.5, $\mu = 0.25$, 25 °C at an enzyme concentration of 1.8×10^{-7} M in the presence of various concentrations of Zn^{2+}.

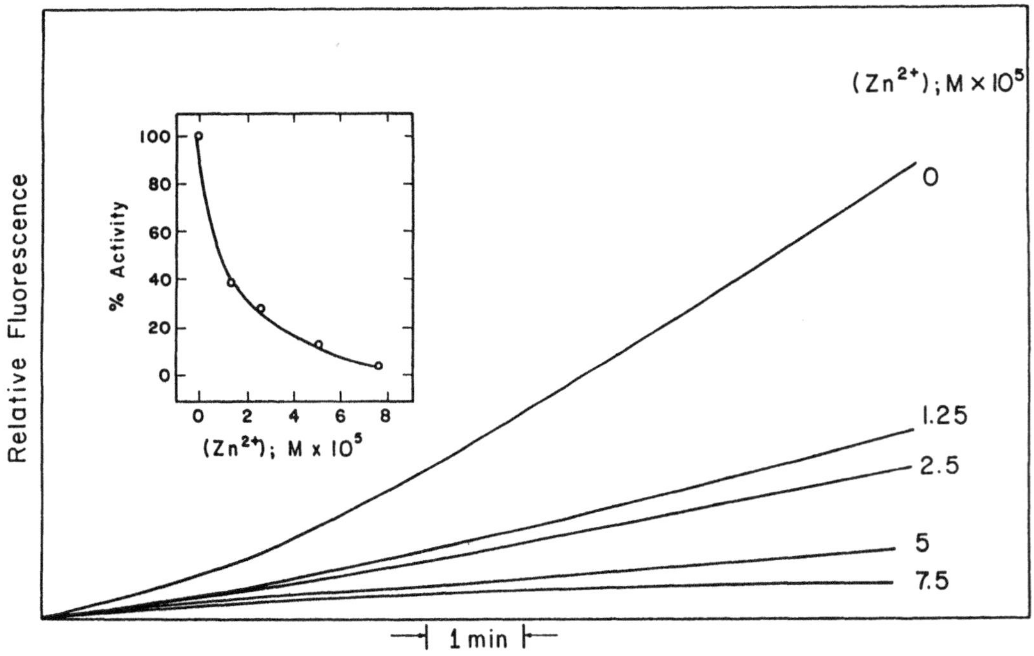

Fig. 12. Inhibition of human fibrinoligase (Factor XIII_a) by Zn^{2+}-ions (85). Amide formation between β-phenylpropionylthiocholine and dansylcadaverine was measured at pH 7.5, $\mu = 0.37$, 25 °C at an enzyme concentration of 1.6×10^{-7} M, in the presence of various concentrations of Zn^{2+}.

similarly to the regulation of phosphofructokinase (86), could block enzyme activity by a disulfide exchange reaction: ESH + RSSR → ESSR + RSH, where ESSR represents a reversibly inactivated form of transglutaminase. An illustration for this type of inactivation is given in Fig. 14 (unpublished work of Svahn, Credo and Lorand).

A remarkable example of *activation by a solute* is shown in Fig. 15 regarding the non-secretory transglutaminase of the guinea pig prostate gland. In the absence of the organic solute tosylglycine:

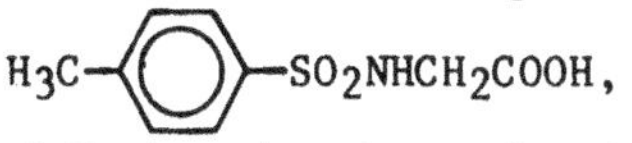

, the enzyme was essentially inactive, but a hundred-fold activation occurred in the presence of the solute. The activity of epidermal transglutaminase could be increased by heating at 56 °C or by treatment with dimethylsulfoxide (87).

In addition to such examples for 'enzyme-level' control, there are indications that transglutaminase activity may also be regulated at the *'substrate level'*, and this may take different forms. The possibility of proteolytic unmasking of the amine incorporating γ-glutamine sites has already been referred to in connection with the finding that fibrin was a much better substrate for Factor XIII_a than the parent protein, fibrinogen. In fact, the removal of only fibrinopeptide A by thrombin: $(\alpha\beta\gamma)_2(A)_2(B)_2 \xrightarrow{\text{thrombin}} (\alpha\beta\gamma)_2(B)_2 + 2A$ is sufficient to enhance the reactivity of the protein substrate toward Factor XIII_a, the specific cross-linking enzyme (but not toward liver transglutaminase!). Such a regulation has a clear implication for coordinating the physiological blood clotting event in an ordered manner (i.e., to give preference to the cross-linking of fibrin over that of fibrinogen; see Fig. 16), and may be taken as a

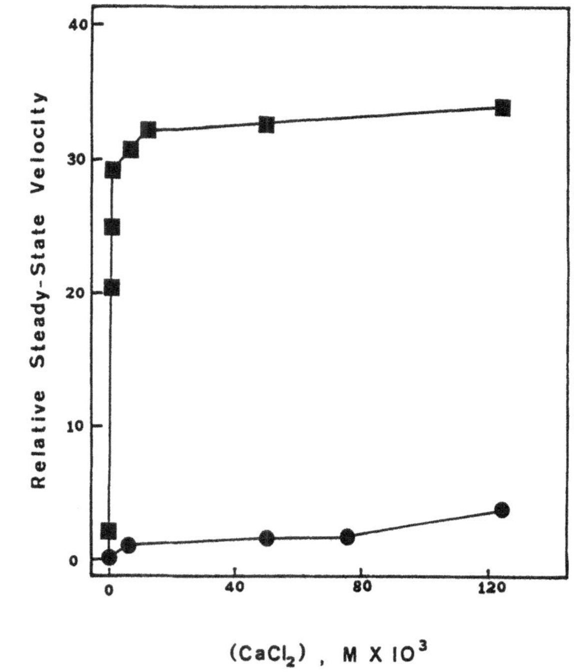

Fig. 15. Effect of tosylglycine on the activity of non-secretory guinea pig prostate transglutaminase (139). Steady-state velocities for amide formation were determined at 30 °C in reaction volumes of 120 μl containing 50 mM Tris-acetate of pH 7.0, 3.5 mM β-phenylpropionylthiocholine, 1.13 mM dansylthiacadaverine, 0.036 mg/ml enzyme, appropriate amounts of sodium chloride to maintain a constant ionic strength of 0.5 and varying concentrations of calcium chloride (●). Formation of the amide product was followed also in the presence of 96 mM p-tosylglycine (■).

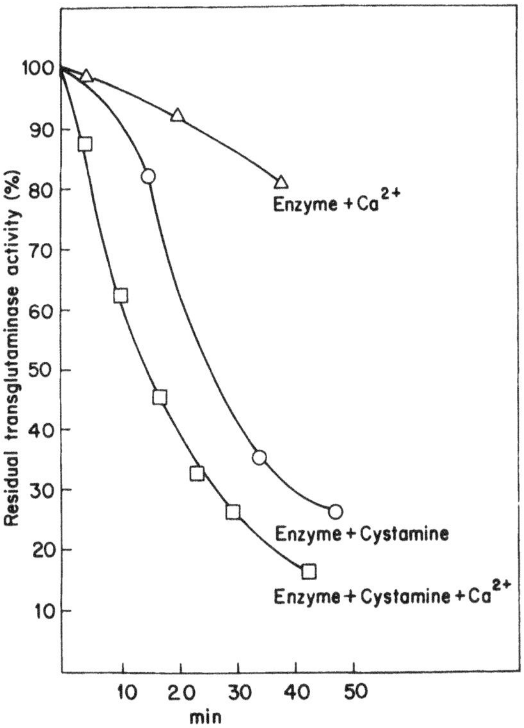

Fig. 14. Inhibition of guinea pig liver transglutaminase by incubation (time on abscissa) with cystamine (10^{-4} M) at pH 7.5 (Svahn and Lorand, unpublished results). When indicated, 17 mM Ca^{2+} was also present during incubation. Residual enzyme activity was measured by formation of the fluorescent amide product in the reaction of β-phenylpropionylthiocholine with dansylcadaverine.

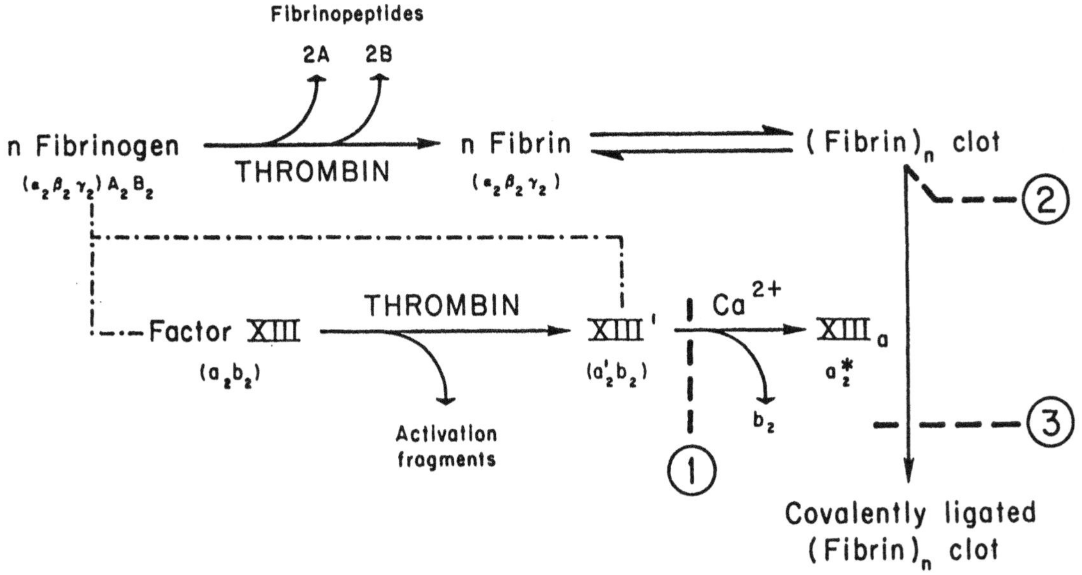

Fig. 16. Outline of the clotting reaction in normal human plasma. Disorders of fibrin stabilization can be genetic, such as Factor XIII-deficiency, or acquired due to the appearance of circulating inhibitors directed against one of the steps in the sequence marked with broken lines (---) as ①, ②, and ③. For a discussion of the various molecular diseases, consult Lorand et al. (7). The dotted lines (-·-·-) refer to the regulation exerted by fibrinogen on the conversion of the Factor XIII zymogen, in regard to enhancing the rate of release of activation fragments by thrombin and the lowering of Ca^{2+} requirement.

sign/of co-evolution between the cross-linking enzyme and its protein substrate (33).

Apart from limited proteolytic activation, there is also a possibility that 'substrate-level' control could be exercised through the *prior covalent modification* of the protein serving as target for transglutaminase action. Phosphorylation (e.g., —**O**—PO_3H_2), methylation (e.g., —**COOCH$_3$**) or acetylation (e.g., —**NHCOCH$_3$**) of various side chains of the substrate (in bold letters) might alter its reactivity towards transglutaminase. It may be more than coincidence that many of the best protein substrates for transglutaminases, i.e., casein, β-lactoglobulin, or histones, are also among the best substrates of other enzymes which post-translationally modify proteins (88, 89).

Specific *interaction between a* particular *solute* (L) and the protein *substrate* (S) might render the latter (L · S) more reactive toward a transglutaminase: $L + S \rightleftharpoons L \cdot S$. This type of regulation was first recognized by Williams-Ashman et al. (90) who showed that the addition of polyanions, by simulating the effect of bulbourethral gland secretion, greatly enhanced the reactivity of seminal vesicle secretion proteins (but not that of casein!) towards the clotting enzyme.

Fluctuations in the *intra-cellular concentrations of polyamines* (spermidine, spermine, putrescine and cystamine) or their monoacetylated forms might play a regulatory role by competing against the cross-linking of proteins, as suggested by the equations in Fig. 1.

Strategies for probing transglutaminase-mediated events in biological systems

When the participation of transglutaminase in a biological reaction is suspected, the first step is often to try to interfere with the phenomenon by the addition of one of the synthetic amine substrates. This approach was introduced in connection with the study of fibrin stabilization during the process of blood coagulation (35) and, in general, the following rules should apply.

(i) Specificity for the *primary amino* group functionality of the inhibitor should be demonstrated. If dansylcadaverine is inhibitory, it is expected that the tertiary amino group containing N-dimethyl analogue[12] (Table 5) would be less so,

[12] Because of widely differing solubilities, dansylamidopentanol is not an appropriate control for dansylcadaverine.

Table 5. Amine substrates for probing transglutaminase activities in biological systems and recommended control compounds.

Substrates	Controls
glycine ethylester $H_2NCH_2COOCH_2CH_3$	sarcosine ethylester $CH_3NHCH_2COOCH_2CH_3$
histamine 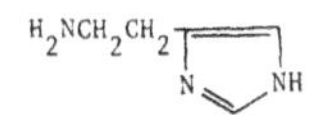	$^{\alpha}$N-dimethylhistamine 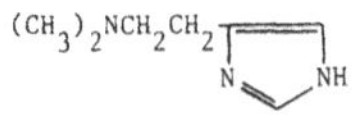
dansylcadaverine 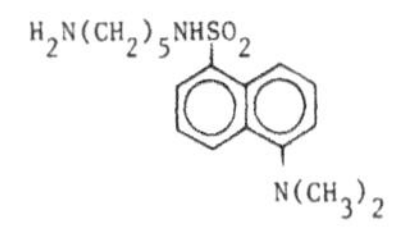	dimethylated dansylcadaverine 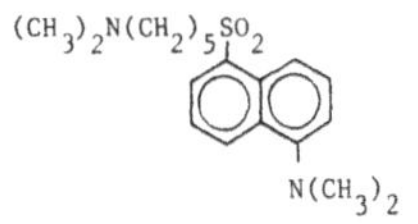

and the same could be said for comparing the effects of histamine and $^{\alpha}$N-dimethylhistamine, or that of glycine ethylester and sarcosine ethylester. Such controls are all the more important in cellular systems, where inhibition might simply be due to interference by the hydrophobic dansylamidoalkyl moiety with some membrane function or to the changing of pH in an intra-cellular compartment (e.g., lysomotropic effect), which could be elicited by primary and tertiary amines equally well.

There are other complications with the use of amine inhibitors in cell biology. Even a compound with as favorable a K_M ($\sim 10^{-5}$ M) for transglutaminase as dansylcadaverine cannot be considered so restricted in its mode of action as to be assumed to interfere with this enzyme exclusively. There is some information already available that dansylcadaverine can inhibit a variety of metabolic processes [ketosteroid reductase (91); glycoprotein synthesis (C. G. Curtis, personal communication); phosphatidylcholine synthesis (92)] and the list is almost certain to grow. In addition, considering the fact that some commercially marketed 'calmodulin inhibitors' (W-5 and W-7 of Rikaken Co., Japan):

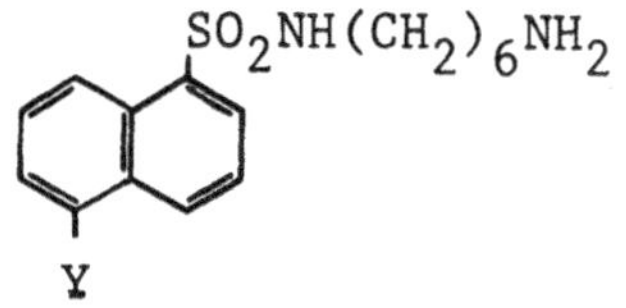

Y = H or Cl

are structurally quite similar to dansylcadaverine and other synthetic substrates of transglutaminase (17), one wonders if dansylcadaverine and its analogues might not also act as calmodulin inhibitors[13].

(ii) *Incorporation of the amine* inhibitor into target proteins must be demonstrable. Under ideal conditions, all the enzyme-reactive γ-glutaminyl acceptor sites of the protein can be modified. Various amine isotopes are available for this purpose, and other synthetic amines may be studied by virtue of their fluorescence or color, or because antibodies are easily obtained against some substituents (e.g., dansyl or dinitrophenyl). Though a necessary requirement, the covalent attachment of the label to proteins by itself is not sufficient for concluding that incorporation was mediated by transglutaminase. Such a difficulty was encountered when work with lymphocyte activation showed that the putrescine-derived radioactive label was converted into hypusine and not, as previously thought, into γ-glutamylspermidine (62, 64).

Transglutaminase-catalyzed reactions between proteins and amines produce amides; thus, acid hydrolysis (5.7 N HCl, 24 hr, 100°) should permit recovery of the amine. Furthermore, total enzymatic digestion, as in the analysis of the γ-glutamyl-ε-lysine cross-linking peptide (46), should produce a labelled γ-glutamyl-amide.

Apart from the hypusine pathway (64), putrescine could also undergo oxidation to aldehyde which, in turn, might react with proteins in a transglutaminase-independent manner. When such a possibility is transglutaminase-independent manner. When such a possibility is suspected, diamine oxidase inhibitors (such as methylglyoxalbis (guanylhydrazone) or aminoguanidine) could be employed. As discussed already, spurious incorporation of an amine label could also arise from the action of myeloperoxidase (67).

(iii) *Other inhibitors* of transglutaminases fall into various categories:

(a) Active site-directed compounds may block the essential cysteine thiol group of the enzyme in various ways. Disulfides, such as cystamine (Fig. 14), participate in an exchange reaction as shown by Svahn, Credo and Lorand (unpublished results)

[13] In cell biology, we are forced to realize that the specificity of an inhibitor is inversely proportional to our knowledge regarding its mode of action.

of the: $RSSR + ESH \rightleftharpoons (RSSR; ESH) \rightarrow RSSE + RSH$ type. Other compounds, e.g., halomethyl ketones (93) or blocked diazooxonorvaline or norleucine derivatives (94), are also inhibitory. However, these are not naturally occurring substances, nor is it known whether they are more selective in reacting with transglutaminase than iodoacetamide or N-ethylmaleimide would be.

(b) As shown in Fig. 12, Zn^{2+} appears to be a good inhibitor of transglutaminase by virtue of competing against Ca^{2+}.

(c) Lorand, Barnes, Michalska, Hawkins, and Bruner-Lorand (unpublished results) recently explored a novel, *non-competitive inhibitor* (95) for the purpose of preventing protein cross-linking reactions in Ca^{2+}-enriched human erythrocytes and in thrombin-activated human platelets:

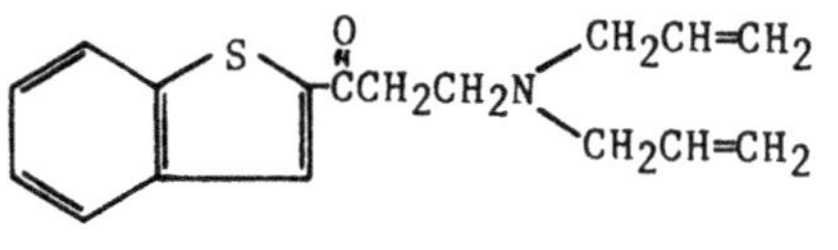

Both reactions were abolished by adding less than 10^{-3} M of the inhibitor to the medium. Though again there are no data regarding selectivity, the approach of using such non-competitive inhibitors may be very useful in the future.

(iv) In addition to the previous approaches, *isolation and identification of γ-glutamyl-ε-lysine* cross-linked polymers is important in studying the participation of transglutaminase in a biological reaction (12, 35, 96). However, one should always keep in mind the possibility that such cross-links may also arise without the involvement of transglutaminase.

The blood clotting paradigm

It is now possible to reconstruct, in molecular terms, the sequence of events leading to the formation of a γ-ε cross-linked (ligated) fibrin clot in normal plasma (Fig. 16) (7). The regulatory aspects of generating the cross-linking enzyme (fibrinoligase, Factor XIII_a) from the circulating zymogen (fibrin stabilizing factor, Factor XIII) offer unique insights into possible control mechanisms regarding transglutaminases in general. Elucidation of the physiological pathway of clot formation in normal plasma, together with the development of ancillary methodologies, also made it possible to diagnose novel molecular hemorrhagic diseases related to various defects of fibrin stabilization (7, 97).

The plasma zymogen comprises two types of subunits (98) of about 80 000 MW each with a presumed molecular weight of ~320 000: a_2b_2. Activation can be effected simply by treating the zymogen with high concentrations of Ca^{2+} (37 °C, pH 7.5, $\mu = 0.3$): $a_2b_2 \rightarrow a_2^{\circ} + b_2$, when the active center cysteine on the catalytic *a* subunits is unmasked coincidentally with dissociation from the carrier *b* subunits. Addition of chaotropic anions (p-toluenesulfonate, thiocyanate, iodide, bromide) can reduce the Ca^{2+}-requirement to about 0.05 M (99, 100).

The physiological pathway of enzyme generation, however, is thrombin dependent. The a_2b_2 zymogen is first modified in a limited proteolytic step (101, 102):

$$a_2b_2 \xrightarrow{\text{thrombin}} a_2'b_2 + \text{activation peptides}$$

to yield an $a_2'b_2$ ensemble which, however, is still devoid of catalytic activity. Nevertheless, the hydrolytic removal of the N-terminal peptide fragment from the *a* subunit leads to a loosening of the heterologous association between *a* and *b* subunits, as seen by the fact that they can be dissociated at the physiological ionic strength of $\mu = 0.15$ merely by the addition of 10–20 mM $CaCl_2$, at pH 7.5, 37 °C in 10 min (80). However, since normal plasma contains only 1.5 mM Ca^{2+}, some other regulatory feature still had to be uncovered and it was found that fibrinogen (ϕ), at the physiological concentration of about 10^{-5} M, exerted a profound effect on the $a_2'b_2 \xrightarrow[\phi]{Ca^{2+}} a_2^{*} + b_2$ transition by lowering the Ca^{2+}-requirement to 1.5 mM (99). A peptide fragment corresponding to a sequence in the mid-portion of the α-chain of fibrinogen (residues 242–424) displayed the same activity as the native protein (103).

In addition to interacting with the hydrolytically modified ($a_2'b_2$) species, fibrinogen was also found to regulate the rate of release of the N-terminal activation peptide from the parent zymogen (104):

$$a_2b_2 \xrightarrow{\text{thrombin}} a_2b_2 + \text{activation peptides (AP)}.$$

Using an HPLC system for measuring AP release, a k_{cat}/K_M of 10 μM^{-1} min^{-1} was obtained (at

$\mu = 0.15$, pH 7.4 in phosphate buffered saline and 37 °C). The addition of human fibrinogen caused a marked increase, in a dose-dependent fashion, in the rate of release of AP. In the absence of fibrinogen, 0.57 nM human α-thrombin, for example, released 50% of AP from 0.1 μM Factor XIII in 150 min, whereas in the presence of 0.45 μM fibrinogen the time required for 50% release was only 4 min.

From an evolutionary point of view, the clotting systems of *Homarus* plasma (9, 25, 105) and of the rodent seminal fluid (10) could be considered to be more primitive, because they depend only upon the encounter of the latent or the sequestered transglutaminase with the protein substrates in the presence of Ca^{2+}.

Transglutaminase and cell aging. The Ca^{2+}-enriched human erythrocyte

Due to the operation of an efficient, outward directed Ca^{2+} pump, the Ca^{2+} concentration in normal human red blood cells is kept below 10^{-6} M. When the pump fails (as in aged cells, sickle cells or in some hereditary metabolic diseases), the intra-cellular Ca^{2+} may rise as high as 2×10^{-4} M (106–109), which would be sufficient to activate transglutaminase to an appreciable extent. The human erythrocyte enriched by Ca^{2+} in the presence of ionophore (A23187) is a suitable experimental model. If Ca^{2+} loading is allowed to proceed for periods longer than a few minutes, the cell suffers a permanent shape change (110) (see also Fig. 17) and a loss of membrane deformability (111). Parallel with these physical manifestations, the proteins of the membrane also undergo profound structural changes.

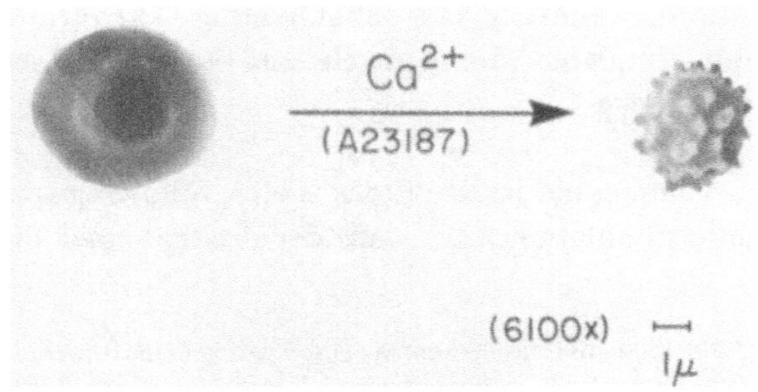

Fig. 17. Irreversible shape change, from discocyte (on left) to spheroechinocyte (on right) in the human erythrocyte, following enrichment with Ca^{2+} in the presence of ionophore A23187. Scanning electronmicrograph (work of M. Michalska).

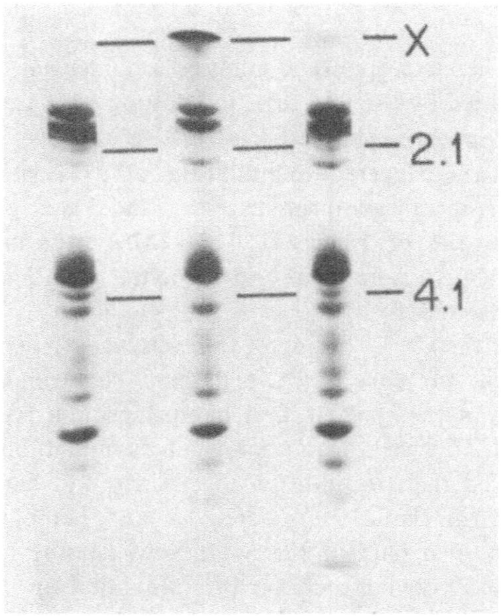

Fig. 18. Cross-linking of membrane proteins in Ca^{2+}-enriched human erythrocytes can be prevented by adding small molecular weight substrates or inhibitors for transglutaminase. Right-hand gel: the presence of 20 mM aminoacetonitrile during incubation of the cells (50% hematocrit, 37 °C, 3 hr, 10 μM ionophore) with 1.5 mM Ca^{2+}, prevents loss of bands 2.1 and 4.1, as well as the formation of membrane protein polymer (X). Center gel: cells withou aminoacetonitrile, but with Ca^{2+}. Left-hand gel: cells with 1.5 mM Mg^{2+} instead of Ca^{2+}.

A prominent feature of the membrane profile of Ca^{2+}-treated cells, as revealed by electrophoretic analysis, is the appearance of high molecular weight polymers (Fig. 18) which can be shown to contain appreciable amounts of γ-glutamyl-ϵ-lysine cross-links. This effect of Ca^{2+} may be prevented if either histamine, cystamine, aminoacetonitrile or a non-competitive inhibitor of transglutaminase is included in the incubation medium

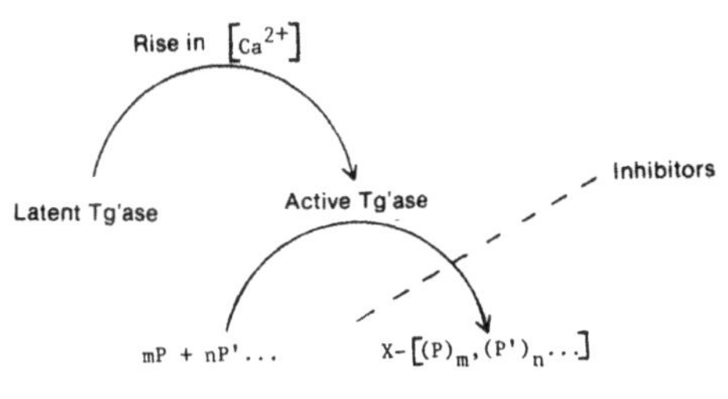

prior to and during Ca^{2+}-loading (11, 29, 96, 112).

The electrophoretic analysis is somewhat complicated by the fact that, in addition to the transglutaminase-catalyzed cross-linking process: a Ca^{2+}-triggered degradation of at least two trans-membrane proteins, i.e., band 3 and glycophorin, takes place (41, 113–115). Pepstatin was shown to inhibit the proteolytic phenomenon rather selectively.

If the Ca^{2+}-loading of the cells was carried out in the presence of ^{14}C-histamine, covalent labelling in the spectrin and band 3 protein regions was observed. Immunological examinations revealed that, in addition to spectrin and band 3, ankyrin (band 2.1), band 4.1 and hemoglobin were also part of the γ-glutamyl-ε-lysine cross-linked polymeric cluster (41; also Bjerrum et al., unpublished results). It appears that the treatment of intact erythrocytes with Ca^{2+} activates an *in situ* covalent fusion of nearest neighbors at the membrane cytoskeletal interface (Fig. 19)[14].

The fusion between the cytoskeletal matrix and membrane proteins by γ-glutamyl-ε-lysine bonds, giving rise to large immobile structures as in the human erythrocyte, might be a general paradigm for cells undergoing terminal differentiation in the

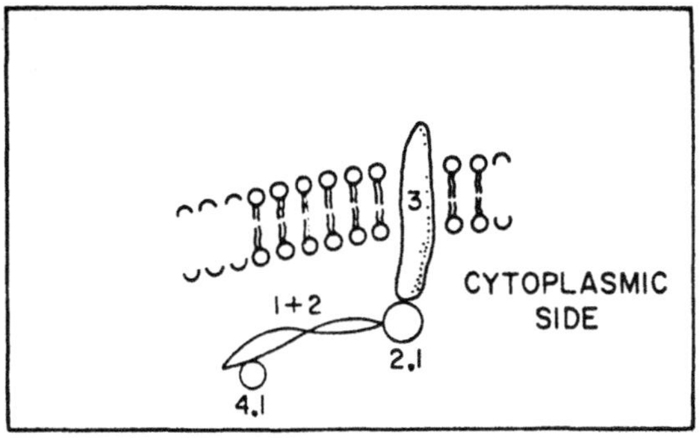

Fig. 19. In intact human erythrocytes the disappearance of monomeric forms of 1 and 2, 2.1, 3 and 4.1 membrane proteins, occurring when intracellular Ca^{2+} is high, can be inhibited by added amines (i.e. by alternate transglutaminase substrates). Radioactive amines label 1 and 2, and 3, indicating that γ-glutamine side chains of these proteins would participate in cross-bridge formation in the absence of added amines.

[14] Catalase activity was also found to be associated with the polymer (116).

process of aging and dying[15]. Transglutaminase-catalyzed cross-linking was shown to play an essential role in the process of cornification in keratinocytes (30, 119). Another system currently under study in our laboratory relates to lens transglutaminase (120–123), with obvious possible overtones for cataract formation[16]. After demonstrating a Ca^{2+}-dependent transglutaminase activity in lens, isolating the enzyme and identifying some of its intrinsic protein substrates, attention was focused on a model system (125). Rabbit lenses were bathed in Ca^{2+}-containing buffer, then solubilized by the addition of urea-dithiothreitol-sodium dodecylsulfate and electrophoresed in 10% polyacrylamide gels, which allowed resolution of protein components below 100 000 chain weight. In comparison to controls where EDTA or Mg^{2+} replaced Ca^{2+}, two unrelated major differences were evident: a new band appeared at about 55 000 MW, and a band disappeared from the 62 000 MW region (Fig. 20). Since the appearance of the 55 000 weight material could be selectively inhibited by histamine, and since this amine would interfere with the γ-glutamyl-ε-lysine type of cross-linking of proteins by the intrinsic transglutaminase, formation of the 55 000 MW protein in the Ca^{2+}-enriched lens is thought to be the result of transglutaminase action. On the other hand, the disappearance of the 62 000 MW protein was attributed to proteolysis, because it could be selectively prevented by leupeptin, an inhibitor of trypsin-like enzymes and of Ca^{2+}-dependent intracellular proteases. The parallel occurrence of a cross-linking and a proteolytic event reminds one of the phenomenon observed in Ca^{2+}-enriched human erythrocytes (114, 115).

Transglutaminase and cell activation. The thrombin-stimulated human platelet and the fertilized sea urchin egg

There could be situations where, following specific stimulation, Ca^{2+}-ions could either enter the

[15] Possibility of transglutaminase participation in the formation of paired helical filaments found in the brains of Alzheimer patients is now under investigation (117, 118).

[16] In addition to γ-glutamyl-ε-lysine bridges (121), the formation of protein bound γ-glutamyl-putrescine was also reported (124) in cataract proteins.

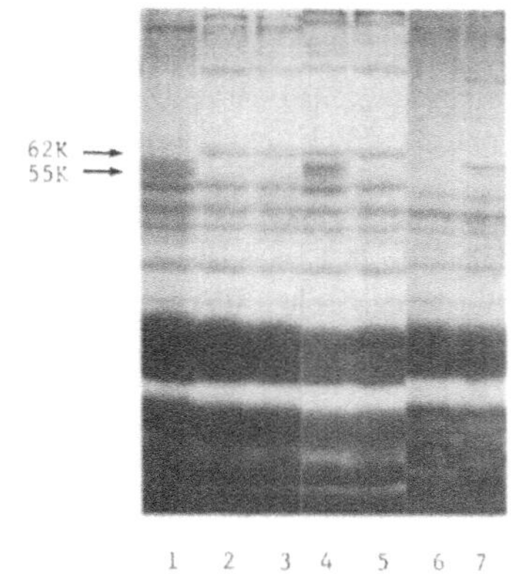

Lane	$CaCl_2$	$MgCl_2$	EDTA	Leupeptin	Histamine
1	+	-	-	-	-
2	-	-	+	-	-
3	-	+	-	-	-
4	+	-	-	+	-
5	-	-	+	+	-
6	+	-	-	-	+
7	-	-	+	-	+

Fig. 20. Cross-linking and proteolysis in Ca^{2+}-enriched rabbit lenses (125). Whole, frozen rabbit lenses were incubated in 500 μl of 50 mM Tris-HCl, 100 mM NaCl, pH 7.5 for 1 hr at 37 °C with the specified inhibitors (1 mM leupeptin or 63 mM histamine) followed by the addition of 100 μl of either $CaCl_2$ (8.3 mM, final concentration), $MgCl_2$ (8.3 mM) or EDTA (1.7 mM) for an additional 3 hr incubation at 37 °C. Following the incubation, the lenses were solubilized in 4 ml of 0.02 M Na_xPO_4, 9 M urea, 40 mM dithiothreitol, 2% sodium dodecylsulfate. Electrophoresis on 10% SDS-polyacrylamide was carried out on 20 μl aliquots of sample using the Laemmli system.

cell from the outside or could become liberated from intra-cellular storage sites in high enough concentration to cause activation of transglutaminase. The phytohemagglutinin-treated lymphocyte might have been an example of this type but, as it was mentioned before, putrescine incorporation in this case was shown to follow a different (hypusine) pathway (62, 64). Considerable effort was focused in recent years on the possible involvement of transglutaminase in receptor-mediated endocytosis with regard to the internalization of ligands such as the epidermal growth factor, α_2-macroglobulin and insulin (126–131; see also chapter by Davies). Activation of transglutaminase in Aplysia nerve may be significant in the axonal transport of proteins (132, 133).

In our own laboratory, two cell activation systems are currently under study. One relates to the formation of γ-glutamyl-ϵ-lysine cross-linked proteins in human platelets following thrombin stimulation and the prevention of this transglutaminase-catalyzed cross-linking process by alternate substrates and inhibitors of the enzyme (134).

The other system is the fertilized egg of the sea urchin *(Arbacia punctulata* or *lixula).* When the eggs are pre-incubated in ^{3}H-putrescine, there is a significant enhancement in the covalent incorporation of the isotope into proteins soon after fertilization (61)[17]. Methylglyoxalbis(guanylhydrazone), which is an inhibitor of S-adenosylmethionine decarboxylase as well as diamine oxidase, had no effect on the incorporation. Emetine, an inhibitor of protein synthesis, also had no effect. Dansylcadaverine, however, was inhibitory. Following acid hydrolysis, the label could be recovered in the form of putrescine and proteolytic degradation of labelled proteins gave rise to γ-glutamylputrescine (Cariello, Wilson and Lorand, unpublished results).

Concluding remarks

It is now clear that Ca^{2+}-dependent transamidases of the endo-γ-glutamine:ϵ-lysine transferase type play important roles in a number of biological systems. We have particularly good understanding about the modes of operation of these enzymes when the cross-linking of protein substrates by γ-ϵ peptide bonds leads to the formation of identifiable polymeric structures in the extra-cellular environment (e.g. as in clot formation). However, the intracellular covalent fusion of membrane and cytoplasmic components by similar bonds is also a rather well established phenomenon in Ca^{2+}-enriched cells, such as human erythrocytes and keratinocytes, serving as a general paradigm for structural changes accompanying cell aging.

Another post-translational role for these enzymes is to modify γ-glutamine residues of specific

[17] Washing the eggs before fertilization and removing them from the isotope-containing medium made no difference. Thus, the observed effect could not be due to increased permeability of the fertilized egg to putrescine.

protein substrates upon cell activation by the incorporation of biological amines (e.g. putrescine). Formation of protein-bound γ-glutamylputrescine by such reaction was shown to occur during the process of egg fertilization in the sea urchin (61) and has been suggested to accompany axonal transport in Aplysia neuron (132, 133). Mouse peritoneal macrophages cultured in mouse serum accumulate very high levels of transglutaminase (1–2% of cellular proteins) due to accelerated synthesis (Davies, this volume).

The covalent bonding of putrescine and spermidine to proteins in the cyst coat of the protozoan *Colpoda steinii* (135) may or may not be due to transglutaminase action. The origin of polyamine (primarily putrescine)-conjugated peptide material in human plasma (136) and of free γ-glutamyl-putrescine in the mammalian brain (137) also awaits elucidation. More than seventy γ-glutamyl derivatives of amino acids and amines have been isolated from plants, many of them from mushrooms (138). Could some of these represent proteolytic breakdown products of cell proteins which became derivatized with the amines in a transglutaminase-catalyzed earlier metabolic step? Or, did they simply arise by conventional biosynthetic reactions requiring either activation by ATP or the participation of more typical γ-glutamyltransferase? Current evidence seems to favor the latter two pathways (138).

Acknowledgments

USPHS Career Award HL-03512, NIH grants HL-02212, AM-25412, and EY-03542. Special thanks are due to Pauline T. Velasco for her kind assistance.

References

1. Lorand, L. and Stenberg, P., 1976. Handbook of Biochemistry and Molecular Biology (Fasman, G. D., ed.), Vol. II, pp. 669–685, CRC Press, Cleveland.
2. Folk, J. E., 1980. Ann. Rev. Biochem. 49: 517–531.
3. Clark, D. D., Mycek, M. J., Neidle, A. and Waelsch, H., 1959. Arch. Biochem. Biophys. 79: 338–354.
4. Neidle, A. and Acs, G., 1961. Fed. Proc. 20: 234.
5. Kornguth, S. E. and Waelsch, H., 1963. Nature 198: 188–189.
6. Myhrman, R. V., 1973. Ph.D. dissertation, Northwestern University, Evanston, Illinois.
7. Lorand, L., Losowsky, M. and Miloszewski, K., 1980. Progress in Hemostasis and Thrombosis (Spaet, T. H., ed.), Vol. 5, pp. 245–290, Grune and Stratton, Inc., New York.
8. Shen, L. and Lorand, L., 1983. J. Clin. Invest. 71: in press.
9. Lorand, L., Doolittle, R. F., Konishi, K. and Riggs, S. K., 1963. Arch. Biochem. Biophys. 102: 171–179.
10. Williams-Ashman, H. G., Notides, A. C., Pabalan, S. S. and Lorand, L., 1972. Proc. Natl. Acad. Sci. USA 69: 2322–2325.
11. Lorand, L., Weissmann, L. B., Epel, D. L. and Bruner-Lorand, J., 1976. Proc. Natl. Acad. Sci. USA 73: 4479–4481.
12. Rice, R. H. and Green, H., 1977. Cell 11: 417–422.
13. Lorand, L., Campbell, L. K. and Robertson, B. J., 1972. Biochemistry 11: 434–438.
14. Lorand, L., Maldonado, N., Fradera, J., Atencio, A., Robertson, B. and Urayama, T., 1972. Brit. J. Haematol. 23: 17–27.
15. Lopaciuk, S., Bykowska, K., McDonagh, J. M., McDonagh, R. P., Yount, W. J., Fuller, C. R., Cooperstein, L., Gray, A. Jr. and Lorand, L., 1978. J. Clin. Invest. 61: 1196–1203.
16. Rosenberg, R. D., Colman, R. W. and Lorand, L., 1974. Brit. J. Haematol. 26: 269–284.
17. Lorand, L., Parameswaran, K. N., Stenberg, P., Tong, Y. S., Velasco, P. T., Jonsson, N. A., Mikiver, L. and Moses, P., 1979. Biochemistry 18: 1756–1765.
18. Russell, D. H., and Womble, J. R., 1982. Life Sci. 30: 1499–1508.
19. Williams-Ashman, H. G. and Canellakis, Z. N., 1979. Perspect. Biol. Med. 22: 421–453.
20. Russell, D. H., 1981. Biochem. Biophys. Res. Commun. 99: 1167–1172.
21. Tyagi, A. K., Tabor, H. and Tabor, C. W., 1982. Biochem. Biophys. Res. Commun. 109: 533–540.
22. Lockridge, O., 1971. Ph.D. dissertation, Northwestern University, Evanston, Illinois.
23. Birckbichler, P. J., Orr, G. R., Conway, E. and Patterson, M. K. Jr., 1977. Cancer Res. 37: 1340–1344.
24. Lorand, L., 1972. Ann. N.Y. Acad. Sci. 202: 6–30.
25. Fuller, G. M. and Doolittle, R. F., 1971. Biochemistry 10: 1311–1315.
26. Mosher, D., Schad, P. E. and Kleinman, H. K., 1979. J. Clin. Invest. 64: 781–787.
27. Ichinose, A. and Aoki, N., 1982. Biochim. Biophys. Acta 706: 158–164.
28. Tamaki, T. and Aoki, N., 1982. J. Biol. Chem. 257: 14767–14772.
29. Lorand, L., Siefring, G. E. Jr. and Lowe-Krentz, L., 1978. J. Supramol. Struct. 9: 427–440.
30. Lorand, L., 1982. In: Membranes and Transport (Martonosi, A., ed.), Vol. 2, pp. 473–478.
31. Lorand, L. and Chenoweth, D., 1969. Proc. Natl. Acad. Sci. USA 63: 1247–1252.
32. Chen, R. and Doolittle, R. F., 1971. Biochemistry 10: 4486–4491.

33. Lorand, L., Chenoweth, D. and Gray, A., 1972. Ann. N.Y. Acad. Sci. 202: 155–171.
34. Fretto, L. J., Ferguson, E. W., Steinman, H. M. and McKee, P. A., 1978. J. Biol. Chem. 253: 2184–2195.
35. Lorand, L., Rule, N. G., Ong, H. H., Furlanetto, R., Jacobsen, A., Downey, J., Oner, N. and Bruner-Lorand, J., 1968. Biochemistry 7: 1214–1223.
36. Jilek, F. and Hormann, H., 1977. Hoppe Seyler's Z. Physiol. Chem. 358: 1165–1168.
37. Mosher, D. F., Schad, P. E. and Vann, J. M., 1980. J. Biol. Chem. 255: 1181–1188.
38. Akiyama, S. K., Yamada, K. M. and Hayashi, M., 1981. J. Supramol. Struct. 17: 263–275.
39. Petersen, T. E., Thorgersen, H. C., Skorstengaard, K., Vibe-Pedersen, K., Sahl, P., Sottrup-Jensen, L. and Magnusson, S., 1983. Proc. Natl. Acad. Sci. USA 80: 137–141.
40. Molnar, J., Lai, M. Z., Siefring, G. E. Jr. and Lorand, L., 1983. Unpublished results.
41. Bjerrum, O. J., Hawkins, M., Swanson, P., Griffin, M. and Lorand, L., 1981. J. Supramol. Struct. Cell. Biochem. 16: 289–301.
42. Campbell-Wilkes, L., 1973. Ph.D. dissertation, Northwestern University, Evanston, Illinois.
43. Lorand, L., Downey, J., Gotoh, T., Jacobsen, A. and Tokura, S., 1968. Biochem. Biophys. Res. Commun. 31: 222–229.
44. Matacic, S. and Loewy, A. G., 1968. Biochem. Biophys. Res. Commun. 30: 356–362.
45. Pisano, J. J., Finlayson, J. S. and Peyton, M. P., 1969. Biochemistry 8: 871–876.
46. Griffin, M., Wilson, J. and Lorand, L., 1982. Anal. Biochem. 124: 406–413.
47. Tack, B. F., Janatova, J., Thomas, M. L., Harrison, R. A. and Hammer, C. H., 1981. Methods in Enzymology (Lorand, L., ed.), Vol. 80, pp. 64–101, Academic Press, New York.
48. Barrett, A. J., 1981. Methods in Enzymology (Lorand, L., ed.), Vol. 80, pp. 737–754, Academic Press, New York.
49. Mortensen, S., Sottrup-Jensen, L., Hansen, H. F., Rider, D., Petersen, T. E. and Magnussen, S., 1981. FEBS Lett. 129: 314–317.
50. Lorand, L., Lockridge, O. M., Campbell, L. K., Myhrman, R. and Bruner-Lorand, J., 1971. Anal. Biochem. 44: 221–231.
51. Curtis, C. G. and Lorand, L., 1976. Methods in Enzymology (Lorand, L., ed.), Vol. 45, pp. 177–191, Academic Press, New York.
52. Gorman, J. J. and Folk, J. E., 1980. J. Biol. Chem. 255: 419–427.
53. Lorand, L., Campbell-Wilkes, L. K. and Cooperstein, L., 1972. Anal. Biochem. 50: 623–631.
54. Lorand, L. and Campbell, L. K., 1971. Anal. Biochem. 44: 207–220.
55. Lorand, L., Ong, H. H., Lipinski, B., Rule, N. G., Downey, J. and Jacobsen, A., 1966. Biochem. Biophys. Res. Commun. 25: 629–637.
56. Lorand, L., Curtis, C. G., Domanik, R. A., Gray, A., Bruner-Lorand, J., Myhrman, R. and Campbell-Wilkes, L. K., 1973. Fed. Proc. 32: 274, abstr. 338.
57. Lorand, L., Parameswaran, K. N., Velasco, P. T., Hsu, L. K.-H. and Siefring, G. E. Jr., 1983. Anal. Biochem., in press.
58. Lorand, L., Gray, A. J., Brown, K., Credo, R. B., Curtis, C. G., Domanik, R. A. and Stenberg, P., 1974. Biochem. Biophys. Res. Commun. 56: 914–922.
59. Lorand, L., Siefring, G. E. Jr., Tong, Y. S., Bruner-Lorand, J. and Gray, A. J. Jr., 1979. Anal. Biochem. 93: 453–458.
60. Folk, J. E., 1982. Methods in Enzymology (Purich, D., ed.), Vol. 87, pp. 36–42, Academic Press, New York.
61. Takeuchi, K., Cariello, L. and Lorand, L., 1982. Fed. Proc. 41: 1453, abstr. 6913.
62. Folk, J. E., Park, M. H., Chung, S. I., Schrode, J., Lester, E. P. and Cooper, H. L., 1980. J. Biol. Chem. 255: 3695–3700.
63. Chen, K. Y. and Liu, A. Y. C., 1981. FEBS Lett. 134: 71–74.
64. Park, M. H., Cooper, H. L. and Folk, J. E., 1981. Proc. Natl. Acad. Sci. USA 78: 2869–2873.
65. Zeller, E. A., 1963. The Enzymes, 2nd ed. (Boyer, P. D., Lardy, H. and Myrback, K., eds.), Vol. 8, pp. 313–335, Academic Press, New York.
66. Bardsley, W. G., James, M., Crabbe, C. and Scott, I. V., 1974. Biochem. J. 139: 169–181.
67. Thomas, E. L., Jefferson, M. M. and Grisham, M. B., 1982. Biochemistry 21: 6299–6308.
68. Folk, J. E. and Cole, P. W., 1965. J. Biol. Chem. 240: 2951–2960.
69. Waelsch, H., 1962. Monoamines et Systeme Nerveux Central, pp. 93–104, Masson, Geneva.
70. Folk, J. E., Cole, P. W. and Mullooly, J. P., 1967. J. Biol. Chem. 242: 2615–2621.
71. Folk, J. E., Cole, P. W. and Mullooly, J. P., 1967. J. Biol. Chem. 242: 4329–4333.
72. Chung, S. I., Shrager, R. I. and Folk, J. E., 1970. J. Biol. Chem. 245: 6424–6435.
73. Gross, M., Whetzel, N. K. and Folk, J. E., 1977. J. Biol. Chem. 252: 3752–3759.
74. Lorand, L., Chou, C-H. J. and Simpson, I., 1972. Proc. Natl. Acad. Sci. USA 69: 2645–2648.
75. Curtis, C. G., Stenberg, P., Brown, K. L., Baron, A., Chen, K., Gray, A., Simpson, I. and Lorand, L., 1974. Biochemistry 13: 3257–3262.
76. Stenberg, P., Curtis, C. G., Wing, D., Tong, Y. S., Credo, R. B., Gray, A. and Lorand, L., 1975. Biochem. J. 147: 155–163.
77. Parameswaran, K. N. and Lorand, L., 1981. Biochemistry 20: 3703–3711.
78. Jencks, W. P., 1968. Handbook of Biochemistry (Sober, H. A., ed.), pp. J-146, The Chemical Rubber Co., Cleveland.
79. Curtis, C. G., Stenberg, P., Chou, C-H. J., Gray, A., Brown, K. L. and Lorand, L., 1973. Biochem. Biophys. Res. Commun. 52: 51–56.
80. Curtis, C. G., Brown, K. L., Credo, R. B., Domanik, R. A., Gray, A., Stenberg, P. and Lorand, L., 1974. Biochemistry 13: 3774–3780.

81. Chung, S. I., Lewis, J. S. and Folk, J. E., 1974. J. Biol. Chem. 249: 940–950.
82. Lorand, L. and Ong, H. H., 1966. Biochemistry 5: 1747–1753.
83. Cheung, W. Y., 1980. Science 207: 19–27.
84. Puszkin, E. G., Raghuraman, V., Jenkins, C. S. P. and Grossman, L., 1982. Fed. Proc. 41: 1235, abstr. 5637.
85. Credo, R. B., Stenberg, P., Tong, Y. S. and Lorand, L., 1976. Fed. Proc. 35: 1630, abstr. 1390.
86. Gilbert, H. F., 1982. J. Biol. Chem. 257: 12086–12091.
87. Ogawa, H. and Goldsmith, L. A., 1976. J. Biol. Chem. 251: 7281–7288.
88. Martelo, O. J., Woo, S. L. C., Reimann, E. M. and Davie, E. W., 1970. Biochemistry 9: 4807–4813.
89. Reimann, E. M., Walsh, D. A. and Krebs, E. G., 1971. J. Biol. Chem. 246: 1986–1995.
90. Williams-Ashman, H. G., Wilson, J., Beil, R. E. and Lorand, L., 1977. Biochem. Biophys. Res. Commun. 79: 1192–1198.
91. Moger, W. H., 1982. Can. J. Physiol. Pharm. 60: 858–861.
92. Mato, J. S., Pencer, D., Vasanthakumar, G., Schiffmann, E. and Pastan, I., 1983. Proc. Natl. Acad. Sci. USA 80: in press.
93. Folk, J. E. and Gross, M., 1971. J. Biol. Chem. 246: 6683–6691.
94. Holcenberg, J. S., Ring, B. J. and Zirneskie, J. D., 1981. Fed. Proc. 40: 1687 (Abstract 852).
95. Samama, M., Soria, J., Soria, C., Maamer, M. and Otero, A. M., 1979. Progress in Chemical Fibrinolysis and Thrombolysis (Davidson, J. F., Cepelak, V., Samama, M. and Desnoyers, P., eds.), Vol. IV, pp. 235–240, Churchill Livingstone, Edinburgh.
96. Siefring, G. E. Jr., Apostol, A. B., Velasco, P. T. and Lorand, L., 1978. Biochemistry 17: 2598–2604.
97. Lorand, L., 1977. Haemostasis: Biochemistry, Physiology and Pathology (Ogston, D. and Bennett, B., eds.), pp. 405–423, Wiley, London.
98. Schwartz, M. L., Pizzo, S. V., Hill, R. L. and McKee, P. A., 1973. J. Biol. Chem. 248: 1395–1407.
99. Credo, R. B., Curtis, C. G. and Lorand, L., 1978. Proc. Natl. Acad. Sci. USA 75: 4234–4237.
100. Lorand, L., Credo, R. B. and Janus, T. J., 1981. Methods in Enzymology (Lorand, L., ed.), Vol. 80, pp. 333–341, Academic Press, New York.
101. Lorand, L. and Konishi, K., 1964. Arch. Biochem. Biophys. 105: 58–67.
102. Takagi, T. and Doolittle, R. F., 1974. Biochemistry 13: 750–756.
103. Credo, R. B., Curtis, C. G. and Lorand, L., 1981. Biochemistry 20: 3770–3778.
104. Janus, T. J., Lewis, S. D., Lorand, L. and Shafer, J. A., 1983. Fed. Proc. 42: 1032, abstr. 4335.
105. Doolittle, R. F. and Lorand, L., 1962. Biol. Bull. 132: 481.
106. LaCelle, P. L., 1971. The Red Cell Membrane (Weed, R. I., Jaffe, E. R. and Miescher, P. A., eds.), pp. 107–123, Grune & Stratton, New York.
107. LaCelle, P. L., Kirkpatrick, F. H., Udkow, M. P. and Arkin, B., 1973. Red Cell Shape (Bessis, M., Weed, R. I. and Leblond, P. F., eds.), pp. 69–78, Springer Verlag, New York.
108. Eaton, J. W., Shelton, T. D., Swofford, H. L., Kolpin, C. E. and Jacob, H. S., 1973. Nature 246: 105–106.
109. Palek, J., 1973. Blood 42: 988.
110. White, J. G., 1976. Sem. Hematol. 13: 121–132.
111. Smith, B. D., LaCelle, P. L, Siefring, G. E. Jr., Lowe-Krentz, L. and Lorand, L., 1981. J. Membr. Biol. 61: 75–80.
112. Lorand, L., Siefring, G. E. Jr. and Lowe-Krentz, L., 1979. Sem. Hematol. 16: 65–74.
113. Lorand, L., Siefring, G. E. Jr. and Lowe-Krentz, L., 1980. Membrane Transport in Erythrocytes (Lassen, U. V., Ussing, H. H. and Wieth, J. O., eds.), pp. 285–299, Munksgaard, Copenhagen.
114. Lorand, L., Bjerrum, O. J., Hawkins, M., Lowe-Krentz, L. and Siefring, G. E. Jr., 1982. Fed. Proc. 41: 974, abstr. 4115.
115. Lorand, L., Bjerrum, O. J., Hawkins, M., Lowe-Krentz, L. and Siefring, G. E. Jr., 1983. J. Biol. Chem., in press.
116. Bjerrum, O. J., Hawkins, M., Michalska, M. and Lorand, L., 1983. Fed. Proc. 42: 604, abstr. 1828.
117. Selkoe, D. J., Ihara, Y. and Salazar, F., 1982. Science 215: 1243–1245.
118. Selkoe, D. J., Abraham, C. and Ihara, Y., 1982. Proc. Natl. Acad. Sci. USA 79: 6070–6074.
119. Rice, R. H. and Green, H., 1978. J. Cell Biol. 76: 705–711.
120. Lorand, L., Siefring, G. E. Jr., Hsu, L., Quraishi, A. and Rafferty, N. S., 1980. Fed. Proc. 39: 1695, abstr. 475.
121. Lorand, L., Hsu, L. K. H., Siefring, G. E. Jr. and Rafferty, N. S., 1981. Proc. Natl. Acad. Sci. USA 78: 1356–1360.
122. Azari, P., 1980. Invest. Ophthalmol. Vis. Sci. Suppl., p. 55, abstr. 12.
123. Azari, P., Rahim, I. and Clarkson, D. P., 1981. Curr. Eye Res. 1: 463–469.
124. Kremzner, L. T., Roy, D. and Spector, A., 1982. Fed. Proc. 41: 676, abstr. 2374.
125. Conrad, S. M., Velasco, P. T., Michalska, M., Hsu, L. and Lorand, L., 1983. Invest. Ophthalmol. Vis. Sci. Suppl. 24: 157 (Abstract 38).
126. Maxfield, F. R., Willingham, M. C., Davies, P. J. A. and Pastan, I., 1979. Nature 277: 661–663.
127. Davies, P. J. A., Davies, D. R., Levitzki, A., Maxfield, F. R., Milhaud, P., Willingham, M. C. and Pastan, I. H., 1980. Nature 283: 162–167.
128. Haigler, H. T., Maxfield, F. R., Willingham, M. C. and Pastan, I., 1980. J. Biol. Chem. 255: 1239–1241.
129. Haigler, H. G., Willingham, M. C. and Pastan, I., 1980. Biochem. Biophys. Res. Commun. 94: 630–637.
130. Chuang, D.-M., 1981. J. Biol. Chem. 256: 8291–8293.
131. Van Leuven, F., Cassiman, J.-J. and Van den Berghe, H., 1980. Cell 20: 37–43.
132. Ambron, R. T. and Kremzner, L. T., 1982. Proc. Natl. Acad. Sci. USA 79: 3442–3446.
133. Kremzner, L. T. and Ambron, R. T., 1982. J. Neurochem. 38: 1719–1727.
134. Bruner-Lorand, J., Bjerrum, O. J. and Lorand, L., 1982. Fed. Proc. 41: 1118, abstr. 4959.
135. Tibbs, J., 1982. Eur. J. Biochem. 122: 535–539.
136. Seale, T. W., Chan, W.-Y., Shulka, J. and Rennert, O. M., 1979. Arch. Biochem. Biophys. 198: 164–174.

137. Nakajima, T., Kakimoto, Y., Tsuji, M. and Konishi, H., 1976. J. Neurochem. 26: 115–118.
138. Kasai, T. and Larsen, P. I., 1980. Progress in the Chemistry of Organic Natural Products (Herz, W., Grisebach, H. and Kirby, G. W., eds.), Vol. 39, pp. 173–285, Springer-Verlag, New York.
139. Tong, Y. S., 1980. Ph. D. dissertation, Northwestern University, Evanston, Illinois.

Received 29 March 1983.

Molecular and Cellular Biochemistry 58, 37–49 (1984).

Detection of ε(γ-glutamyl) lysine

M. Griffin[a] and J. Wilson[b]
[a] *Department of Life Sciences, Trent Polytechnic, Nottingham, NG1 4BU, England;* [b] *Department of Biochemistry, Molecular and Cell Biology, Northwestern University, Evanston, IL 60201, U.S.A.*

Summary

Detection of ε(γ-glutamyl) lysine crosslinks is not only necessary for establishing the importance of the dipeptide as a post-translational modification of proteins, but provides information as to the importance of the transglutaminase enzyme in a biological system. The crosslink may be detected using both indirect and direct methodology. Indirect methods for its detection include measurement of 'masked lysines' within a protein, detection of polymer formation by gel-electrophoresis and the inhibition of crosslinking by the incorporation of small molecular weight amines into the substrate protein. Direct methods for the detection of ε(γ-glutamyl) lysine require the actual isolation of the dipeptide following its release from the sample protein by exhaustive proteolytic digestion. Separation of the dipeptide from other components of the digest may be achieved by either ion-exchange chromatography or gel filtration and its qualitative identification achieved by techniques such as paper-electrophoresis or thin layer chromatography. Quantitative estimation of ε(γ-glutamyl) lysine normally involves its further separation by ion-exchange chromatography and its post-column detection following derivatisation with ninhydrin. More recent techniques include pre-column derivatisation of the dipeptide with fluorogenic reagents such as σ-pthalaldehyde and separation by reverse phase HPLC. With the recent advances in liquid chromatography resulting in the improved resolution of amino acids, increased sensitivity, rapid analysis times, and small sample sizes, it appears likely that direct quantitation of ε(γ-glutamyl) lysine will be the preferred method for the future.

Introduction

The ability of transglutaminase enzymes to catalyse ε(γ-glutamyl) lysine crosslinks in proteins has resulted in this product becoming a means of assessing the enzyme's importance in a biological system.

Detection of ε(γ-glutamyl) lysine within a system is therefore not only merited by the dipeptide's importance as a post-translational protein crosslink, but serves as an alternative means of estimating the activity of transamidating enzymes of the endo-γ-glutamine:ε-lysine transferase type, of which transglutaminase is an example.

Measurement of ε(γ-glutamyl) lysine in crosslinked fibrin (1, 2, 3), hair proteins (4), epidermal proteins (5), and the clotted seminal vesicle protein of guinea pig (6) has enabled physiological functions to be ascribed to the transglutaminase enzymes which catalyse these reactions.

Transglutaminase enzymes still without a proven physiological function include the tissue enzyme (7) and the factor XIIIa enzyme found in platelets and in the placenta (8). Isolation of ε(γ-glutamyl) lysine has now been achieved from a wide variety of biologcial sources (Table 1), but like the tissue transglutaminase, no definite physiological function can always be ascribed to its presence.

To establish the cellular importance of the ε(γ-glutamyl) lysine crosslink and ultimately the enzymes which catalyse its formation, continued investigations into the dipeptide's presence in tissues and cells of both normal and pathological states are

Table 1. ε(γ-glutamyl) lysine in selected proteins and biological materials.

Protein or biological material	Source	Amount of ε(γ-glutamyl) lysine (moles per 10^6 g of protein)	Reference
Clotted vesicular protein	Seminal clot (guinea pig)	370	6
Fibrin	Human plasma	16–20	28, 29
Keratin	Wool (Merino sheep)	15	19
Fibrous protein fraction	Stratum corneum (human)	9	9
Polymeric lens protein	Cataract lens (human)	4–5	11
Plasma membrane	L-cells (mouse)	1.7	20
Endoplasmic reticulum	L-cells (mouse)	0.67	20
Skeletal myofibrils	Tissue-cultured embryonic chick	0.29–0.33	30
Skeletal myofibrils	Mouse	0.65–0.78	30
E. coli B		0.11–0.15	30

still necessary. It is therefore important that techniques of the required sensitivity and accuracy be readily available for the quantitative assessment of this important crosslink.

Direct analysis of ε(γ-glutamyl) lysine

Indirect as well as direct approaches have been used for the quantitative assessment of ε(γ-glutamyl) lysine but the direct approach of isolating the dipeptide from crosslinked protein still remains the ultimate method for its analysis. The protocol used in this procedure has changed little from that used in the early investigations upon crosslinked fibrin (1, 2, 3). The flow diagram (Fig. 1) illustrates the number of steps which are generally required before final analysis of the dipeptide can be made.

Not included in the protocol is the actual isolation of the sample material, a procedure which is

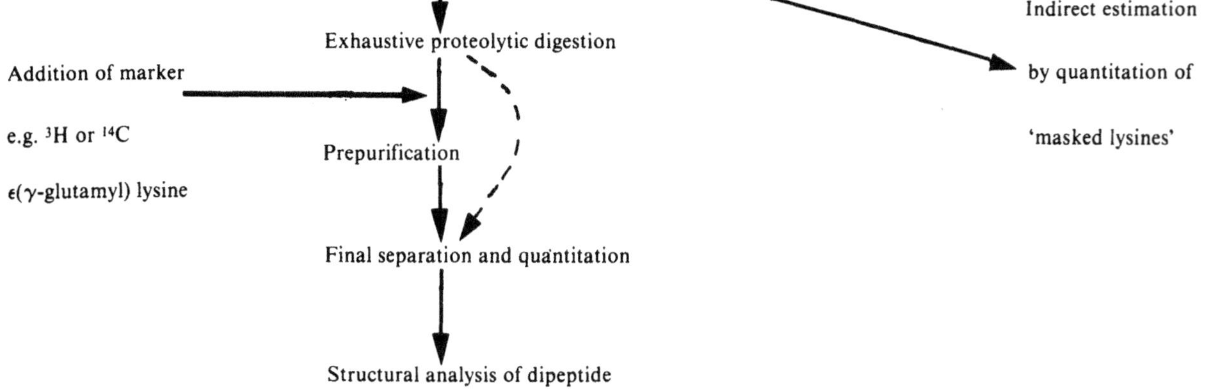

Fig. 1. General procedures for the direct estimation of ε(γ-glutamyl) lysine.

largely dependent upon its source of origin and the degree of purity required. These techniques may simply involve precipitation of total protein with agents such as cold aqueous trichloroacetic acid or more complex procedures involving selective solubilisation of protein by detergents and/or separation by column chromatography (9, 10, 11).

Expression of the final amount of crosslink detected in the sample in terms of protein will depend upon determination of the amount of protein present in the original material. Acid hydrolysis of a known weight of material followed by determination of its amino acid content provides an accurate method of protein estimation (14).

A further technique which is dependent upon the type and source of protein to be analysed is the reduction and derivatisation of disulphide linkages. If present in large amounts in the sample material then reduction and blocking of these linkages should be performed before commencement of proteolytic digestion. The derivatisation of keratin (9) and more recently, protein polymers from cataract lens (12) by S-carboxymethylation (13) are such examples.

Exhaustive proteolytic digestion

Release of the dipeptide from the sample protein is achieved by exhaustive digestion of the protein with proteolytic enzymes. The importance of this procedure is well demonstrated by the work of Schweet (15, 16) who was unable to isolate $\epsilon(\gamma$-glutamyl) lysine following incorporation of ^{14}C-lysine into the soluble fraction of guinea pig liver. This author used a combination of proteolysis and partial acid hydrolysis and it was not until later that Korngruth et al. (17) demonstrated the unsuitability of partial acid hydrolysis for the isolation of the crosslink. These authors suggested the use of proteolytic enzymes as an alternative and made the crucial observation that the dipeptide was resistant to papain, chymotrypsin, trypsin, leucine aminopeptidase and pronase. Since this finding many different combinations of proteolytic enzymes have been reported for the release of the dipeptide from sample material.

The enzymatic digestion of crosslinked human fibrin included the sequential use of trypsin, pronase twice, leucine aminopeptidase and prolidase (18). The digestion of wool (19) included the sequential use of pepsin, pronase, aminopeptidase M and prolidase, while the digestion of cell membrane proteins required only pronase followed by leucine aminopeptidase (20). A combination of proteolytic enzymes found effective for the digestion of the Ca^{2+}-mediated polymer of red blood cell membranes (10), protein polymers from cataract lens (11, 12) and the clotted seminal vesicle protein from guinea pig (21) included subtilisin, pronase, leucine aminopeptidase, prolidase and either carboxypeptidase A or Y.

In the majority of cases and in the examples quoted, proteolytic digestion is performed in the first instance with proteases such as pronase, subtilisin or trypsin which hydrolyse the protein into a mixture of small peptides. The digestion is then followed by a series of exopeptidases which hydrolyse the peptides into their constitutive amino acids.

Pronase which is a common component of most digestion protocols contains both endo- and exopeptidases of broad specificity (22). Other proteases of broad specificity include both subtilisin and papain which serve as good alternatives to trypsin, pepsin and chymotrypsin (23, 24).

The choice of exopeptidase may include leucine aminopeptidase, aminopeptidase M, carboxypeptidase A or carboxypeptidase Y. Each of these enzymes has a broad substrate specificity. Of the N-terminal exopeptidases leucine aminopeptidase requires preactivation with either Mg^{2+} or Mn^{2+} and needs Mg^{2+} for its stabilisation during digestion (25). Apart from carboxypeptidase Y these exopeptidases show little or no hydrolysis of peptide bonds containing the imino group of proline. Hydrolysis of the X-Pro- (X refers to any other amino acid) bond can be achieved by the inclusion of prolidase (26) in the digestion mixture. Like leucine aminopeptidase this enzyme requires preactivation with Mn^{2+}.

Mixtures containing the proteolytic enzymes, with and without the addition of standard $\epsilon(\gamma$-glutamyl) lysine, should be included as controls. A useful indication of enzymic digestion may be obtained by measurement of the ninhydrin positive material that is released during hydrolysis. This sample may then be compared to a duplicate sample which has undergone acid hydrolysis (27). It should be remembered that incomplete digestion can lead to a conclusion that the dipeptide is not present, or can produce low values for its estimation.

Table 2. Ion-exchange systems for the isolation and quantitation of ε(γ-glutamyl) lysine.

Biological material	Amino acid analyser	Ion-exchange column	Elution buffers	Method of detection and quantitation	Reference
Fibrin (human)	Spinco 120C	Column 0.9 × 52 cm packed with PA-28 resin	Isocratic elution with 0.2 M sodium citrate buffer, pH 3.83	Ninhydrin/isotope dilution	1
Clotted vesicular protein (guinea pig) Ca^{2+}-mediated polymer from erythrocyte membranes (human)	Beckman 120C	Column 0.9 × 55 cm packed with Custom AA-15 resin	Isocratic elution with 0.2 M sodium citrate buffer, pH 3.31 at 55 °C	Ninhydrin	6, 10
Fibrin (human)	Spinco 120C	Column 0.9 × 58 cm packed with UR-30 resin	Isocratic elution with 0.2 M sodium citrate buffer, pH 4.12 at 32 °C	Ninhydrin/isotope dilution	28
Epidermal protein (rat)	Hitachi KLA-3B	Column 0.9 × 52 cm packed with Custom 2612 resin	Isocratic elution with 0.2 M sodium citrate buffer, pH 3.31 at 55 °C	Ninhydrin	31
Fibrin (bovine) Membrane proteins (mouse L-cells)	Technicon	Column 0.6 × 130 cm packed with Type A resin	Buffer gradient of sodium citrate; 0.2 M, pH 2.9 to 0.8 M, pH 5.0 delivered from a nine-chamber varigrad	Ninhydrin/isotope dilution	2, 20
Citrulline containing medulla protein from hair (guinea pig)	Technicon	Column 0.6 × 130 cm packed with Type A resin	Buffer gradient of sodium citrate, pH 2.96 to pH 10.0 at 60 °C, delivered from a nine-chamber varigrad	Ninhydrin	39
Normal and transformed fibroblasts (human) Crosslinked transglutaminase (guinea pig)	Glenco Model 100AS	Column 0.2 × 60 cm packed with Dionex DC-4A resin (Durrum Chemical Co.)	Buffer gradient with sodium citrate buffers (Pico II buffer system, Pierce Chemical Co.) excluding the additive from buffer B	σ-Pthalaldehyde/isotope dilution	38, 32
Clotted vesicular protein (guinea pig) Protein polymers from cataract lens	Glenco Model MM Series	Column 0.325 × 30 cm packed with Dionex DC-4A resin (Durrum Chemical Co.)	Buffer gradient with lithium citrate buffers. (Pico IV buffer system, Pierce Chemical Co.)	Ninhydrin	21, 12

Addition of markers to the digested material

The addition of a known amount of radiolabelled ε(γ-glutamyl) lysine to the digested material serves as a useful marker during the isolation of the dipeptide. It may also be used to quantitate the yield of dipeptide in the final analysis by the method of isotope dilution. Reference to Table 2 illustrates that isotope dilution is a method frequently used for quantitation of ε(γ-glutamyl) lysine.

Radiolabelled dipeptide may be prepared enzymically as reported by Matacic and Loewy (2) or by chemical synthesis as reported by Pisano et al. (1, 28). Enzymatic synthesis of either ^{14}C- or ^{3}H-ε(γ-glutamyl) lysine uses factor XIIIa to incorporate the appropriately labelled lysine into casein. The incorporation is normally very slow and requires incubation periods of up to 48 hours. The radiolabelled dipeptide is isolated following the sequential proteolytic digestion of the casein with pronase and leucine aminopeptidase.

Isolation of ε(γ-glutamyl) lysine

Direct analysis of the digested material for ε(γ-glutamyl) lysine without the need for further purification procedures would be the preferred technique, but in many cases this is practically impossible.

The amount of dipeptide found in proteins is very often a small percentage of the total amino acid content (Table 1). In cellular proteins this value may be as low as 0.11–1.7 moles of crosslink per 10^6 g of protein (20, 30). Procedures which eliminate a large percentage of the amino acids and interfering peptides are therefore often required before final analysis of the ε(γ-glutamyl) lysine can be performed. The extent of this purification procedure is also dependent upon the resolution, sensitivity and loading capacity of the final analytical procedure to be used.

Examples of such prepurification procedures have included column chromatography based upon separation by size, with column packings like Biogel P_2 (Biorad Labs), used in the analysis of insoluble epidermal proteins of rat (31) and Sephadex G-10 used in the analysis of wool proteins (19). By far the greater number of prepurification procedures have dependend upon ion-exchange chromatography, with the use of a variety of resin types and elution buffers. Such examples include the use of Biorad AG-50 × 4 or HPB-80 (Hamilton Co.) resins eluted with a gradient of pyridine-acetate buffers (9, 32) used in the analysis of keratin and cellular proteins respectively. The resin Dowex 50 X-8 eluted with a gradient of sodium citrate buffers has also been used in the investigations upon cross-linked fibrin (2).

Perhaps the most interesting technique reported is the adsorption of excess amino acids by a species of *Bacillus*, a procedure used in the enzymic preparation of radiolabelled ε(γ-glutamyl) lysine (2). Whatever the technique of choice the presence of radiolabelled ε(γ-glutamyl) lysine in the sample material serves as a useful marker and may also be used as a measure of dipeptide recovery. When radiolabelled dipeptide is not available, alternative procedures have used the addition of radiolabelled amino acids known to co-elute with or immediately precede the elution of the dipeptide in the purification procedure chosen. This procedure has been used in the purification of ε(γ-glutamyl) lysine from the red blood cell membrane polymer isolated from a patient with sickle cell anemia (Lorand, personal communication). In this instance the amino acid leucine immediately preceded the elution of the dipeptide during ion-exchange chromatography (conditions as those in Fig. 2), thus enabling radiolabelled leucine to be used as a useful marker when ε(γ-glutamyl) lysine needed to be collected.

Qualitative techniques for identification of ε(γ-glutamyl) lysine

Qualitative techniques for the identification of ε(γ-glutamyl) lysine provide useful evidence for the unambiguous identification of this dipeptide. Earlier techniques used high voltage paper electrophoresis and descending paper chromatography (17).

A more recent method (33) uses the dansyl derivative of the dipeptide. This derivative may be separated from a mixture of dansylated amino acids by two-dimensional thin layer chromatography upon polyamide sheets. The possibility of quantitating this procedure utilising radiolabelled dansyl chloride was also reported in this method.

Derivatisation of the α-amino groups of ε(γ-glutamyl) lysine with acid stable derivatives such as the dansyl group have also been used in the structural

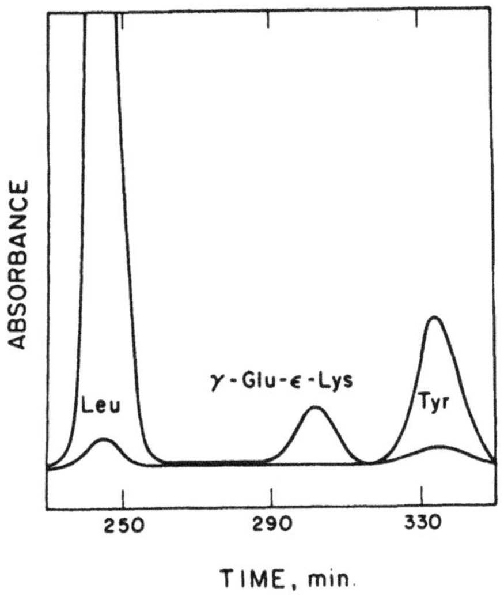

Fig. 2. Isolation of γ-glutamyl-ε-lysine isodipeptide from enzymatic digest of the polymer obtained from the membranes of Ca^{2+}-loaded erythrocytes. The analysis was performed on a Beckman 120C amino acid analyser using a Custom AA-15 resin column (0.9 × 55 cm) eluted with 0.2 M citrate, pH 3.31 at a flow rate of 75 ml/hr at 55 °C. The lower tracing represents the pattern obtained when the enzymes used for digestion were incubated without addition of polymer. Reprinted from Siefring et al. (10) with permission. Copyright ©, 1978, American Chemical Society.

analysis of the dipeptide. This provides further identification of the ε-amino:γ-glutamyl linkage (9). Hydrolysis of the dansylated ε(γ-glutamyl) lysine will lead to production of the corresponding monodansyl α-glutamic acid and monodansyl α-lysine which may be identified by thin layer chromatography (34) or by flat bed paper electrophoresis (35).

Quantitative analysis of ε(γ-glutamyl) lysine by ion-exchange chromatography

Ion-exchange resins have been used for the separation of complex mixtures of amino acids since the late 1940s, a property which has made them the natural selection for the separation and identification of ε(γ-glutamyl) lysine. Factors which affect the separation of amino acids upon ion-exchange resins have been well documented (36, 37).

Various amino acid analyser programmes for separating and detecting ε(γ-glutamyl) lysine have now been reported. A good representative of these are listed in Table 2. Some of the programmes rely upon modification of an existing sodium citrate or lithium citrate buffer gradient used for amino acid analysis. Others use simple isocratic elution utilising a citrate buffer of constant molarity and constant pH. Although normally providing good resolution the main disadvantage of the isocratic programme is the broadening of the solute band width during column elution. This effect can lead to a drastic loss in sensitivity which is often critical to an investigation.

In the majority of programmes the dipeptide elutes with the neutral amino acids in the vicinity of leucine, isoleucine, tyrosine or methionine.

A high affinity microparticulate resin (8–9 μm), which has proved useful for the separation of the dipeptide in more recent reports, is Dionex DC-4A (Durrum Chem. Corp.). Separation of ε(γ-glutamyl) lysine has been obtained with this resin using eluting buffers of either sodium citrate (Pico buffer system II, Pierce Chemical Co.) (37: L. Saidel, personal communication) or lithium citrate (Pico buffer system IV) (21).

Elution programmes with sodium citrate buffers cause the dipeptide to elute in the valley between methionine and isoleucine (Fig. 3). With a lithium citrate buffer programme the dipeptide is eluted in a large trough between leucine and tyrosine (Fig. 4). A programme producing good resolution without loss of sensitivity is important in the separation of the dipeptide, because this minimises the number of prepurification steps which need to be performed before its final analysis.

Good resolution of ε(γ-glutamyl) lysine is also necessary if the dipeptide is to be collected for further analysis. When ion-exchange chromatography is the only method available for identification of the dipeptide, then its collection, further hydrolysis and identification of the constitutive amino acids is a necessary procedure for unambiguous identification.

In the majority of systems listed in Table 2, detection and quantitation of the dipeptide is by post-column derivatisation with ninhydrin, although post-column derivatisation with the fluorogenic reagent σ-pthalaldehyde has been used in the analysis of cellular proteins (32). Both of these methods may be used in combination with isotope dilution for quantitation of the dipeptide.

There seems little doubt that ion-exchange systems for the analysis of ε(γ-glutamyl) lysine will

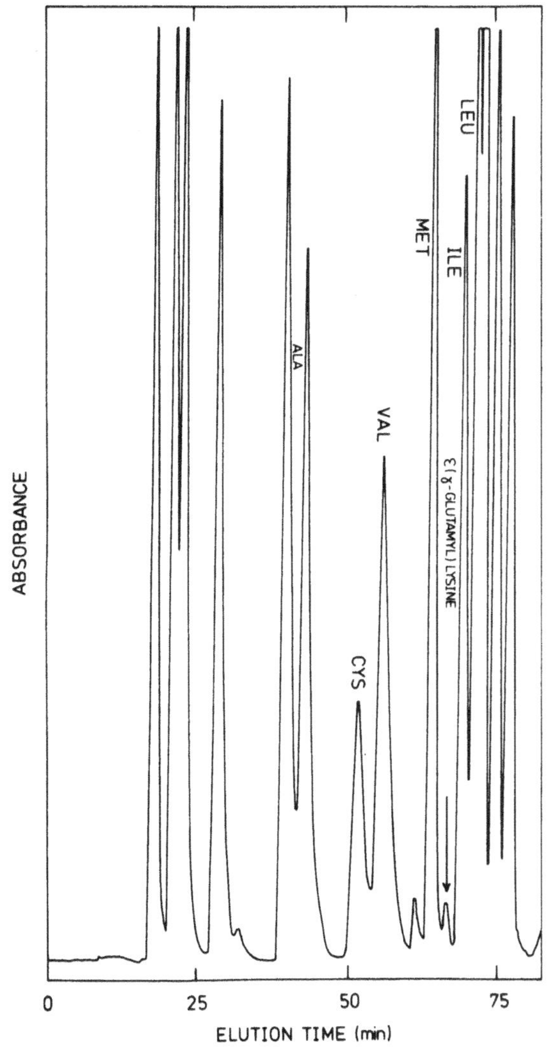

Fig. 3. Separation of ε(γ-glutamyl) lysine upon a Glenco amino acid analyser using a Dionex DC-4A resin column (0.325 × 30 cm) eluted with sodium citrate buffers (Pico buffer system II) at a flow rate of 12 ml/hr. The column was eluted with Buffer A for 40 min at 50 °C, Buffer B (without additives) for 13 min at 70 °C and Buffer C for 50 min at 70 °C. The column was loaded with 10 nmol of each amino acid and 0.2 nmol of ε(γ-glutamyl) lysine. With permission of L. Saidel, Department of Biochemistry, University of Health Sciences, the Chicago Medical School, Chicago, IL 60064, U.S.A.

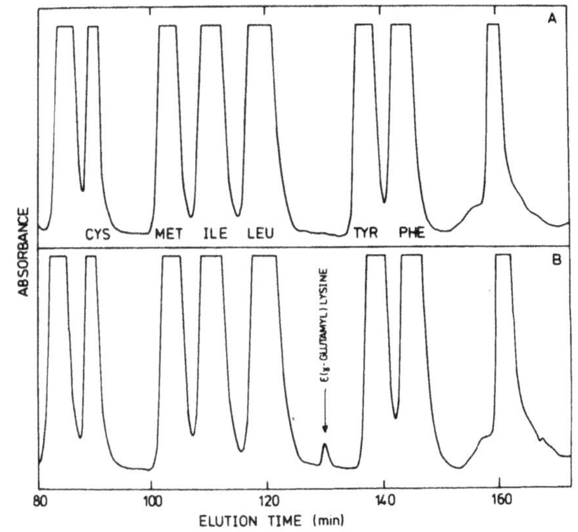

Fig. 4. Separation of ε(γ-glutamyl) lysine upon a Glenco (Model MM) amino acid analyser using a Dionex DC-4A resin column (0.325 × 30 cm) eluted with lithium citrate buffers (Pico buffer system IV) at a flow rate of 12 ml/hr. The column was eluted with Buffer A, pH 2.9 for 30 min at 40 °C, Buffer B, pH 3.04 for 22 min at 40 °C, Buffer C, pH 2.95 for 23 min at 40 °C, and 20 min at 65 °C, Buffer D, pH 3.39 for 30 min at 65 °C and Buffer E, pH 3.44 for 60 min at 65 °C. The column was loaded with 125 nmol of each amino acid and 1 nmol of ε(γ-glutamyl) lysine, B and amino acids only in A.

continue to improve with the introduction of the modern amino acid analyser. The majority of the apparatus now on the commercial market use microbore columns containing small particle high affinity resins like Dionex DC-4A. This type of analyser normally uses microbore detection equipment with high signal to noise ratios and provides the added choice of either colourimetric or fluorescence detection if increased sensitivity is required.

Quantitative analysis of ε(γ-glutamyl) lysine by reverse phase HPLC

The introduction of high performance or high pressure liquid chromatography (HPLC) over the last decade has enabled complex mixtures of thermally labile, nonvolatile and ionic compounds to be readily separated and analysed quantitatively (40, 41).

The basic apparatus used in HPLC for gradient elution with two different solvents is illustrated in Fig. 5. Simple isocratic elution would require only one high pressure pump without the need for the gradient controller.

Unlike the amino acid analyser the apparatus is extremely versatile and lends itself to separation of the mixtures previously listed by a change of either column packing or eluting solvent(s).

One particular technique which has proved extremely successful using HPLC is reverse phase chromatography. This technique utilises high effi-

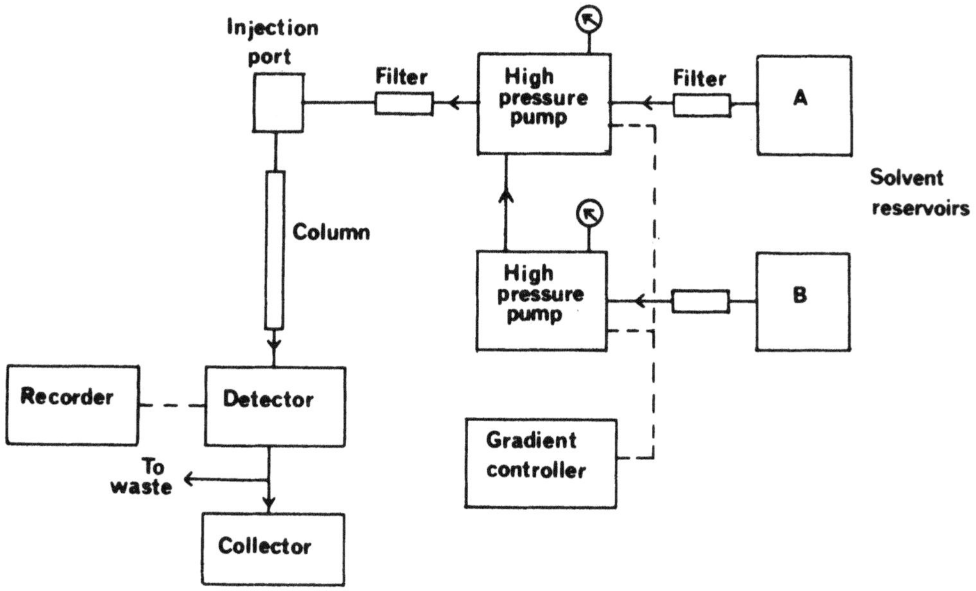

Fig. 5. Basic apparatus used for HPLC utilising two different solvents.

ciency microbore columns packed with microparticulate (3–10 μm) octadecyl-, octyl- of phenyl-silica. The term reverse phase simply refers to the use of a nonpolar stationary phase and a polar mobile phase instead of the once more common method of eluting a polar stationary phase with nonpolar solvents. The advantages of using this system and its application to Biochemical and Biomedical research have been recently reviewed (42).

Separation of amino acids by reverse phase HPLC has now achieved a reasonable measure of success. The majority of the separations rely upon pre-column derivatisation of the amino acids. The derivative is selected in such a way so as to provide

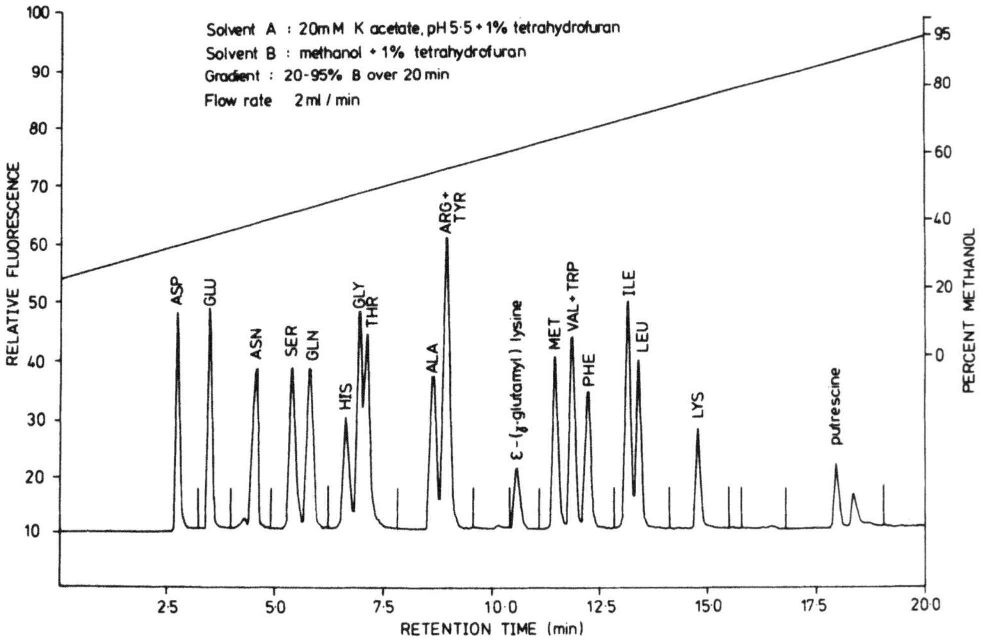

Fig. 6. Separation of ε(γ-glutamyl) lysine by reverse phase HPLC. The column (Zorbax C8, 25 cm × 4.6 mm) was loaded with 568 pmol of dipeptide and 1.14 nmol of each amino acid and putrescine.

detection of the amino acid and to provide the hydrophobicity required for its increased retention upon the reverse phase column. Such derivatives have included phenylthiohydantoin derivatives (43, 44), dansyl derivatives (45, 46) and σ-pthalaldehyde/thiol derivatives (47, 48, 49). Of these methods it is the σ-pthalaldehyde/thiol derivative that provides the greatest sensitivity in terms of detection and is the most quickly prepared.

Reaction of ε(γ-glutamyl) lysine with σ-pthalaldehyde/2-mercaptoethanol in the presence of methanol gives a derivative which is relatively stable and which may be separated from a mixture of amino acids by reverse phase HPLC (Fig. 6) (21). The separation procedure uses a Zorbax C_8 Column (DuPont Inst.) which is eluted with a linear gradient made between methanol:tetrahydrofuran (99:1 v/v) and 0.02 M potassium acetate buffer, pH 5.5:tetrahydrofuran (99:1 v/v).

The general application of this methodology for detection of ε(γ-glutamyl) lysine has been tested with proteolytic digests of protein polymers from human cataract lens (12) and with the transglutaminase mediated clot obtained from the seminal vesicle secretion of guinea pig (21). The elution profiles obtained from digested seminal vesicle secretions are illustrated in Fig. 7.

The advantages of using HPLC methodology for detection of ε(γ-glutamyl) lysine are the small sample size, rapidity of analysis (14 minutes analysis time including derivatisation) and the increased sensitivity it affords (less than 50 pmol dipeptide may be detected). The technique is not only useful as an alternative analytical system for quantitation of ε(γ-glutamyl) lysine, but provides a rapid screening procedure for its detection.

When reverse phase chromatography is used in combination with ion-exchange chromatography, the two different chromatographic procedures offer a possible alternative to the tedious collection, derivatisation and further hydrolysis of the dipeptide (9) for its unambiguous identification.

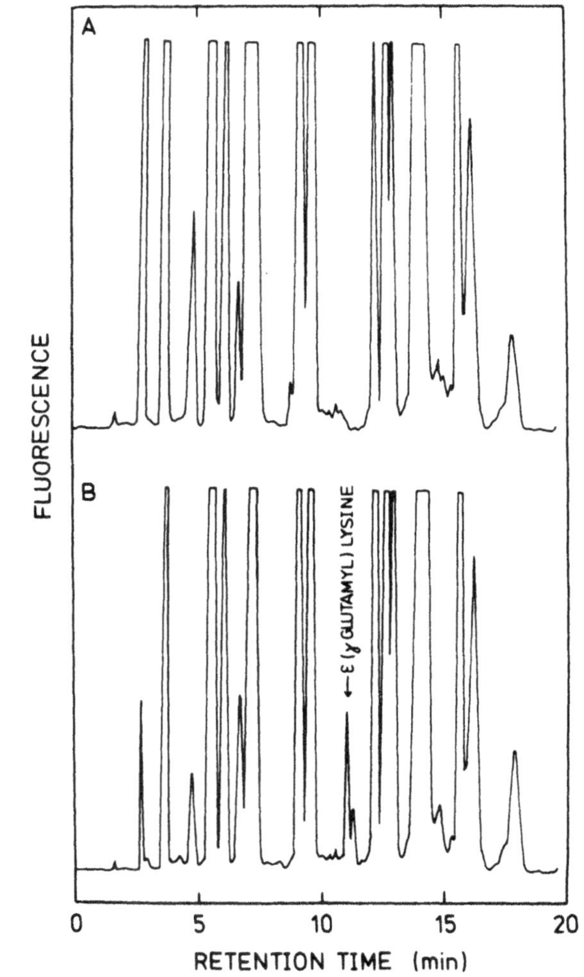

Fig. 7. HPLC measurement of ε(γ-glutamyl) lysine crosslinks in the proteolytic digests of the seminal vesicle protein from guinea pig. Panel A, digest of the acetone powder from the seminal vesicle secretion, Panel B digest of the seminal vesicle secretion clotted with the transglutaminase enzyme from the guinea pig prostate. Chromatographic conditions were the same as those listed in Fig. 6.

Indirect approaches for the detection of ε(γ-glutamyl) lysine crosslinks

Indirect approaches for the detection of ε(γ-glutamyl) lysine crosslinks rely upon a number of techniques which do not require the actual isolation and subsequent identification of the dipeptide. These techniques are useful in providing additional data for the quantitative assessment of the dipeptide as well as providing information about its formation between individual polypeptide chains.

One particular approach relys upon the measurement of all lysine residues within a protein that are involved in ε-lysine linkages (masked lysines). The method involves the blocking of the free amino groups in the protein with acid stable derivatives. The modified protein is then subjected to acid hydrolysis and the underivatised lysine residues (those previously involved in ϵ-lysine linkages) measured by standard amino acid analysis.

Traditional methods commonly used in the food industry employ fluorodinitrobenzene (50, 51) or trinitrobenzene-sulfonic acid (52) for the derivatising agent.

Alternative methods which have been used for the indirect measurement of ϵ(γ-glutamyl) lysine in biological systems, have relied upon either deamination of the ϵ-lysine (53), or its cyanoethylation with acrylonitrile (54). To allow easier access of the acrylonitrile to the reactive lysines the protein may be predigested with trypsin (54) or subtilisin (10).

When the quantitative assessment of total 'masked lysines' is used in combination with the direct estimation of ϵ(γ-glutamyl) lysine, evidence may be gained for the presence of other ϵ-lysine linkages existing in the protein. If the value of the total masked lysines exceeds that of the total amount of ϵ(γ-glutamyl) lysine then other ϵ-lysine linkages may be present.

A further indirect approach for the detection of protein crosslinking by ϵ(γ-glutamyl) lysine crossbridges uses the technique of SDS-gel electrophoresis for which many buffer systems have been reported (55, 56, 57). This approach relys upon and can also confirm the ability of transglutaminase enzymes to crosslink polypeptides into protein aggregates. Electrophoretic patterns of SDS–polyacrylamide or SDS–polyacrylamide/agarose gels can reveal the extent of protein crosslinking and can also indicate which polypeptide chains are involved.

A classic example is the crosslinking of fibrin by Factor XIIIa, hair follicle and liver transglutaminase (7). Comparison of the electrophoretic profiles of crosslinked fibrin (Fig. 8) indicated that each of these enzymes could crosslink fibrin into polymeric aggregates. The significance of this finding was that each of these transglutaminase enzymes had a different preference for the individual chains of fibrin which were crosslinked. Other examples where SDS–gel electrophoresis has been used to indicate protein crosslinking by transglutaminase enzymes, includes the Ca^{2+}-mediated crosslinking of human erythrocyte membranes (10, 58), the crosslinking of collagen (59), the crosslinking of myosin (60), the crosslinking of fibronectin (61), the crosslinking of cold insoluble globulin (62), the crosslinking of fibronectin to collagen (63), the crosslinking of human B_2-microglobulin (64), and the crosslinking of seminal vesicle proteins of the rat (65).

An additional technique which is frequently used in combination with the demonstration of protein crosslinking, is the inhibition of this reaction by small molecular weight amine substrates (NH_2R in II).

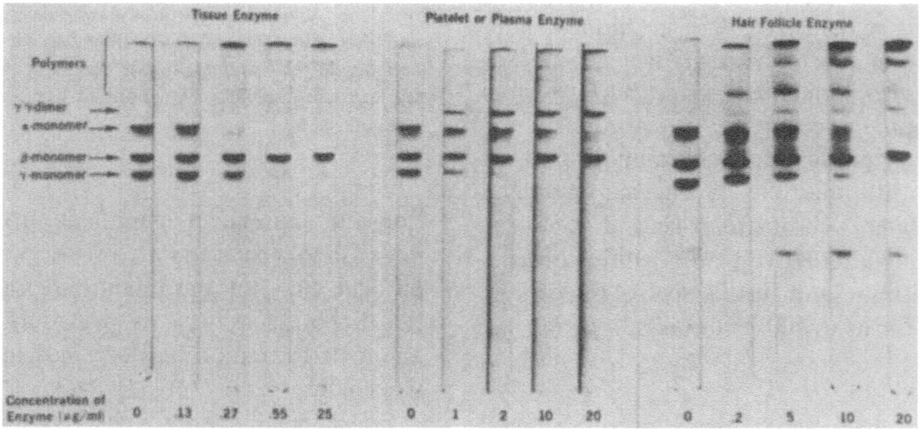

Fig. 8. Sodium dodecyl sulphate–polyacrylamide gel electrophoresis of fibrin stabilised by transglutaminases. Gel electrophoresis was carried out by the method of Weber and Osborn (55) except 6 M urea was included in the gel buffer. Reprinted from Chung (7) with permission.

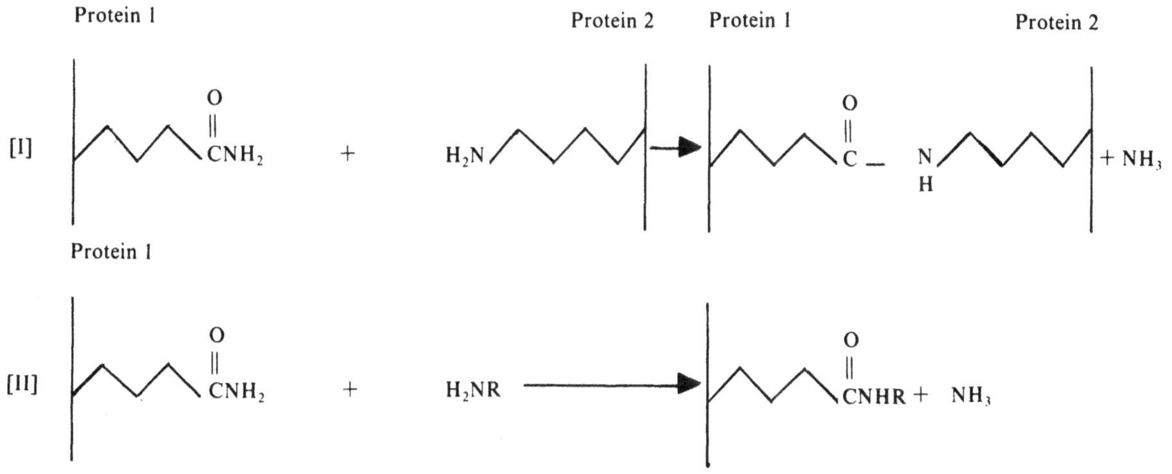

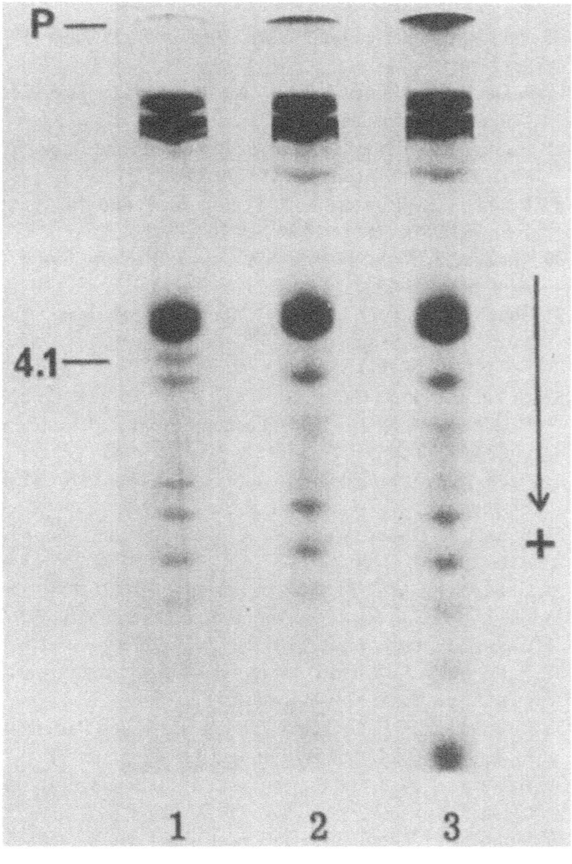

Fig. 9. Inhibition of membrane protein polymerisation by histamine. Fresh human erythrocytes were incubated for 3 hr at 37 °C with 20 μM A23187 ionophore and 1.5 mM $MgCl_2$ (gel 1), 1.5 mM $CaCl_2$ (gel 2) in the presence of 10 mM histamine or 1.5 mM $CaCl_2$ (gel 3). Following incubation, membranes were isolated, solubilised in 1% SDS–40 mM DTT and subjected to SDS-polyacrylamide gel (4.5%) electrophoresis according to the method of Steck and Yu (67). Polymer is denoted as P.

Covalent incorporation of amine substrates into the γ-glutamine groups of the accepter protein, with the concomitant inhibition of protein crosslinking, provides further evidence for the formation of ϵ(γ-glutamyl) lysine in the crosslinking mechanism. Such an example is the Ca^{2+}-mediated crosslinking of human erythrocyte membrane proteins. The inhibition of this process by amines such as histamine and cystamine (Fig. 9) (66), provided good evidence for the involvement of the erythrocyte transglutaminase in polymer formation before the isolation and quantitation of ϵ(γ-glutamyl) lysine was finally achieved (10).

This technique can also provide information regarding the availability of γ-glutamyl acceptor groups in the protein, if the small molecular weight amines are suitably labelled. Examples include the fluorescently labelled amine monodansylcadaverine or radiolabelled amines like ^{14}C-glycine methylester or ^{14}C-putrescine which have provided useful information about the γ-glutamine crosslinking sites in fibrin (68, 69), myosin (60) and cell membrane proteins (70).

Conclusion

This chapter has indicated that direct and indirect techniques for detecting ϵ(γ-glutamyl) lysine in a biological system are available. Which of these techniques are chosen will depend upon the requirements of a particular investigation and in many cases the availability of specific pieces of apparatus and instrumentation.

Looking to the future there seems little doubt that with the advances in liquid chromatography technology and resin technology the direct analysis of ϵ(γ-glutamyl) lysine will continue to improve. HPLC has not only found success in its application of reverse phase chromatography for the analysis of amino acids, but is now an integral part of most modern ion-exchange analysers. The advantages of these systems are better resolution of amino acids, rapid analysis, increased sensitivity, and the capability of handling small sample sizes. Factors which will no doubt prove important in establishing the cellular importance of ϵ(γ-glutamyl) lysine.

Acknowledgement

We would like to thank Professor Laszlo Lorand for his support and helpful discussions which led to the development of some of the new techniques presented in this chapter.

References

1. Pisano, J. J., Finlayson, J. S. and Peyton, M. P., 1968. Science 160: 892–893.
2. Matacic, S. and Loewy, A. G., 1968. Biochem. Biophys. Res. Commun. 30: 356–362.
3. Lorand, L., Downey, J., Gotoh, T., Jacobsen, A. and Tokura, S., 1968. Biochem. Biophys. Res. Commun. 31: 222–230.
4. Harding, H. W. J. and Rogers, G. E., 1972. Biochim. Biophys. Acta 257: 37–39.
5. Rice, R. H. and Green, H., 1977. Cell 11: 417–422.
6. Williams-Ashman, H. G., Notides, A. C., Pabalan, S. S. and Lorand, L., 1972. Proc. Nat. Acad. Sci. 69: 2322–2325.
7. Chung, S. I., 1972. Ann. N.Y. Acad. Sci. 202: 240–255.
8. Folk, J. E. and Chung, S. I., 1973. Adv. Enzymol. 38: 109–191.
9. Abernethy, J. L., Hill, R. L. and Goldsmith, L. A., 1977. J. Biol. Chem. 252: 1837–1839.
10. Siefring, G. E. Jr., Apostol, A. B., Velasco, P. T. and Lorand, L., 1978. Biochemistry 17: 2598–2604.
11. Lorand, L., Hsu, L. K. H., Siefring, G. E. and Rafferty, N. S., 1981. Proc. Nat. Acad. Sci. 78: 1356–1360.
12. Hsu, L. K. H., Griffin, M., Wilson, J. and Lorand, L., 1981. Abstract. Fed. Proc. 40: (3).
13. Hirs, C. H. W., 1967. Methods in Enzymol. 11: 325–329.
14. Walsh, K. A. and Brown, J. R., 1962. Biochim. Biophys. Acta 58: 596–598.
15. Schweet, R., 1955. Fed. Proc., Fed. Am. Soc. Exp. Biol. 14: 277–278.
16. Schweet, R., 1956. Fed. Proc., Fed. Am. Soc. Exp. Biol. 15: 350–351.
17. Korngruth, M. L., Neidle, A. and Waelsch, H., 1963. Biochemistry 2: 740–745.
18. Pisano, J. J., Finlayson, J. S. and Peyton, M. P., 1969. Biochemistry 8: 871–876.
19. Asquith, R. S., Otterburn, M. S., Buchanan, J. H., Cole, M., Fletcher, J. C. and Gardner, K. L., 1970. Biochim. Biophys. Acta. 207: 342–348.
20. Birkbichler, P. J., Dowben, R. M., Matacic, S. and Loewy, A. G., 1973. Biochem. Biophys. Acta. 291: 149–155.
21. Griffin, M., Wilson, J. and Lorand, L., 1982. Anal. Biochem. 124: 406–414.
22. Narahashi, Y., 1970. Methods in Enzymol. 19: 651–664.
23. Markland, F. S. and Smith, E. L., 1971. In: The Enzymes (Boyer, P. D., ed.), 3rd ed., Vol. 3, pp. 561–608, New York and London: Academic Press.
24. Konigsberg, W., Goldstein, J. and Hill, R. J., 1963. J. Biol. Chem. 238: 2028–2033.
25. DeLange, R. J. and Smith, E. L., 1971. In: The Enzymes (Boyer, P. D., ed.), 3rd ed., Vol. 3, pp. 81–119, New York and London: Academic Press.
26. Hill, R. L. and Schmidt, W. R., 1962. J. Biol. Chem. 237: 389–396.
27. Matacic, S. and Loewy, A. G., 1966. Biochem. Biophys. Res. Commun. 24: 858–866.
28. Pisano, J. J., Finlayson, J. S., Peyton, M. P. and Nagai, Y., 1971. Proc. Nat. Acad. Sci. 68: 770–772.
29. Pisano, J. J., Bronzert, T. J., Peyton, M. P. and Finlayson, J. S., 1972. Ann. N.Y. Acad. Sci. 202: 98–113.
30. Matacic, S. S. and Loewy, A. G., 1979. Biochim. Biophys. Acta. 576: 263–268.
31. Sugawara, K., 1977. In: Bioch. Cutaneous Epidermal Differ. Proc. Jpn-US Semin. (Seijii, M. and Bernstein, I. A., eds.), pp. 387–396. Baltimore: Univ. PK. Press.
32. Birkbichler, P. J., Carter, H. A., Orr, G. R., Conway, E. and Patterson, M. K. Jr., 1978. Biochem. Biophys. Res. Commun. 84: 232–237.
33. Venn, R. F., Basford, J. M. and Curtis, C. G., 1978. Anal. Biochem. 87: 278–282.
34. Woods, K. R. and Wang, K. T., 1967. Biochim. Biophys. Acta. 133: 369–370.
35. Gray, W. R., 1972. Methods in Enzymol. 25: 121–138.
36. Blackburn, S. (ed.), 1978. Amino Acid Determination, 367 pages. New York: Marcel Dekker.
37. Rattenbury, J. M. (ed.), 1981. Amino Acid Analysis, 370 pages. New York: Ellis-Horwood.
38. Birkbichler, P. J., Orr, G. R., Carter, H. A. and Patterson, M. K. Jr., 1977. Biochem. Biophys. Res. Commun. 78: 1–7.
39. Harding, H. W. J. and Rogers, G. E., 1971. Biochemistry 10: 624–629.
40. Simpson, C. F. (ed.), 1976. Practical High Performance Liquid Chromatography. New York: Heyden Press.
41. Snyder, L. R. and Kirkland, J. J. (ed.), 1979. Introduction to Modern Liquid Chromatography, 2nd ed. New York: Wiley-Interscience.
42. Brown, P. R. and Krstulovic, A. M., 1979. Anal. Biochem. 99: 1–21.
43. Haag, A. and Langer, K., 1974. Chromatographia. 7: 659–662.
44. Annan, W. D., 1981. In: Amino Acid Analysis (Rattenburg, J. M., ed.), pp. 66–70. Chichester: Ellis-Horwood.

45. Engelhart, H., Asshauer, J., Neue, U. and Weigand, N., 1974. Anal. Chem. 46: 336–340.
46. Bayer, E., Grom, E., Kaltenegger, B. and Uhman, R., 1976. Anal. Chem. 48: 1106–1109.
47. Hill, D. W., Walters, F. H., Wilson, T. D. and Stuart, J. D., 1979. Anal. Chem. 51: 1338–1341.
48. Jones, B. N., Pääbo, S. and Stein, S., 1981. J. Liquid Chromat. 4: 565–586.
49. Griffin, M., Price, S. J. and Palmer, T., 1982. Clin. Chemica. Acta 125: 89–95.
50. Carpenter, K. J., 1960. Biochem. J. 77: 604–610.
51. Horwitz, W. (ed.), 1980. Official Methods of Analysis, 13th ed., pp. 776–777, No 43.224. Washington D.C.: Association of Official Analytical Chemists.
52. Kakade, M. L. and Leiner, I. H., 1969. Anal. Biochem. 27: 273–280.
53. Lorand, L., Ong, H. H., Lipinski, B., Rule, N. G., Downey, J. and Jacobsen, A., 1966. Biochem. Biophys. Res. Commun. 25: 629–637.
54. Pisano, J. J., Finlayson, J. S. and Peyton, M. P., 1969. Biochemistry 8: 871–876.
55. Weber, K. and Osborn, M., 1969. J. Biol. Chem. 244: 4406–4412.
56. Fairbanks, G., Steck, T. L. and Wallach, D. F. H., 1971. Biochemistry 10: 2606–2617.
57. Laemmli, U. K., 1970. Nature (London) 227: 680–685.
58. Bjerrum, O. J., Hawkins, M., Swanson, P., Griffin, M. and Lorand, L., 1981. J. Supramol. Struct. Cell. Biochem. 16: 289–301.
59. Soria, A., Soria, C. and Boulard, C., 1975. Experientia. 31: 1355–1357.
60. Cohen, I., Young-Bandala, L., Blankenberg, T. A., Siefring, G. R. and Bruner-Lorand, J., 1979. Arch. Biochem. Biophys. 192: 100–111.
61. Birkbichler, P. J. and Patterson, M. K., 1978. Ann. N.Y. Acad. Sci. 3: 54–65.
62. Mosher, D. F., 1975. J. Biol. Chem. 250: 6614–6621.
63. Mosher, D. F., Schad, P. E. and Kleinman, H. K., 1979. J. Clin. Invest. 64: 781–787.
64. Fesus, L., Falus, A., Erdei, A. and Laki, K., 1981. J. Cell Biology 89: 706–710.
65. Williams-Ashman, H. G., Beil, R. E., Wilson, J. H., Hawkins, M., Grayhack, J., Zunamon, A. and Weinstein, N. K., 1980. Advances in Enzyme Regulation 18: 239–258.
66. Lorand, L., Weissmann, L. B., Epel, D. L. and Bruner-Lorand, J., 1976. Proc. Nat. Acad. Sci. 73: 4479–4481.
67. Steck, T. L. and Yu, J., 1973. J. Supramol. Struct. 1: 220–232.
68. Lorand, L. and Ong, H. H., 1976. Biochemistry 5: 1747–1753.
69. Lorand, L., Chenoweth, D. and Gray, A., 1972. Ann. N.Y. Acad. Sci. 202: 155–171.
70. Dutton, A. and Singer, J. J., 1975. Proc. Nat. Acad. Sci. 72: 2568–2571.

Received 28 July 1982.

Molecular and Cellular Biochemistry 58, 51–61 (1984).

Transglutaminases and the clotting of mammalian seminal fluids

H. Guy Williams-Ashman
Ben May Laboratory for Cancer Research, and Dept. of Pharmacological and Physiological Sciences, University of Chicago, Chicago, IL 60637, U.S.A.

Summary

Rapid coagulation of seminal fluid in rats, guinea pigs, and several other mammalian species including certain non-human primates is responsible for the post-coital formation of copulatory plugs in the vagina. The clotting of rodent seminal plasma results from coagulation of certain proteins derived from the seminal vesicles by enzymes secreted mainly by the coagulating (anterior prostate) gland. Several lines of evidence indicate that the clotting enzymes of coagulating gland secretions are transglutaminases, and that the extreme insolubility of the seminal clot in rodents is due to transglutaminase-catalyzed formation of ε(γ-glutamyl)lysine cross-links between polypeptide chains. Various features of the apparently unique forms of transglutaminases produced by rat coagulating gland and the actions of these enzymes on vesicular secretory and other proteins are discussed. The aliphatic polyamines spermidine and spermine are incorporated covalently into the proteins of the clot as the corresponding N-mono-ε-(γ-glutamyl)- and N,N-*bis*(γ-glutamyl)-adducts during the enzymatic coagulation process. At the greater than millimolar concentrations at which cross-spermidine and spermine are present in normal rat seminal plasma, these polyamines attenuate the formation of hard clots in reconstituted rat semen coagulation systems, seemingly by competing with lysyl residues in vesicular secretion proteins as transglutaminase amine donor substrates, and thus preventing formation of ε-(γ-glutamyl)lysine cross-bridges. It is proposed that in those species such as the rat and man in which seminal plasma contains large amounts of spermidine and(or) spermine of prostatic origin, the seminal polyamines may serve to stop blockage of the urethra by preventing too explosive a rate of seminal clot formation during the ejaculatory process.

Introduction

Transglutaminase-catalyzed formation of ε(γ-glutamyl)lysine cross-bridges between polypeptide chains is involved in the rigidification of fibrin clots, and of certain hair and epidermal proteins (1–6). Another type of insolubilization of proteins via establishment of ε-(γ-glutamyl)lysine cross-links occurs during the coagulation of semen in certain mammalian species that results from polymerization of seminal vesicle secretion proteins by transglutaminases of prostatic origin (7–9). Biochemical studies on the clotting of seminal fluid have uncovered novel forms of transglutaminases for which certain proteins secreted by the seminal vesicles are particularly active substrates. Investigations in this area have also provided new insights into the roles of natural aliphatic amines as substrates and modulators of transglutaminases. A brief overview of the phenomenology and possible physiological significance of semen coagulation may serve to introduce discussion of the chemical basis of this process.

Semen clotting in mammals

The seminal flids that mammals emit at ejaculation consist of spermatozoa originating from the cauda region of the epididymis swimming around in the seminal plasma that is largely made up of secretions of the male accessory glands of reproduction (seminal vesicles and the prostate and bulbourethral glands) plus components derived from rete testis fluid and the epithelia of the epididymis and vas deferens (10–12). In most mammalian species, the seminal plasma represents a very large portion of the total volume of seminal fluid. In some animals, such as bulls, and also dogs, cats and other carnivores, ejaculated semen when kept at room or body temperature either *in vitro* or in the vagina remains permanently in the liquid state. But in a substantial number of other species, the semen swiftly clots on its own accord. In some cases the coagulated semen consists of a hard, rubbery lump, as in many myomorphic rodents, and certain moles, hedgehogs, marsupials and non-human primates. When semen from such animals is deposited into the vagina at coitus, the clotting process results in formation of a copulatory plug or *bouchon vaginal* which occludes the vaginal barrel close to the uterine cervix and may stay in that location for many hours or even days. In the majority of species in which a vaginal plug is evident, the coagulation of the semen reflects clotting of seminal plasma proteins secreted by the seminal vesicles (12). However, it is believed that the copulatory plug produced in the opossum (13) and certain bats (14) is formed as a result of coagulation of female secretions by material in seminal plasma. Rodent copulatory plugs eventually undergo dissolution in the vagina, probably because of infiltration by leucocytes and the actions of their proteinases (15). There are other species in which the semen normally coagulates into a soft gel quickly after ejaculation (domesticated rabbit, stallion, man), or produces a gel rather slowly on standing at tepid temperatures, as in the boar. In the latter animals, the protein substrates for seminal gelation appear to come from the seminal vesicles, and in the boar (but apparently not in the stallion and man) the vesicular fluid contribution to seminal plasma forms a gel when it comes into contact with bulbourethral gland secretion (11, 12). In the normal human male, the seminal clot gradually liquefies at 25–37 °C over a period of 15 to 30 min *in vitro* (and apparently faster when the semen is in the vagina [16] as a result of the actions of proteinases and peptidases in seminal plasma (11, 12, 16).

What is the physiological utility of semen clotting in mammals? Surgical excision of the seminal vesicles and the adjacent coagulating (anterior prostate) glands, which prevents formation of a vaginal plug, has a very deleterious effect on fertility in certain rodent species (11, 12, 17–19). Scrutiny of guinea pig copulatory plugs with an electron microscope revealed a tubular structure composed of a dense network of short, flattened protein fibers in a sponge like arrangement (20). Curiously enough, few spermatozoa were found inside or on the surface of the coagulum. Comparably, the rhesus monkey copulatory plug (which undergoes limited liquefaction on its own accord) appeared as a disorganized array of protein fibers with relatively small numbers of spermatozoa inside (20). During ejaculation in many mammalian species the seminal vesicle secretion is the last of the male accessory gland fluids to enter the urethra following discharge of prostate and bulbourethral gland secretions and the contents of the vas deferens containing sperm cells. Since the protein substrates for semen coagulation are mainly derived from vesicular secretion, it is thus not altogether surprising that in some species the vaginal plug contains few spermatozoa, although in man the seminal gel enmeshes the majority of the spermatozoa until liquefaction is well advanced (20). On the basis of the aforementioned and other findings, it has been proposed that semen clotting and any attendant formation of a vaginal plug might to varying degrees: (a) hinder backflow of semen out of the vagina after coitus; (b) enhance sperm transport from the upper vagina into the cervix and hence the uterus, or even stimulate transport of spermatozoa via the uterine lumen towards the oviducts, if it is assumed that the vaginal plug stimulates myometrial contractions; (c) act as a reservoir for gradual release of free-swimming sperms from solid material deposited into the vicinity of the uterine cervix. Water-tight evidence in favor of all of these eventualities is lacking, and the first proposition seems unlikely. This is because in various rodents a substantial proportion of spermatozoa in the ejaculate are catapulted through the cervix into the uterus within seconds after ejaculation has taken place in the vagina; and thus prevention of semen outflow from the vagina post-coitally is an improbable function of the vaginal plug in such animals. Indeed, removal of the copulatory plug from female guinea pigs immediately after service by males did not prevent most of the females from becoming pregnant (21).

Martan and Shepherd (21) observed that when guinea pig females in estrus were copulated first with albino males and then with colored males, all of the offspring had albino coat colors. Yet in other experiments in which vaginal plugs produced by insemination with albino males were removed and the females subsequently engaged in coitus with colored males, the litters that followed were apparently sired either by colored males exclusively, or by males of both hues. It was concluded that the copulatory plug in guinea pigs might be of adaptive significance because only the progeny of the first male to engage in coitus (e.g. a behaviorally dominant stud) will be sired if the resultant vaginal plug remains in place during intromissions and ejaculations by subsequently mounting males. (If these notions are valid, the copulatory plug consisting of coagulated seminal plasma proteins might be regarded as a sort of enzymatically fabricated chastity belt.)

Enzymatic coagulation of rodent seminal plasma

As reviewed elsewhere (10-12, 22), early studies established that semen clotting in the rat and guinea pig is due to coagulation of certain proteins in vesicular secretion by heat-labile enzymes from the anterior prostate (coagulating) gland. Gotterer et al. (23, 24) developed reconstituted rodent semen coagulation systems in which the clotting of stable preparations of seminal vesicle secretion proteins by partially purified coagulating gland enzymes could be measured quantitatively by optical methods. The coagulation processes were shown to require Ca^{2+}, which could be replaced by Mn^{2+} but not by Mg^{2+} or various other divalent metal cations. Production of a coagulum entailed a lag period that depended on pH, ionic strength, temperature, and the concentrations of vesicular secretion protein substrates and clotting enzymes. Thrombin in the absence or presence of exogenous Ca^{2+} did not coagulate the vesicular secretion proteins and also did not influence their clotting by anterior prostate enzymes. No proteins resembling fibrinogen could be detected in seminal vesicle secretions. The enzymatically-induced clots did not resemble fibrin in appearance or consistency, and were insoluble at 37 °C or lower temperatures in 0.15 or 2 M NaCl; 8 M urea; 6 M guanidium chloride; or water, even if 50 mM dithiothreitol or 2-mercaptoethanol was added with or without 1% sodium dodecyl sulfate. The clots were quickly solubilized by treatment with trypsin at close to neutral pH and even faster by trypsin plus chymotrypsin. Attempts to demonstrate during the clotting reactions any liberation of fairly low molecular weight peptides comparable to those formed by action of thrombin on fibrinogen were unsuccessful.

The bulk proteins of vesicular secretion, and also the transglutaminases in soluble extracts of the coagulating gland or its secretion, are quite dissimilar from those of the rat, as are many of the factors that influence the course of the coagulation processes. Details of transglutaminase-mediated clotting of seminal fluid in these two species will therefore be considered separately.

Clotting of guinea pig semen

The major proteins of guinea pig seminal vesicle secretion dissolved by 150 mM NaCl at neutral pH have been separated by ion-exchange chromatography, gel filtration, and gel electrophoresis under both non-denaturing (7) and denaturing conditions (7, 25–28). Prior treatment of these secretory proteins with thiols has little effect on the profiles of protein bands separable by one-dimensional gel electrophoresis in the presence of dodecyl sulfate and (or) concentrated urea (J. Wilson and H. G. Williams-Ashman, unpublished studies) – a finding that contrasts with the striking effects of thiols on protein band profiles separable by similar gel electrophoretic techniques from rat vesicular secretion (22). Notides and Williams-Ashman (7) suggested that a basic protein in guinea pig vesicular fluid of apparent M_r 17 900 which lacks proline, cyst(e)ine and tyrosine was the main precursor of the coagulum produced by action of coagulating gland enzymes. The amino acid composition of the purified basic protein determined after hydrolysis with hot 6 N HCl was very close to that of the washed enzymatically-induced clot, except that the undigested coagulum contained considerably less amide nitrogen. Free ammonia was liberated during the clotting reaction. Acid hydrolysates of the unclotted basic protein (which was partially sequenced (27)) and of the coagulum were similarly rich in glutamate and lysine (7). On the basis of the latter findings, as well as the requirement for Ca^{2+} for the coagulation process and also the extreme insolubility of the clot, it was proposed (7) that transglutaminase(s) in coagulating gland secretion polymerized the basic protein in vesicular secretion via production of ϵ-(γ-glutamyl)lysine cross-links between polypeptide chains.

Subsequent studies by Williams-Ashman et al. (8) showed that the enzymatically produced coagulum after exhaustive washing with concentrated urea and then water contained 6 to 7 mol of ϵ-(γ-glutamyl)lysine isodipeptide liberated by exhaustive proteolytic hydrolysis per mol equivalent (17 900 daltons) of the putative basic vesicular secretion precursor protein. Unclotted guinea pig vesicular secretion did not contain detectable ϵ-(γ-glutamyl)lysine (27). The values obtained by direct measurement of the enzymatically liberated isodipeptide agreed well with determinations of the lysine residues resistant to cyanoethylation by treatment with acrylonitrile prior to acid hydrolysis according to the method of Pisano et al. (29).

Although several investigators have described various transglutaminase reactions catalyzed by extracts of guinea pig prostate, the relationship of apparently multiple forms of transglutaminases in this organ to semen coagulation remains unclear. According to Chung (30) both the human and the guinea pig prostate (which lobe(s) of the latter gland unspecified) contain two transglutaminases. One of these enzymes was obtained in the zymogen form and exhibited properties similar to platelet Factor XIII; the other transglutaminase was stated (cf. 4) to be immunologically cross-reactive with guinea pig liver transglutaminase. These conclusions are at variance with the investigations of Wing (31) and Tong (32) in Lorand's laboratory indicating the existence of two forms of guinea pig coagulating gland transglutaminases whose reactivities towards various substrates, activators and inhibitors are quite different from those of either the liver enzyme or plasma fibrinoligase. One of these forms of transglutaminase which is strongly activated by p-tosyl-glycine (an inhibitor of liver transglutaminase and of fibrinoligase) was extensively purified and studied by Tong (32). This particular enzyme was present in coagulating gland tissue extracts but not in coagulating gland secretion, the latter fluid containing two additional forms of transglutaminase that were separable from fibrinoligase and the liver enzyme by electrophoresis on agar gels and detected by an enzyme activity stain that utilizes monodansylcadaverine as a fluorescent substrate (33). Further work will be necessary to clarify the relationships of these different coagulating gland transglutaminases to semen clotting in the guinea pig, and to determine whether or not they may represent post-synthetically modified forms of transglutaminase found in other tissues.

Semen coagulation in the rat

More is known about the mechanisms and control of seminal fluid clotting in the rat than in the guinea pig. An advantage of working with rats in this context is that in this species, the coagulating glands empty their secretions into the urethra through a single terminal duct, a circumstance that permits pure rat coagulating gland secretion to be easily obtained. Before considering the coagulation of vesicular secretory proteins by coagulating gland transglutaminases in the rat, two aspects of male genital secretions in this species must be mentioned.

Although rat coagulating gland secretion readily clots homologous vesicular secretion proteins into hard lumps that resemble material in copulatory plugs, several lines of evidence (34–36) indicate that the secretion of bulbourethral (Cowper's) glands in this species accelerate seminal fluid clotting both *in vivo* and in reconstituted systems *in vitro*. It appears that the material in bulbourethral gland secretion that enhances rat semen clotting consists of acidic oglycoproteins which under appropriate conditions interact non-enzymatically with certain vesicular secretion proteins so as to cause their precipitation and increased susceptibility to enzymatic coagulation (36).

The abundant proteins in rat seminal vesicle fluid are quite different from their guinea pig counterparts (in both species and as in many other mammals, most of the vesicular secretory proteins do not originate from blood plasma but rather are synthesized *de novo* in the gland). When rat seminal vesicle secretion proteins are separated by one-dimensional gel electrophoresis under strongly denaturing conditions at neutral or mildly alkaline pH, the relative migrations of a substantial proportion of many of the higher molecular weight (> 35 000) protein bands depends on whether or not the secretion is exposed to thiols. Thus some of the slower migrating protein bands separated by standard dodecyl sulfate gel electrophoresis in the presence of 2-mercaptoethanol seem to represent peptide chains that in the fresh secretion may be linked via disulfide bridges as parts of larger protein molecules. By contrast, certain other protein bands including 3 that migrate as if their molecular weights are in the 10 to 16 kilodalton range, and also a yellow flavoprotein sulfhydryl oxidase of M_r 66 000 (37, 38), are not affected by incubation with thiols.

The only proteins in rat vesicular secretion that as yet have been purified to homogeneity are the latter flavoprotein (which does not participate in seminal clot formation), and a small protein often denoted as rat SVS-IV (39) of 9 758 daltons that has been completely sequenced (40), and apparently is polymerized enzymatically into seminal clots only sluggishly by virtue of acting as an amine donor but not as an amine acceptor substrate for transglutaminases (H. G. Williams-Ashman, unpublished observations).

When fresh rat vesicular secretion, or the proteins therein that are soluble in 150 mM NaCl buffered at pH 7.4, are incubated with crude or semipurified coagulating gland soluble enzyme preparations at optimal Ca^{2+} concentrations (> 3 mM), many though not all of the larger vesicular secretory proteins are polymerized rapidly to extremely insoluble products that do not penetrate 10–20% polyacrylamide gels. The coagula formed by treatment of fresh vesicular secretion with coagulating gland enzymes under arbitrary conditions were found to contain more than 9 mol of ε-(γ-glutamyl)lysine isodipeptide per 100 000 g of protein that was liberated by exhaustive proteolytic digestion (J. Wilson, L. Lorand and H. G. Williams-Ashman, unpublished studies), and to contain substantial quantities of lysine residues resistant to reaction with acrylonitrile (41). No ε-(γ-glutamyl)-lysine could be detected in unclotted rat vesicular secretion proteins. The estimations of the isodipeptide cross-bridges in the seminal clot may not represent the maximal possible values that might be obtained as a result of enzymatic coagulation under other conditions. Our findings indicating that semen clotting in the rat involves transglutaminase-mediated cross-linking of several different vesicular secretion proteins (9, 41) – which contrast with an apparently homopolymeric system in the guinea pig – agree with conclusions drawn by Higgins et al. (42) and Bradshaw and Wolfe (43) from their independent studies on insolubilization of rat vesicular secretory proteins by coagulating gland enzyme preparations that were undertaken under rather different conditions.

In line with the aforementioned evidence for transglutaminase catalysis of the clotting reactions was the demonstration that several types of low molecular weight amines that are known to be highly effective amine donor substrates for mammalian transglutaminases (putrescine, spermidine, spermine, glycine ethyl ester, methylamine), when added in the [^{14}C]-labeled form in millimolar concentrations to the reconstituted coagulation systems, became covalently attached with varying kinetics to proteins of the clot and the clot liquor (9, 41). There was no detectable 'self-incorporation' of the labeled low molecular weight amines into semipurified coagulating gland enzyme preparations of a type analogous to the incorporation of polyamines observed with high concentrations of homogeneous guinea pig liver transglutaminase (44).

Germane to the aforementioned enhancement of rat seminal fluid coagulation by bulbourethral gland secretion was the finding that the latter material at low concentrations (0.1 mg protein per ml) enhanced the initial rate (but not the final extent) of clotting of rat vesicular secretion proteins by coagulating gland enzymes, and accelerated the incorporation of [^{14}C]-putrescine into the clotted proteins. These effects could not be ascribed to the negligible transglutaminase activities of the amounts of bulbourethral gland secretion employed, but on several grounds seemed to be related to non-covalent interactions of glycoproteins in bulbourethral gland secretions with clottable vesicular secretion proteins resulting in (conformational ?) changes in the vesicular secretion proteins that might enhance their reactivity of amine acceptor substrates for transglutaminases. In accord with such notions was our finding 9, 41) that low concentrations of certain macromolecular polyanions that are not transglutaminase substrates (e.g. poly-L-glutamate, poly-

L-aspartate or heparin at pH 7.4, and poly(A) that was particularly active at pH 6.5), and under circumstances that did not cause non-enzymatic precipitation of the vesicular secretion proteins, markedly enhanced coagulating gland transglutaminase-mediated coagulation, and also increased the covalent incorporation of [^{14}C]-polyamines into the proteins of the clot and clot liquor.

Poly-L-glutamate and other active macromolecular polyanions likewise increased the covalent incorporation of putrescine into N,N-dimethylated vesicular secretion proteins but had no effect on putrescine incorporation into casein or N,N-dimethylcasein.

The fluorescent amine monodansyl-cadaverine (i.e. N-[5-aminopentyl]-5-dimethylamino-l-naphthalene-sulfonamide), a substance which Lorand et al. (45) introduced as a very potent substrate for various mammalian transglutaminases, was observed to be extensively incorporated covalently into vesicular secretion proteins during the clotting process, and to inhibit the formation of insoluble protein at concentrations greater than 0.5 mM. Presumably monodansyl-cadaverine competes with lysine residues in some of these proteins for reaction that act as amine donor substrates for reaction with glutaminyl residues in those vesicular secretory proteins that serve as amine acceptor substrates. Clots formed in the presence of monodansylcadaverine at low concentrations in reconstituted rat seminal fluid coagulation systems, after successive washing with concentrated urea, water, and acetone and subsequent desiccation, were found to exhibit an intense fluorescence when the dried material was suspended in isotonic sodium chloride and viewed with an ultraviolet lamp.

The rat coagulating gland cytosol and the secretion of this organ appear to contain at least two forms of transglutaminases that have not been detected in many other rat organs except the adjacent dorsolateral (but not the ventral) lobe of the rat prostate. Using assays based on incorporation of [^{14}C]-putrescine into either N,N-dimethylcasein or N,N-dimethylated rat vesicular secretion proteins as substrates – thus obviating formation of ϵ-(γ-glutamyl)lysine cross-links – it was observed that the transglutaminase activity of cytosolic extracts of rat coagulating gland was much greater than that of a variety of other rat tissues (9, 41). Moreover, crude and partially refined preparations of these coagulating gland enzymes utilized N,N-dimethylated vesicular secretion proteins more effectively than N,N-dimethylcasein as amine acceptor substrates, whereas soluble transglutaminases from various other rat tissues reacted with saturating levels of the aforementioned modified proteins to roughly the same degree. However, the transglutaminase activities of dialyzed 0–50% saturation ammonium sulfate fractions of coagulating gland cytosol toward N,N-dimethylcasein were more labile to storage at 2 °C, and were depressed to a greater extent by heating to 65 °C for 5 min at pH 7.4 in the presence of 2 mM EDTA, than were the activities with N,N-dimethylated vesicular secretion proteins (J. Wilson and H. G. Williams-Ashman, unpublished observations). Comparably, raising the ionic strength of the enzyme reaction mixtures by addition of 0.6 M NaCl increased the activity of such coagulating gland extracts with modified secretory proteins as substrates but markedly decreased the reactivity towards N,N-dimethylcasein (9). When the ammonium sulfate fractions from rat coagulating gland cytosol were subject to agarose gel electrophoresis in two directions at pH 7.4 and then processed for visualization of bands of transglutaminase activity using a procedure based on reactivity of the fluorescent monodansyl-cadaverine as amine donor substrate (33), with N,N-dimethylated derivatives of either casein or vesicular secretion proteins as substrates, two enzyme activity bands that migrated towards the cathode were evident. Liver transglutaminase or thrombin-activated human Factor XIII electrophoresed under the same conditions traveled to separate single bands towards the anode. Coagulating gland secretion contained two forms of transglutaminase that migrated to the same positions towards the cathode as the transglutaminases in the cytosol fraction of homogenates of this organ (9). The latter transglutaminases were hardly retained on DEAE-cellulose at pH 7.4, and could be purified further by chromatography on carboxylmethyl-cellulose and gel filtration on Bio-Gel P-100, but only in poor yield (J. Wilson and H. G. Williams-Ashman, unpublished experiments). Which of the two enzyme activity bands separated by agarose gel electrophoresis represent the form of rat coagulating gland transglutaminases that reacts preferentially with vesicular secretion proteins remains unsettled. But it is is clear that the transglutaminases in this organ and its secretion exhibit different properties including their apparent net charge at pH 7.4 than those of liver transglutaminase, plasma fibrinoligase, and also the transglutaminases found in guinea pig coagulating gland and its secretion; whether this reflects some post-translational modifications of transglutaminases found in other tissues by addition of adventitious groups or limited proteolysis is not known, although no evidence was obtained for catalytically inactive zymogen forms of guinea pig coagulating gland transglutaminases that could be activated by thrombin and a number of other proteinases at neutral pH. It is noteworthy that dimethylsulfoxide, which can activate transglutaminases from epidermis (46), cultured human prostate cells (47), and guinea pig coagulating gland (32), did not directly enhance the activity of rat coagulating gland transglutaminases in our experience.

Effects of aliphatic polyamines on clotting of rat seminal fluid

Two considerations prompted more detailed examination of effects of polyamines on reconstituted rat semen clotting systems. First, polyamines are present in high concentrations in the seminal plasmas of several though not all mammalian species. For example, human semen contains about 3 mM spermine alongside much lesser amounts of spermidine which largely originate from prostatic secretion. In the rat, the ventral and to a smaller extent the dorsal lobes of the prostate contribute spermidine plus spermine in levels of several mM to the ejaculate. By contrast, rat seminal vesicle secretion is essentially devoid of polyamines (22, 48–52). The levels of these polyamines in human and rat semens are more than three orders of magnitude greater than the levels of spermidine and spermine in normal blood plasma. Thus it was of interest to examine whether polyamines either modulate, or participate in, the transglutaminase-mediated clotting of seminal fluid. As already mentioned labeled putrescine, spermidine and spermine become covalently attached to vesicular secretion proteins during the coagulation process. Since these natural polyamines contain primary amino groups attached to methylene moieties at both ends, another question that arose was whether once a diamine or polyamine incorporated into an amine acceptor substrate as the corresponding peptide-bound mono-N-γ-glutamyl derivatives might react in a further transglutaminase-catalyzed reaction with a glutaminyl residue on another peptide chain to produce the corresponding N,N-*bis*(γ-glutamyl) derivative that might cross link two polypeptide chains. This eventuality in comparison with the transglutaminase-mediated formation of ε-(γ-glutamyl)lysine crossbridges may be represented by the following equations:

Experiments in the author's laboratory (9, 53, 54) indicated that when [^{14}C]-putrescine was incorporated into N,N-dimethylated rat vesicular secretion proteins as catalyzed by coagulating gland transglutaminases over brief incubation periods, the radioactivity covalently attached to proteins was, after liberation of the products by exhaustive proteolytic digestion with subtilisin, carboxypeptidase and aminopeptidase at pH 7, almost exclusively in

$$\begin{array}{c} \text{CONH}_2 \\ | \\ \text{-Glu-} \end{array} + \begin{array}{c} \text{NH}_2 \\ | \\ \text{-Lys-} \end{array} \longrightarrow \text{NH}_3 + \begin{array}{cc} \text{CO—} & \text{NH} \\ | & | \\ \text{-Glu-} & \text{-Lys-} \end{array}$$

Formation of ε-(γ-glutamyl)lysine

$$\begin{array}{c} \text{CONH}_2 \\ | \\ \text{-Glu-} \end{array} + \text{H}_2\text{N-R-NH}_2 \longrightarrow \text{NH}_3 + \begin{array}{l} \text{CONH-R-NH}_2 \\ | \\ \text{-Glu-} \end{array}$$

$$\begin{array}{l} \text{CONH-R-NH}_2 \\ | \\ \text{-Glu-} \end{array} + \begin{array}{c} \text{CONH}_2 \\ | \\ \text{-Glu-} \end{array} \longrightarrow \text{NH}_3 + \begin{array}{lr} \text{CONH-R-NHOC} & \\ | & | \\ \text{-Glu-} & \text{-Glu-} \end{array}$$

Successive production of N-mono-γ-glutamyl and N,N-*bis* (γ-glutamyl) derivatives of a diamine or polyamine.

the form of mono-N-γ-glutamyl-putrescine. But when the reactions were conducted with higher levels of enzyme for longer time periods, a second radioactive product could be separated by paper chromatography from the proteolytic digests (9). This material was identified as the N,N-*bis* (γ-glutamyl) putrescine adduct (R. E. Beil and H. G. Williams-Ashman, unpublished studies) in experiments for which the availability of authentic N,N-*bis*(γ-glutamyl)-putrescine generously provided by Dr J. E. Folk was crucial. Elegant model studies by Schrode and Folk (55) utilizing various protein or synthetic glutaminyl peptide amine acceptor substrates for fibrinoligase and guinea pig liver transglutaminase established more thoroughly and independently that aliphatic diamines and polyamines via the enzymatic synthesis of the corresponding N,N-*bis*(γ-glutamyl)-polyamine cross-bridges could link peptide chains together. By a combination of enzymatic and chemical procedures, the synthesis of free isodipeptides representing all of the possible N-mono-and N,N-*bis*-(γ-L-glutamyl) derivatives of putrescine, spermidine and spermine was accomplished in Folk's laboratory (50, 55), with parallel development of methods for the rapid separation of these substances by high performance liquid chromatography.

Using a reconstituted rat semen coagulation system comparable to those developed in the author's laboratory (9, 41), Folk et al. (50) found that when exogenous [^{14}C]-spermidine plus -spermine at roughly 0.09 mM of each were incubated with rat vesicular secretion proteins and coagulating gland extracts, the following radioactive substances could be isolated from proteolytic digests of the clotted proteins: N^1- and N^8-(γ-glutamyl) spermidines, N^1, N^8-*bis*(γ-glutamyl) spermidine; N^1-(γ-glutamyl) spermine; and N^1, N^{12}-*bis* (γ-glutamyl)spermine. A very substantial proportion of the total spermidine and spermine added to the semen coagulation system was converted into these protein-bound adducts. Moreover, several proteins in both the clot and the clot liquor were found to have polyamines bound to them covalently.

Virtually all of the spermidine and spermine in rat seminal plasma originates from the secretions of the dorsal and especially ventral lobes of the prostate gland which commingle with vesicular and coagulating gland fluids at the time of ejaculation. There are many reasons to believe that the extreme insolubility of the proteinaceous seminal clot that forms a copulatory plug in the normal course of coital events is due primarily to transglutaminase-mediated production of ε(γ-glutamyl)lysine cross-bridges between polypeptide chains, rather than to cross-links involving polyamines. This is because coagula exhibiting properties that closely resemble those of vaginal plugs are formed by action of coagulating gland transglutaminases on refined vesicular secretion proteins in the complete absence of polyamines (9, 41) (although in the rat polyamine-dependent cross-links between these highly insoluble seminal clot proteins may also be formed under natural circumstances) and also because transglutaminase-catalyzed incorporation of labeled spermidine and spermine into N,N-dimethylated rat vesicular secretion proteins does not result in rapid formation of extremely insoluble clots, even though the modified proteins may well undergo polymerization to a limited extent via *bis* (γ-glutamyl)-polyamine adduct formation. In fact, at concentrations of roughly 2–3 mM or higher, spermidine and spermine progressively attenuate the coagulation of vesicular secretion proteins by coagulating gland secretion (9, 54) in reconstituted rat semen clotting systems. Williams-Ashman and Canellakis (53, 54) have postulated that inhibition of semen coagulation by polyamines in those species such as rat and man in which large amounts of spermine and(or) spermidine are added to seminal plasma from porstatic secretions during the ejaculatory process may be of some physiological utility by preventing too rapid formation of very insoluble protein during the time that seminal constituents traverse the penile urethra. For it is easy to imagine that blockage of the male urethra with seminal clots could evoke disastrous damage to the male genitourinary tract with retention of urine. Noteworthy in this context is the suggestion of Vulpé et al. (56) that the soft calculus occasionally evident in the urinary bladders of male but not female rats is produced by retrograde flow of vesicular and coagulating gland secretions into the bladder. A comparable phenomenon has been noticed in adrenalectomized guinea pigs (57). Furthermore, D. S. Coffey (personal communication) observed that electroejaculation of rats with the aid of a rectal probe faradic stimulator can sometimes result in total blockage of the urethra by coagulated semen with massive retention of urine. Such an artificial mode of seminal emission could perhaps involve emptying of prostatic and vesicular secretions into the urethra in successions or extents that might differ from that occurring during the normal ejaculatory process (in most mammals the vesicular secretion typically comprises the last fraction of 'split' ejaculates).

The enzymatic formation of mono-N-(γ-glutamyl)polyamine and N,N-*bis*(γ-glutamyl)polyamine adducts of proteins was first demonstrated in physiological contexts in the aforementioned studies on the extracellular process of seminal fluid coagulation. Considerations of the affinity of polyamines for fibrinoligase (2, 3) suggests that, at the truly minuscule concentrations at which putrescine, spermidine and spermine are circulating in normal blood plasma, these amines are unlikely to become attached extensively to fibrin or other proteins in plasma during normal blood clotting. There is growing evidence that intracellular proteins in some animal tissues may be post-translationally modified via transglutaminase-catalyzed attachment of putrescine and spermidine as the corresponding mono-N-(γ-glutamyl) adducts but probably not as N, N-bis(γ-glutamyl) derivatives (54, 58–62). (Note that conversion of putrescine via spermidine to protein-bound hypusine (63–65) can masquerade the formation of protein-bound γ-glutamyl spermidines in certain analytical systems.) At present, the physiological importance of the intracellular formation of peptide-bound γ-glutamyl adducts of proteins inside mammalian cells – which might depend heavily on the ambient concentrations of Ca^{2+} required for transglutaminase activity in appropriate locations inside cells – remains far from clear.

Fate of amine adducts of proteins produced by transglutaminase action during semen coagulation

Cross-linking of proteins of the seminal clots of rodents clearly involves transglutaminase-mediated formation of ε-(γ-glutamyl)lysine cross-bridges and the modulation of these reactions by polyamines is largely if not exclusively due to the low molecular weight amines competing with certain lysine residues of proteins as amine donor substrates for the clotting enzymes. No known proteinase or peptidase can sunder the isopeptide bonds of ε-(γ-glutamyl)lysine or γ-glutamyl polyamine derivatives. However, Fink et al. (66) discovered an enzyme in kidney and several other rat tissues that catalyzes cleavage of free γ-glutamylamines including ε-(γ-glutamyl)lysine and various N-mono- and N,N-*bis* (γ-glutamyl) derivatives of polyamines. The products of the action of this enzyme, denoted as γ-glutamylamine cyclotransferase, were shown to be the parent amines plus 5-oxo-L-proline (i.e. pyrrolidone-2-carboxylate, or pyroglutamate). This enzyme (which is quite distinct from the γ-glutamyl amino acid cyclotransferase involved in glutathione metabolism) does not act on γ-glutamyl-ε-lysine or γ-glutamylpolyamine derivatives in which the α-carboxyl and α-amino groups of the glutamyl-moiety are in peptide linkage, but rather attacks only the correspondig free (i.e. non-peptide bound) isodipeptides. Whether γ-glutamylamine cyclotransferase together with various proteinases and peptidases participates in the dissolution of copulatory plugs in the vagina remains to be worked out.

Gelation of rat coagulating gland secretion

According to Joshi et al. (67). when rat coagulating gland secretion expressed from the single duct of the gland is diluted with buffers of physiological ionic strength at pH 7.4, a translucent gel gradually forms at 37 °C when sodium bicarbonate is added in concentrations between 5 and 50 mM. No evidence was obtained that this effect of bicarbonate ions could be accounted for by attendant alteratios in pH, or mimicked by a variety of other inorganic anions. Gel formation was stated to be critically dependent on the concentration of protein in the diluted coagulating gland secretion. In unpublished studies, R. E. Beil and H. G. Williams-Ashman have readily confirmed these observations. The gelation proteins accelerated by bicarbonate was observed not only with diluted secretion from the gland, but also with cytosol extracts of the organ and dialyzed 0–50% saturation ammonium sulfate fractions of proteins obtained therefrom. Bicarbonate-induced gelation was considerably faster at 37 °C as compared with 25 °C and extremely slow at 0 °C. The gels produced under these conditions bore no resemblance to the clots of vesicular secretion proteins formed by the action of coagulating gland transglutaminases. It appeared conceivable that the gels of soluble coagulating gland proteins might result from cross-linking via ε-(γ-glutamyl)-lysine linkages catalyzed by the very powerful transglutaminases in the preparations. However, this appeared very unlikely, because the gelation at 37 °C induced by $NaHCO_3$ was not inhibited nor enhanced by addition of EGTA (5 mM). $CaCl_2$ (5 mM) or monodansylcadaverine (5 mM), all of which profoundly influence the transglutaminase-mediated coagulation of vesicular secretion proteins. Moreover, [^{14}C]-putrescine at 1 mM was not incorporated into trichloroacetic-acid precipitable proteins of diluted coagulating gland secretion after prolonged incubation at pH 7.4 and 37 °C in the presence of 5 mM $CaCl_2$, in either the absence or presence of 50 mM $NaHCO_3$. However, the bicarbonate-induced gels readily dissolved on addition of 10 mM dithiothreitol and the same concentration of the latter thiol prevented gel formation when it was added before bicarbonate to coagulating gland secretion diluted at pH 7.4 and incubated at either 25 °C or 37 °C. 2-Mercaptoethanol at 10 mM also caused rapid dissolution of the bicarbonate-induced gels although not so rapidly as dithiothreitol. These results hint the gelation of guinea pig coagulation gland secretion – which Joshi et al. (67) believe also takes place in the uterine lumen after coitus in the rat – is at least in part due to establishment of disulfide or perhaps dityrosine bridges between certain secretory protein chains rather than to transglutaminase-catalyzed cross-linking.

Acknowledgement

The author is deeply indebted to Professor Laszlo Lorand of Northwestern University for his unfailingly wise counsel, and for his active experimental collaboration in many aspects of the work presented.

References

1. Lorand, L., 1972. Ann. N.Y. Acad. Sci. 202: 6–30.
2. Lorand, L. and Stenberg, P., 1976. In: CRC Handbook on Biochemistry and Molecular Bilogy, Vol. II Proteins (Fasman, G. D., ed.), 3rd ed. pp. 669–685, Chemical Rubber Publ. Co., Cleveland.
3. Curtis, C. G. and Lorand, L., 1977. In: Haemostasis: Biochemistry, Physiology and Pathology (Ogston, D. and Bennett, B., eds.), Chapter 8, pp. 186–201, Wiley, Chichester.
4. Folk, J. E. and Finlayson, J. S., 1977. Adv. Protein Chem. 31: 1–133.
5. Rice, R. H. and Green, H., 1977. Cell 11: 417–422.
6. Goldsmith, L. A. and Martin, C. M., 1975. J. Invest. Dermatol. 64: 316–321.
7. Notides, A. C. and Williams-Ashman, H. G., 1967. Proc. Natl. Acad. Sci. U.S.A. 58: 1991–1995.
8. Williams-Ashman, H. G., Notides, A. C., Pabalan, S. S. and Lorand, L., 1972. Proc. Natl. Acad. Sci. U.S.A. 69: 2322–2325.
9. Williams-Ashman, H. G., Beil, R. E., Wilson, J., Hawkins, M., Grayhack, J., Zunamon, A. and Winstein, N. K., 1980. Adv. Enzyme Regul. 18: 239–258.
10. Price, D. and Williams-Ashman, H. G., 1961. In: Sex and Internal Secretions (Young, W. C., ed.), 3rd ed, Vol. I, Chapter 6, pp. 366–448, Williams and Wilkins, Baltimore.
11. Mann, T., 1964. The Biochemistry of Semen and of the Male Reproductive Tract. Wiley, New York.
12. Mann, T. and Lutwak-Mann, C., 1981. Male Reproductive Function and Semen. Springer, New York.
13. Hartman, C. G., 1924. Anat. Rec. 27: 293–304.
14. Eadie, W. R., 1948. Anat. Rec. 102: 259–267.
15. Stockard, C. R. and Papanicolaou, G. N., 1919. Biol. Bull. 37: 222–245.
16. Tauber, P. F. and Zaneveld, L. J. D., 1976. In: Human Semen and Fertility Regulation in Men (Hafez, E. S. E., ed.), Chapter 14, pp. 153–166, Mosby, St. Louis.
17. Walker, G., 1911. Johns Hopkins Hosp. Rep. 16: 223–226.
18. Blandau, R. L., 1945. Am. J. Anat. 77: 253–272.
19. Pang, S. F., Chow, P. H. and Wong, T. M., 1979. J. Reprod. Fertil. 56: 129–132.
20. Zaneveld, L. J. D., Tauber, P. F., Port, C., Propping, D. and Schumacher, G. B. F., 1974. J. Reprod. Fertil. 40: 223–230.
21. Martan, J. and Shepherd, B. A., 1976. J. Ex. Zool. 196: 79–84.
22. Williams-Ashman, H. G., 1983. Current Topics in Cellular Regulation 23: 201–275.
23. Gotterer, G., Ginsburg, D., Schulman, T., Banks, J. and Williams-Ashman, H. G., 1955. Nature 176: 1209–1211.
24. Gotterer, G. and Williams-Ashman, H. G., 1957. Proc. Soc. Exp. Biol. Med. 94: 60–64.
25. Ballard, P. and Williams-Ashman, H. G., 1964. Invest. Urol. 2: 38–46.
26. Veneziale, C. M. and Deering, N. G., 1976. Andrologia 8: 73–82.
27. Pabalan, S. S., 1975. Guinea Pig Seminal Vesicle Basic Secretory Protein. Ph. D. Thesis, Northwestern University, Evanston, Illinois.
28. Veneziale, C. M., 1977. Biochem. J. 166: 155–166.
29. Pisano, T. J., Finlayson, J. S. and Peyton, M. P., 1969. Biochemistry 8: 871–876.
30. Chung, S. I., 1977. In: 'Isozymes' (Markert, C. L., ed.), Vol. I, pp. 259–274, Academic Press, New York.
31. Wing, D. A., 1977. Guinea Pig Coagulating Gland Transamidase. Ph. D. Thesis, Northwestern University, Evanston, Illinois.
32. Tong, Y. S., 1980. Guinea Pig Prostate Gland Transamidases. Ph. D. Thesis, Northwestern University, Evanston, Illinois.
33. Lorand, L., Siefring, Jr., G. E., Tong, Y. S., Bruner-Lorand, J., and Gray, Jr., A. J., 1979. Anal. Biochem. 93: 453–458.

34. Hart, R. G., 1968. J. Reprod. Fertil. 17: 223–226.
35. Hart, R. G., 1970. Biol. Reprod. 3: 347–352.
36. Beil, R. E. and Hart, R. G., 1973. Biol. Reprod. 8: 613–617.
37. Ostrowski, M., Kistler, W. S. and Williams-Ashman, H. G., 1979. Biochem. Biophys. Res. Commun. 87: 171–176.
38. Ostrowski, M. C. and Kistler, W. S., 1980. Biochemistry 19: 2639–2645.
39. Ostrowski, M. C., Kistler, M. K. and Kistler, W. S., 1979. J. Biol. Chem. 254: 383–390.
40. Pan, Y.-C. E., Silverberg, A. G., Harris, S. E. and Li, S. S.-I., 1980. Int. J. Peptide Protein Res. 16: 143–146.
41. Williams-Ashman, H. G., Wilson, J., Beil, R. E., and Lorand, L., 1977. Biochem. Biophys. Res. Commun. 79: 1192–1198.
42. Higgins, S. J., Burchell, J. M. and Mainwaring, W.I.P., 1976. Biochem. J. 160: 43–48.
43. Bradshaw, B. S. and Wolfe, H. G., 1977. Biol. Reprod. 16: 292–297.
44. Birckbichler, P. J., Orr, G. R., Cartes, H. A. and Patterson, M. K., 1979. Biochem. Biophys Res. Commun. 78: 1–7.
45. Lorand, L., Rule, N. G., Ong, H. H., Furlanetto, R., Jacobson, A., Downey, J., Öner, N. and Brunes-Lorand, J., 1968. Biochemistry 7: 1214–1223.
46. Buxman, M. M., Lopitz, C. J. and Wuepper, K. D., 1980. J. Biol. Chem. 255: 1200–1203.
47. Bures, D. M., Goldsmith, L. A. and Stone, K. R., 1980. Invest. Urol. 17: 298–302.
48. Rhodes, J. B. and Williams-Ashman, H. G., 1964. Med. Exp. 10: 281–285.
49. Williams-Ashman, H. G. and Lockwood, D. H., 1970. Ann. N.Y. Acad. Sci. 171: 882–894.
50. Folk, J. L., Park, M. H., Chung, S. I., Schrode, J., Lester, E. P. and Cooper, H. L., 1980. J. Biol. Chem. 255: 3695–3700.
51. Williams-Ashman, H. G., 1981. Adv. Polyamine Res. 3: 55–64.
52. Williams-Ashman, H. G., Jänne, J., Coppoc, G. L., Geroch, M. E. and Schenone, A., 1972. Adv. Enzyme Regul. 10: 225–245.
53. Williams-Ashman, H. G. and Canellakis, Z. N., 1979. Perspect. Biol. Med. 22: 421–453.
54. Williams-Ashman, H. G. and Canellakis, Z. N., 1980. Physiol. Chem. Phys. 12: 457–472.
55. Schrode, J. and Folk, J. E., 1978. J. Biol. Chem. 253: 4837–4840.
56. Vulpé, M., Usher, D. and Leblond, C. P., 1956. Endocrinology 59: 257–259.
57. Clayton, B. E., Hammant, J. E. and Hawkins, J. B., 1956. Nature 178: 1296–1297.
58. Haddox, M. K. and Russell, D. H., 1981. J. Cell. Physiol. 109: 447–452.
59. Ambron, R. T. and Kremzner, L., 1982. Proc. Natl. Acad. Sci. U.S.A. 79: 3442–3446.
60. Patterson, M. K., Maxwell, M. D., Birchbichler, P. J., Conway, E. and Carter, H. A., 1982. Cell. Biol. Int. Rep. 6: 461–470.
61. Canellakis, Z. N., Lande, L. A., and Bondy, P. K., 1981. Biochem. Biophys. Res. Commun. 100: 675–680.
62. Chan, W. Y., Griesmann, G. and Rennert, O. M., 1981. Adv. Polyamine Res. 3: 213–223.
63. Park, M. H., Cooper, H. L. and Folk, J. E., 1981. Proc. Natl. Acad. Sci. U.S.A. 79: 2869–2873.
64. Park, M. H., Cooper, H. L. and Folk, J. E., 1982. J. Biol. Chem 257: 7217–7222.
65. Cooper, H. L., Park, M. H. and Folk, J. E., 1982. Cell. 29: 791–797.
66. Fink, M. L., Chung, S. I. and Folk, J. E., 1980. Proc. Natl. Acad. Sci. U.S.A. 77: 4564–4568.
67. Joshi, M. S., Yaron, A., and Lindner, H. R., 1972. J. Reprod. Fertil. 30: 27–37.

Received 17 November 1982.

Molecular and Cellular Biochemistry 58, 63–68 (1984).

Cross-linking of fibronectin to collagenous proteins

Deane F. Mosher
Dept. of Medicine, University of Wisconsin, Madison, WI 53706, U.S.A.

Summary

Attempts were made to cross-link several collagenous proteins to fibronectin with Factor XIII_a (plasma transglutaminase). Cross-linking was demonstrated with type I collagen, type II collagen, type III collagen, type V or AB collagen, and α1(I)-CB7 and α1(I)-CB8 cyanogen bromide fragments of type I collagen. Cross-linking was not demonstrated with type IV collagen, Clq, and cyanogen bromide fragment α1(I)-CB6. The pH optimum for cross-linking of α1(I)-CB7 to fibronectin was 8.5 to 9.6. Cross-linking of α1(I)-CB7 to fibronectin was somewhat enhanced at lower than physiological ionic strength.

Introduction

Fibronectin is one of only four proteins in clotting plasma which are substrates for Factor XIII_a (plasma transglutaminase). Other proteins are fibrin, α_2-macroglobulin (1,2), and α_2-plasmin inhibitor (3,4). Fibronectin is also a substrate for guinea pig liver transglutaminase (5). The principal site of transglutamination of fibronectin by Factor XIII_a is very near the amino-terminus (6,7). There is a secondary site closer to the carboxyl terminus (8). Fluorescence polarization studies of dansylcadaverine incorporated into the amino-terminal site suggest that this portion of fibronectin has a good deal of flexibility (9).

Plasma fibronectin can be cross-linked by Factor XIII_a to the α-chain of fibrin (4,10,11), to *Staphylococcus aureus* (12), and, if self-aggregation is induced with sulfhydryl-containing reagents, to itself (11,14). Fibronectin convently incorporated into fibrin clots changes the mechanical properties of the clots (14) and allows cell adherence (15). The non-covalent interactions between fibronectin and fibrin are very weak at 37 °C (1,16), and, thus, covalent cross-linking is required if these modifications are to occur. In contrast, there may be little biologic significance in the fact that Factor XIII_a mediates cross-linking of fibronectin to *S. aureus*, because fibronectin binds irreversibly to *S. aureus* even without covalent cross-linking (17). The biologic relevance of the cross-linking of reduced fibronectin to itself is also unclear; the concentrations of sulfhydryl-containing reagents needed to support this reaction probably never occur in extracellular spaces (13, 18).

Fibronectin in the collagen-containing, non-periodic, fibrillar extracellular matrix of cultured cells is cross-linked by Factor XIII_a into a large complex (19). The protein pairs involved in such cross-linking have not been identified. It seems reasonable to suppose that much of the cross-linking is among polymerized fibronectin molecules. The importance of such fibronectin–fibronectin cross-linking, however, may be slight because fibronectin, which is a disulfide-bonded dimer in the circulation (10,20), forms disulfide-bonded multimers once it is deposited in extracellular matrix (21–25).

Fibronectin can be shown to interact with different types and forms of collagen. Although there are conflicts in the literature, several generalizations seem true (26). First, at 4 °C and 20 °C, fibronectin interacts better with denatured collagen and collagen fragments than with native collagen. Second, at 37 °C, the interaction of fibronectin with native type I collagen is only at the site of cleavage of collagen by vertebrate collagenase. Because collagenase cleaves the $\alpha 1$(I) chain of type I collagen in a region that lacks proline and hydroxyproline over a span of 12 residues, there may be local uncoiling of the collagen helix in this region at 37 °C (27). Third, cyanogen bromide fragments of collagen that contain the -GLY-ILE-ALA- sequence recognized by vertebrate collagenase ($\alpha 1$(I)-CB7, $\alpha 1$(I)-CB8, α_2-CB5, $\alpha 1$(II)-CB10) interact better with fibronectin than fragments which lack the sequence. We have measured the binding of iodinated collagen peptides to fibronectin-coated microbeads (D. F. Mosher and S. K. Larsen, unpublished observation, 1979); ^{125}I-$\alpha 1$(I)-CB7 and ^{125}I-$\alpha 1$(I)-CB8 bind reversibly to fibronectin with a dissociation constant of $2\text{–}5 \times 10^{-9}$ M. ^{125}I-$\alpha 1$(I)-CB6 binds less strongly to the fibronectin-coated beads. Fragments that lack the -GLY-ILE-ALA- sequence, however, do interact with fibronectin (27–29). Fourth, interstitial collagens interact better with fibronectin than do basement membrane collagens (types IV and AB or V). Finally, the interaction between fibronectin and collagen is enhanced by low ionic strength (30).

Fibronectin can be cross-linked to Types I and III collagen and to $\alpha 1$(I)-CB7 by Factor XIII$_a$ (31). The glutaminyl residues of fibronectin which participate in Factor XIII$_a$-mediated cross-linking to collagen are in the 27 kilodalton (kd) amino-terminal fragment, whereas the site for strong binding of fibronectin to collagen is in the adjacent 30 to 50 kd region (26,32–34). Present evidence suggests that $\alpha 1$(I)-CB7 (and other forms and types of collagen contributes only lysyl residues to the ϵ-(γ-glytamyl)-lysine linkages (6). The nearest lysyl residues are 20 residues from the -GLY-ILE-ALA- sequence of $\alpha 1$(I)-CB7 or 5.3 nm away if the polypeptide chain is in a collagen helix (31). Thus, it seems likely that fibronectin has an extended binding site for collagen which contains two subsites: one which interacts strongly with uncoiled collagen and particularly prefers the sequence around the collagenase cleavage point, and a second which brings the Factor XIII$_a$-reactive glutaminyl residue of fibronectin and lysyl residues of collagen together in a manner that favors cross-linking. Binding of $\alpha 1$(I)-CB7 to fibronectin causes an inappropriately large decrease in the diffusion coefficient and an increase in the mobility of dansylcadaverine incorporated in the amino-terminal transglutamination site (9). These findings indicate that the collagen fragment causes a partial unfolding and extension of fibronectin (9).

Indirect evidence suggests that Factor XIII$_a$ catalyzes the cross-linking of fibronectin to the collagenous 'tail' of asymmetric acetylcholinesterase (35). Cross-linking could not be shown, however, to *Ascaris* collagen (31). In the present communication, I describe our attempts to cross-link fibronectin to several other collagenous proteins and discuss the possible significance of such cross-linking.

Materials and methods

The following were gifts in 1978 or 1979: types I, II, III, and V collagens and cyanogen bromide fragments $\alpha 1$(I)-CB6, $\alpha 1$(I)-CB7, and $\alpha 1$(I)-CB8 of the $\alpha 1$(I) chain of type I collagen, Dr Hynda Kleinman, National Institutes of Dental Research, Bethesda, Maryland; types IV and V collagens, Dr Helene Sage, University of Washington, Seattle, Washington; Clq, Dr David Bing, The Blood Center, Boston, Massachusets; and thrombin, Dr John Fenton, II, New York Department of Health, Albany, New York. Factor XIII and fibronectin were purified as described previously (6). Cross-linking reactions were usually performed at 37 °C in Tris-buffered saline (10 mM Tris, 150 mM sodium chloride, pH 7.4) containing 10 mM calcium chloride. Lyophilized collagen was routinely prepared by solubilization in 100 mM acetic acid followed by dialysis against 0.4 M sodium chloride at 4 °C (31). Samples of the reaction mixtures were denatured in 2% sodium dodecyl sulfate, 2% 2-mercaptoethanol and analyzed by discontinuous polyacrylamide slab gel electrophoresis (31). Protein bands were visualized with Coomassie brilliant blue, and radioactive bands were detected by autoradiography (31).

Cyanogen bromide fragments were labeled with ^{125}I-Bolton-Hunter reagent, purchased from New England Nuclear, Boston, Massachusetts. The lyophilized collagen fragment was dissolved in 0.1 M sodium borate, pH 8.5, at a concentration of 3.2×10^{-5} M. Ice-cold solution, 400 μl, was added to 1 mCi of Bolton-Hunter reagent, previously dried from benzene in the reaction vial as recommended by the manufacturer. After 15 min at 0 °C, the reaction was terminated by adding 1 ml of 0.2 M glycine in borate buffer, pH 8.5. Iodinated fragments were separated from unbound reagent by chromatography on Sephadex G-25 previously equilibrated with Tris-buffered saline. Iodinated fragments were stored at −70 °C in Tris-buffered saline which contained 1 mg/ml bovine albumin. Approximately 0.01 mol of ^{125}I-reagent was incorporated per mol of cyanogen bromide fragment.

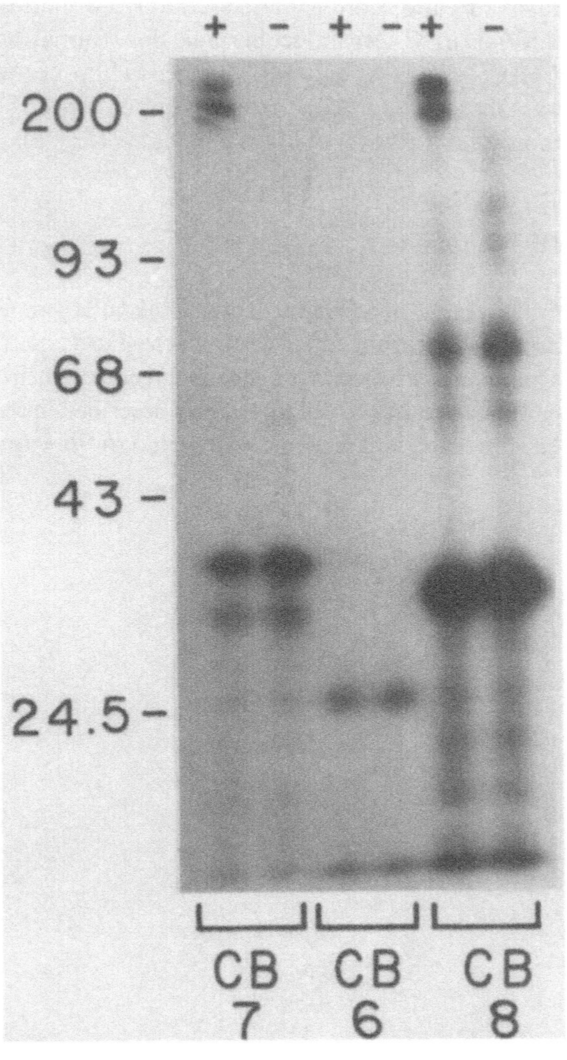

Fig. 1. Cross-linking of ^{125}I-α1(I)-CB7, ^{125}I-α1(I)-CB6, and ^{125}I-α1(I)-CB8 to fibronectin. Cross-linking mixtures contained the following final concentrations of reactants: fibronectin, 300 μg/ml; Factor XIII, 23 μg/ml; thrombin, 0.5 μg/ml; ^{125}I-fragment, approximately 3 μg/ml; and either 10 mM calcium chloride (+) or 10 mM EDTA (−). Thrombin was added last. Incubations were for 2 hr at 37 °C. Samples were then denatured and analyzed by electrophoresis on 10% polyacrylamide gels. Shown is a picture of the autoradiogram. Size markers (on the left) were reduced fibronectin (200 kd), phosphorylase (93 kd), albumin (68 kd), ovalbumin (43 kd), and chymotrypsinogen (24.5 kd).

Results

Electrophoresis through polyacrylamide gels of proteins denatured and reduced in 2% sodium dodecyl sulfate-2% 2-mercaptoethanol was used to ascertain whether Factor XIII$_a$-dependent cross-linking took place. Movement of radiolabel or protein to bands of larger size was taken as evidence of cross-linking.

Factor XIII$_a$ catalyzed the cross-linking of both ^{125}I-α1(I)-CB7 and ^{125}I-α1(I)-CB8 to fibronectin (Fig. 1). In contrast, barely detectable amounts of ^{125}I-α1(I)-CB6 were cross-linked (Fig. 1).

The pH optimum for cross-linking of ^{125}I-α1(I)-CB7 to fibronectin was 8.5 to 9.6 (Fig. 2). Cross-linking was somewhat enhanced at lower than physiological ionic strength (Fig. 3).

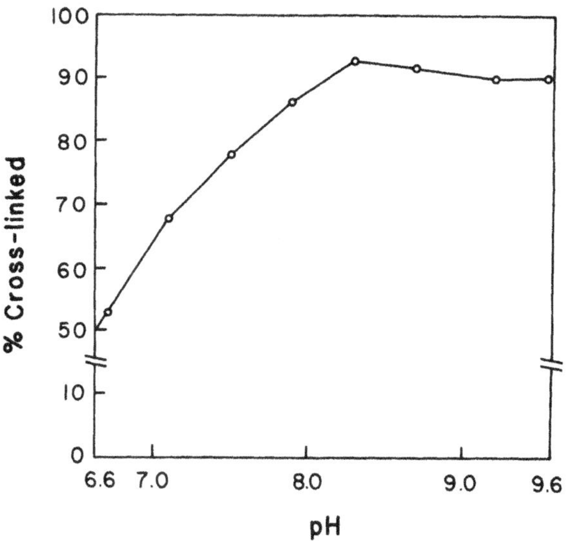

Fig. 2. Effect of pH on the cross-linking of ^{125}I-α1(I)-CB7 to fibronectin. Cross-linking mixtures contained the following final concentrations of reactants: fibronectin, 500 μg/ml; Factor XIII, 25 μg/ml; thrombin, 0.5 μg/ml; ^{125}I-α1(I)-CB7, approximately 20 μg/ml; and 10 mM calcium chloride. The concentration of sodium chloride was 150 mM, and pH was controlled with 20 mM Tris chloride. Cross-linking proceeded for 2 hr at 37 °C. Samples were analyzed as in Fig. 1. The autoradiogram was scanned in a densitometer, and the percent of ^{125}I-α1(I)-CB7 in bands of >200 kd was calculated.

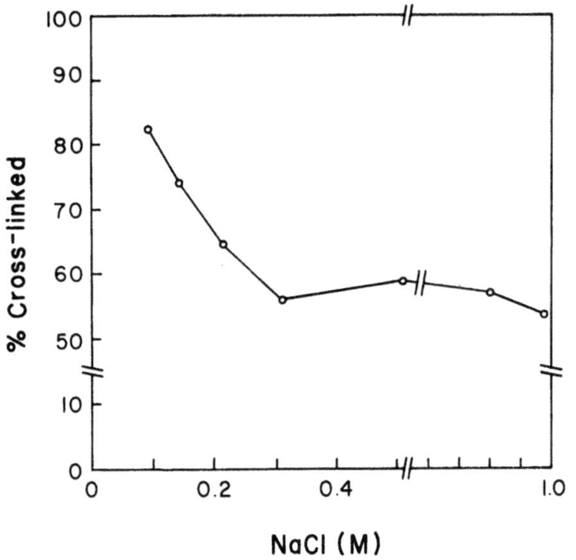

Fig. 3. Effect of ionic strength on the cross-linking of ^{125}I-α1(I)-CB7 to fibronectin. The experiment was similar to that shown in Fig. 2 except that pH was 7.4, and the concentration of sodium chloride (NaCl) was varied.

In contrast to experiments with types I and III collagens, in which these collagens could be cross-linked to fibronectin at 37 °C and not 22 °C or 0 °C, cross-linking was not observed with types IV or V (AB) collagens (Fig. 4). In another experiment, a second preparation of type V collagen did cross-link to fibronectin, but only when dialyzed at 4 °C against 10 mM sodium phosphate, 140 mM sodium chloride, pH 7.4, to induce periodic fibril formation (36). Cross-linking of type II collagen to fibronectin was observed (Fig. 5). Cross-linking of Clq to fibronectin, if present at all, was minimal (Fig. 5).

Discussion

The results of attempts to cross-link collagenous proteins to fibronectin are summarized in Table 1. The results with collagen and collagen fragments paralleled results of binding assays described in the introduction, i.e., proteins which bind to fibronec-

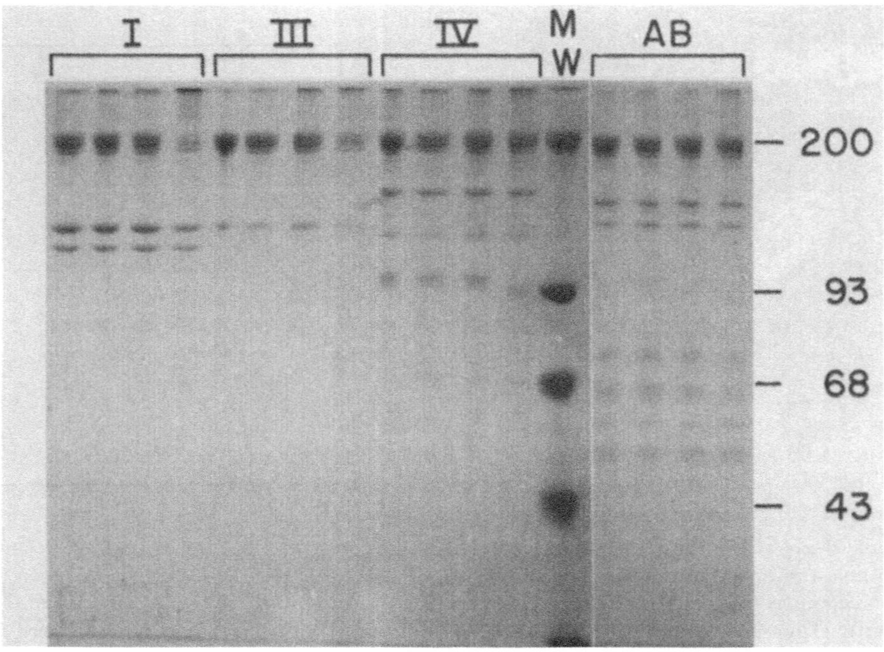

Fig. 4. Cross-linking of collagen types I, III, IV and V (or AB) to fibronectin. Cross-linking mixtures contained the following concentrations of reactants: fibronectin, 650 μg/ml; Factor XIII, 26 μg/ml; thrombin, 0.5 μg/ml; collagen, 250 μg/ml; and 10 mM calcium chloride. In each set of four lanes, the following samples were analyzed: first lane, 10 mM EDTA substituted for calcium chloride; second lane, incubation at 0 °C; third lane, incubation at 22 °C; and fourth lane, incubation at 37 °C. Incubations were for 2 hr. Samples were then denatured and analyzed by electrophoresis on 8% polyacrylamide gels. Shown is the pattern of protein staining. The 3.5% stacking gel, which contained cross-linked proteins, was lost during staining. Size (MW) markers are described in Fig. 1.

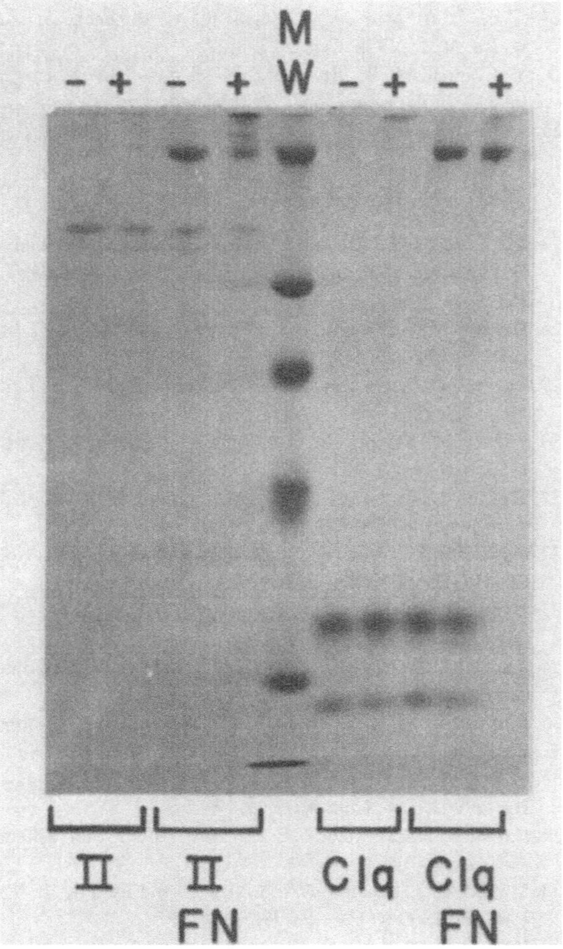

Fig. 5. Cross-linking of type II collagen and Clq to fibronectin (FN). Cross-linking mixtures contained the following final concentration of reactants: fibronectin, 250 μg/ml; Factor XIII, 26 μg/ml; thrombin, 0.5 μg/ml; type II collagen, 250 μg/ml, or Clq, 420 μg/ml; and 10 mM calcium chloride (+) or 10 mM EDTA (-). Incubations were for 2 hr at 37 °C. Samples were then denatured and analyzed by electrophoresis on 10% polyacrylamide gels as described in Fig. 4.

tin can be cross-linked to fibronectin. In addition, the effects of pH and ionic strength on the cross-linking of α1(I)-CB7 to fibronectin paralleled the effects of pH and ionic strength on the binding of fibronectin to gelatin (30). Cross-linking of Clq to fibronectin was not demonstrated, although Clq is in part collagenous (37), and Clq and fibronectin do interact if one of the pair is insolubilized (38–40). It should be noted, however, that cross-linking was not attempted at the low ionic strength which has been shown to favor the Clq-fibronectin interactions (39).

Table 1. Summary of cross-linking results.

Collagenous protein	Cross-linking?
Collagen types	
I	yes[a,b]
II	yes[b]
III	yes[a,b]
IV	no[b]
V (AB)	
amorphous fibrils	no[b]
native fibrils	yes[b]
Ascaris collagen	no[a]
Cyanogen bromide fragments	
α1(I)-CB6	slight[b]
α1(I)-CB7	yes[a,b]
α1(I)-CB8	yes[b]
Other	
asymmetric acetylcholinesterase	yes[c]
Clq	no[b]

[a] Ref. 31.
[b] Present communication.
[c] Ref. 35.

Further experiments are needed to establish if and how transglutaminase-mediated cross-linking between fibronectin and collagen is important. Ultra-structural localization of fibronectin and collagen at various times after subculture of human skin fibroblasts indicates that the early matrix of 2- to 10-day-old cells consists of polymerized fibronectin to which nonhelical collagen is bound and that the matrix of 4- to 8-week-old cells consists of periodic collagen fibrils to which fibronectin is bound (41,42). Fibronectin has been noted around new fibroblasts of scar tissue (43–45). Considerable amounts of Factor XIII are activated during normal blood coagulation, and inhibitors of the active enzyme have not been described. Failure of Factor XIII_a to cross-link fibronectin to collagen during wound healing and tissue remodeling may explain the observations that patients deficient in Factor XIII often suffer from poor wound healing and have difficulty carrying pregnancies to term (46).

Acknowledgements

The experimental work described herein was supported by National Institutes of Health Grant

HL 21644 and was performed during the tenure of an Established Investigatorship from the American Heart Association and its Wisconsin Affiliate.

References

1. Mosher, D. F., 1976. J. Biol. Chem. 251: 1639–1645.
2. Van Leuven, F., Cassiman, J.-J. and Van den Berghe, H., 1981. J. Biol. Chem. 256: 9016–9022.
3. Sakata, Y. and Aoki, N., 1980. J. Clin. Invest. 65: 290–297.
4. Tamaka, T. and Aoki, N., 1981. Biochim. Biophys. Acta 661: 280–286.
5. Birckbichler, P. J. and Patterson, M. K., Jr., 1978. Ann. N.Y. Acad. Sci. 312: 354–365.
6. Mosher, D. F., Schad, P. E. and Vann, J. M., 1980. J. Biol. Chem. 255: 1181–1188.
7. McDonagh, R. P., McDonagh, J., Petersen, T. E., Thøgersen, H. C., Skorstengaard, K., Sottrup-Jensen, L., Magnusson, S., Dell, A. and Morris, H. R., 1981. FEBS Lett. 127: 174–178.
8. Richter, H., Seidl, M. and Hörmann H., 1981. Hoppe-Seyler's Z. Physiol. Chem. 362: 399–408.
9. Williams, E. C., Janmey, P. A., Ferry, J. D. and Mosher, D. F., 1983. J. Biol. Chem. 257: 14973–14978.
10. Mosher, D. F., 1975. J. Biol. Chem. 250: 6614–6621.
11. Iwanaga, S., Suzuki, K. and Hashimoto, S., 1978. Ann. N. Y. Acad. Sci. 312: 56–73.
12. Mosher, D. F. and Proctor, R. A., 1980. Science 209: 927–929.
13 Williams, E. C., Janmey, P. A., Johnson, R. B. and Mosher, D. F., 1983. J. Biol. Chem. (in press).
14. Kamykowski, G. W., Mosher, D. F., Lorand, L. and Ferry, J. D., 1981. Biophys. Chem. 13: 25–28.
15. Grinnell, F., Feld, M. and Minter, D., 1980. Cell 19: 517–525.
16. Ruoslahti, E. and Vaheri, A., 1975. J. Exp. Med. 141: 497–501.
17. Proctor, R. A., Mosher, D. F. and Olbrantz, P. J., 1983. J. Biol. Chem. 257: 14788–14794.
18. Bannai, S. and Ishii, T., 1980. J. Cellular Physiol. 104: 215–223.
19. Keski-Oja, J., Mosher, D. F. and Vaheri, A., 1976. Cell 9: 29–35.
20. Mosesson, M. W., Chen, A. B. and Huseby, R. M., 1975. Biochim. Biophys. Acta 386: 509–524.
21. Hynes, R. O. and Destree, A., 1977. Proc. Natl. Acad. Sci. USA 74, 2855–2859.
22. Keski-Oja, J., Mosher, D. F. and Vaheri, A., 1977. Biochem. Biophys. Res. Commun. 74: 699–706.
23. Chen, L. B., Murray, A., Segal, R. A., Bushnell, A. and Walsh, M. L., 1978. Cell 14: 377–391.
24. McConnell, M. R., Blumberg, P. M. and Rossow, P. W., 1978. J. Biol. Chem 253: 7522–7530.
25. McKeown-Longo, P. J. and Mosher, D. F., 1983. J. Cell Biol. (in press).
26. Mosher, D. F., 1980. Prog. Hemostas. Thrombosis 5: 111–151.
27. Kleinman, H. K., McGoodwin, E. B., Martin, G. R., Klebe, R. J., Fietzek, P. P. and Woolley, D. E., 1978. J. Biol. Chem. 253: 5642–5646.
28. Dessau, W., Adelmann, B. C., Timpl. R. and Martin, G. R., 1978. Biochem. J. 169: 55–59.
29. Engvall, E., Ruoslahti, E. and Miller, E. J., 1978. J. Exp. Med. 147: 1584–1595.
30. Vuento, M., Salonen, E., Österlund, K. and Stenman, U.-H., 1982. Biochem. J. 201: 1–8.
31. Mosher, D. F., Schad, P. E. and Kleinman, H. K., 1979. J. Clin. Invest. 64: 781–787.
32. Pearlstein, E., Gold, L. I. and Garcia-Pardo, A., 1980. Mol. Cell. Biochem. 29: 103–128.
33. Hynes, R. O., 1981. Cell Biology of Extracellular Matrix (Hay, E. D., ed.) pp. 295–333, Plenum Press, New York.
34. Ruoslahti, E., Engvall, E. and Hayman, E. G., 1981. Collagen Res. 1: 95–128.
35. Emmerling, M. R., Johnson, C. D., Mosher, D. F., Lipton, B. H. and Lilien, J. E., 1981. J. Biochem. 20: 3242–3247.
36. Chiang, T. M., Mainardi, C. L., Seyer, J. M. and Kang, A. H., 1979. J. Lab. Clin. Med. 95: 99–107.
37. Porter, R. R. and Reid, K. B. M., 1979. Adv. Protein Chem. 33: 1–71.
38. Menzel, E. J., Smolen, J. S., Liotta, L. and Reid, K. B. M., 1981. FEBS Lett. 129: 188–192.
39. Bing, D. H., Almeda, S., Isliker, H., Lahav, J. and Hynes, R. O., 1982. Proc. Natl. Acad. Sci. USA 79: 4198–4201.
40. Pearlstein, E., Sorvillo, J. and Gigli, I., 1982. J. Immunol. 128: 2036–2039.
41. Furcht, L. T., Smith, D., Wendelschafer-Crabb, G., Mosher, D. F. and Foidart, J. M., 1980. J. Histochem. Cytochem. 28, 1319–1333.
42. Furcht, L. T., Wendelschafer-Crabb, G., Mosher, D. F. and Foidart, J. M., 1980. J. Supramol. Struct. 13: 15–33.
43. Matsuda, M., Yoshida, N., Aoki, N. and Wakabayashi, K., 1978. Ann. N.Y. Acad. Sci. 312: 56–73.
44. Kurkinen, M., Vaheri, A., Roberts, P. J. and Stenman, S., 1980. Lab. Invest. 43: 47–51.
45. Grinnell, F., Billingham, R. E. and Burgess, L., 1981. J. Invest Dermat. 76: 181–189.
46. Duckert, F., 1972. Ann. N.Y. Acad. Sci. 202: 190–199.

Received 3 November 1982.

Molecular and Cellular Biochemistry 58, 69–77 (1984).

Transglutaminase and receptor-mediated endocytosis in macrophages and cultured fibroblasts

Peter J. A. Davies and Michael P. Murtaugh
Dept. of Pharmacology and Division of Endocrinology, Dept. of Medicine, University of Texas Medical School at Houston, Houston, TX 77025, U.S.A.

Summary

The function of intracellular transglutaminases remains to be clarified. In fibroblasts the links between the activity of this enzyme and receptor-mediated endocytosis are complex and open to interpretation. However, the issue cannot be firmly laid to rest until the structural specificity of the alkylamine inhibitors of endocytosis is explained. In macrophages, there is substantial evidence that the enzyme plays some role in receptor-mediated phagocytosis, but what this role is and how it might relate to endocytosis in other types of cells is at present an unresolved issue.

Introduction

Enzymes that catalyze the covalent incorporation of amines into protein-bound glutamine residues, transglutaminases, are widely distributed in both the tissues and extracellular compartments of higher animals. Several of these enzymes, such as the transglutaminase of plasma, Factor XIII, or epidermal transglutaminase, appear to be responsible for the physiologically important cross-linking of structural proteins such as fibrin (1, 2) or involucrin (3). The functions of other transglutaminases, particularly the ubiquitous, intracellular transglutaminase, tissue transglutaminase, are much less well understood. Recently it has been observed that many of the compounds that interfere with the receptor-mediated endocytosis of certain ligands by cultured fibroblasts also are either substrates or inhibitors of cellular transglutaminases (4–7). This has raised the possibility that transglutaminases participate in receptor-mediated endocytosis. Studies to test this hypothesis have been carried out both in cultured fibroblasts and in macrophages and a variety of interesting and sometimes conflicting results have been observed. To clarify this situation we would like to review those studies that have directly examined the role of transglutaminase in endocytosis and to discuss their various interpretations. It is our aim to keep the focus of this review quite narrow and not to digress into a more general discussion of the process of receptor-mediated endocytosis. The reader interested in a more comprehensive discussion of receptor-mediated endocytosis is referred to any of several excellent reviews of the topic that have recently been published (8–11).

Receptor-mediated phagocytosis in macrophages and a potential role for transglutaminase

Although, as will be discussed in a following section, the initial proposal that transglutaminase might participate in endocytosis developed from studies of cultured fibroblasts, recent work from several laboratories has suggested the enzyme may be important to the function of macrophages. Because the evidence in macrophages is relatively direct and concerns a single ligand binding system, we will discuss these studies first and then deal with the more complex issues of fibroblast endocytosis.

Macrophages represent a cell uniquely specialized for endocytosis (12). Not only do they bind and internalize a variety of soluble ligands through receptor-mediated endocytosis but they also phagocytose large insoluble particles and cells. Macrophages can phagocytose inert particles such as latex beads, but they are much more active in internalizing particles coated with ligands for which they express specific cell surface receptors. On the macrophage surface distinct receptors for the Fc region of immunoglobulin (13), for the C3b component of complement (14), and for terminal mannosyl glycoconjugates (15) appear to mediate phagocytosis through separate, specific pathways. Though the presence of different mechanisms for internalization via these three receptors is clear (16–18), the reasons for maintaining separate morphological pathways for particle ingestion are not well understood.

Recognition of the potential importance of transglutaminase in macrophage function has developed from the finding that macrophages activated by a variety of stimuli show marked increases in transglutaminase activity (19–21). Schroff et al. (19) and Leu et al. (20) have reported that macrophages elicited by intraperitoneal injections of several agents that induce an inflammatory response contain much higher levels of transglutaminase activity than resident, unstimulated peritoneal macrophages. We have recently shown that this increase in enzyme activity can be reproduced *in vitro* by culture of resident peritoneal macrophages in homologous serum and is due to a rapid increase in the rate of synthesis and accumulation of tissue transglutaminase (21). The accumulation of tissue transglutaminase in these cells begins within ninety minutes of exposure of macrophages to serum and reaches 1–2% of total cellular protein within 48 hours (21).

It is striking that macrophages possess such a dynamic capability to synthesize and accumulate a specific enzyme. Several studies have suggested that this accumulation of transglutaminase in macrophages is linked to an enhanced capacity for receptor-mediated endocytosis (19–22). Macrophages elicited by inflammatory agents show modest increases in their capacity to endocytose IgG-coated particles (23) and a marked increase in their capacity for C3b-mediated endocytosis (24). Using a fluorescent stain for transglutaminase activity, Schroff et al. showed a correlation between the level of transglutaminase activity in individual macrophages and the extent to which they were capable of phagocytosing Ig-coated erythrocytes (19). Kannagi et al. have also reported that the differentiation of Ml murine myeloblastic leukemia cells induced by lymphokine is associated with a large increase in transglutaminase activity (25). The induction of the enzyme matches the appearance of both functional Fc receptors and phagocytosis in these cells.

Compounds which interfere with both transglutaminase activity and receptor-mediated endocytosis have also been used to support a role for transglutaminase in macrophage phagocytosis. Cystamine directly inhibits transglutaminase activity *in vitro* and in intact cells (26). Concentrations of cystamine that inhibit macrophage transglutaminase also block Fc receptor-mediated endocytosis of opsonized red cells by guinea pig peritoneal macrophages (20). Methylamine and dansylcadaverine are substrates of transglutaminase that inhibit protein cross-linking by competing with endogenous substrates for transglutaminase-catalyzed glutamine transamidation. Both methylamine and dansylcadaverine inhibit Fc receptor-mediated endocytosis by macrophages and dansylcadaverine is one to two orders of magnitude more potent than methylamine in blocking the process (20, 22). Furthermore, pretreating cells with methylamine reduces the binding and rosetting of opsonized erythrocytes (22). While the mechanism of this effect is not clear, Fesus has shown that transglutaminase can polymerize a low affinity Fc receptor and convert it to an aggregated form that has an apparent high affinity for opsonized erythrocytes (27). These studies suggest that transglutaminase might, in some unspecified way, be capable of processing Fc receptors from a low to a high affinity state and thereby promoting ligand binding and phagocytosis.

A link between ligand binding and transglutaminase activity has been suggested by two studies. Lectin binding to lymphocyte membranes has been reported to produce a marked activation of cellular transglutaminase activity (28). The binding of either aggregated IgG or opsonized cells to the Fc receptor of macrophages also can produce marked and immediate activation of transglutaminase (22). Furthermore in cells pre-incubated with radioactive methylamine, this activation produces increased incorporation of the amine into membrane proteins (22). These studies suggest that transglutaminase activity may promote the formation of high affinity Fc receptors and these receptors in turn may produce increased transglutaminase activity and membrane protein cross-linking. It is possible that such an amplifying cycle may facilitate the formation of large and stable attachment sites that may be required to internalize ligands as large as whole cells.

Receptor-mediated endocytosis in fibroblasts

The proposition that transglutaminase might participate in receptor-mediated endocytosis was an outgrowth of studies on the ability of certain drugs to interfere with endocytosis in cultured fibroblasts. To provide a framework for discussion of these studies we will present a brief description of the major steps in receptor-mediated endocytosis of ligands by fibroblasts. A more detailed description of the experimental basis of this description can be found in the review articles referred to previously (8–11).

The distribution of unoccupied cell surface receptors appears to depend on the characteristics of the particular receptor molecule itself. Many types of receptors when unoccupied are uniformly distributed on the cell surface and freely mobile. Other receptors do not appear to be stable in the membrane as monomeric species but spontaneously precluster in coated pits in the membrane. When ligands bind to diffusely-distributed receptors the ligand–receptor complexes seem to self-associate and to accumulate in large aggregates or clusters on the cell surface. These aggregates then rapidly collect in coated pits. If the receptor is already preclustered in the coated pits, then ligand–receptor complexes form there without redistribution. Once ligand–receptor complexes are formed or accumulate in coated pits, they are rapidly endocytosed. The precise nature of this endocytotic event is the subject of some dispute, but it ultimately produces membrane-limited vesicles, receptosomes, in the cytoplasm that contain both ligand and receptor (29). These receptosomes seem to undergo rapid acidification that is presumed to produce dissociation of most ligands from their receptors (30). Once internalized the endocytic vesicles appear to travel to a variety of different targets, depending primarily on the ligand and the cell type (11). One major pathway is fusion with elements of the GERL apparatus and ultimate lysosomal destruction of the ligand. The fate of the receptor is less clear cut. There is substantial evidence that the receptor may be returned to the membrane, a recycling step that conserves the pool of cell surface receptors. Alternatively, some of the receptors may be degraded by lysosomes resulting in a depletion of cell surface receptors and a ligand-induced down-regulation of surface receptor number.

Because receptor-mediated endocytosis is a potential control point for the regulation of cell surface receptor number and hence cellular responsiveness to hormones, it was of interest to study the biochemical mechanisms involved. Two ligands, $\alpha_2$macroglobulin and epidermal growth factor, have proved to be particularly useful for characterizing the various steps in the pathway of receptor-mediated endocytosis. In the following two sections we will discuss separately the studies that have reported the inhibition of endocytosis of these two proteins and then consider other examples where drugs have been used to inhibit ligand internalization.

Transglutaminase and the endocytosis of $\alpha_2$macroglobulin

α_2Macroglobulin is a serum protease inhibitor that binds to specific receptors on the surface of both fibroblasts and macrophages (31–34). The physiological ligand is apparently the complex that forms between α_2M and proteases, but purified α_2M alone will also bind the same receptor (35). Following binding, α_2M is rapidly internalized via a coated pit-type pathway typical of receptor-mediated endocytosis (36, 37). It is worth noting that α_2M contains substantial regions of homology to both the C3 and C4 components of complement (38, 39) and also that the protein is an excellent substrate of transglutaminase (40, 41).

The distribution of α_2M receptors on the cell surface is heterogeneous. Morphological studies have indicated that a fraction of the receptors are distributed diffusely on the cell surface, but many unoccupied α_2M receptors pre-cluster in coated pits (42). Radioligand binding studies also are compatible with two pools of α_2M receptors, one smaller pool with a very high affinity for the protein and a larger pool with slightly lower affinity (43). The complexes of α_2M and its receptor are rapidly internalized; the α_2M is delivered to and degraded by lysosomes (44) and the receptors rapidly recycle to the cell surface. There is no evidence of down-regulation of α_2M receptors.

A survey of compounds that might interfere with the receptor-mediated endocytosis of $\alpha_2$macroglobulin identified several compounds that not only blocked endocytosis but also were substrates or inhibitors of the transglutaminase activity of the cells (4–6, 45). Alkylamines in particular were effective inhibitors of α_2M endocytosis (4–6, 45). The first compound identified, methylamine, inhibited clustering of α_2M on fibroblasts at concentrations of 10–20 mM (45). Straight chain alkylamines with longer alkyl chains were more potent than methylamine in inhibiting endocytosis and branched chain amines were less effective than the straight chain homologues (4, 5, 46). This structural specificity for blocking endocytosis matched the specificity for inhibiting fibroblasts transglutaminase activity (1, 4, 5). Long chain alkylamines have a higher affinity for the enzyme than methylamine and branched chain amines have a lower affinity than their straight chain isomers. Dansylcadaverine, a substituted diamine, was both a particularly potent inhibitor of endocytosis and an excellent substrate of transglutaminase (2, 47). Concentrations as low as 15–20 μM dansylcadaverine produced marked inhibition of the formation of fluorescent clusters of α_2M in cells treated with rhodamine-conjugated α_2M (4, 5, 6, 46). Furthermore the effects of dansylcadaverine were quite specific. Dansylcadaverinol and N,N-dimethyldansylcadaverine, two structural analogues with similar physical properties but which are inactive as inhibitors of transglutaminase activity, were also inactive as inhibitors of endocytosis (46).

Although the initial observations on the effects of amines as inhibitors of α_2M endocytosis were based on visual observation of clustering of fluorescent ligand on the cell surface (4, 5, 45), subsequent studies of endocytosis using radiolabelled α_2M have confirmed the observation (6, 46). However it is notable that the concentrations of amines, either methylamine or dansylcadaverine, required to block visible clustering of α_2M have been consistently lower than those required to block completely internalization.

Only amines that can diffuse across the cell membrane block α_2M endocytosis. Impermeant amines such as putrescine, histamine, glycinamide and cystamine were only effective as inhibitors of endocytosis in cells that were permeabilized by treatment with 10% DMSO (4, 5). Furthermore, methylamine and dansylcadaverine were only effective when cells were preincubated in the compound at 37°C for at least 30 minutes (4, 6).

Several possibilities to account for the inhibitory action of the amines on receptor-mediated endocytosis have been ruled out. Neither methylamine nor dansylcadaverine directly interfere with the binding of α_2M to its receptor (46). The amines also do not alter the mobility of the ligand–receptor complexes in the membrane (48). The effects of the alkylamines are reversible, cells pretreated with dansylcadaverine rapidly recover the capacity to cluster and internalize α_2M (4, 45). Also α_2M bound to cells whose clustering has been inhibited by the presence of the amine is still competent to cluster and be internalized following removal of the inhibitor (49).

Two studies have examined the kinetics of α_2M binding and internalization in amine-treated cells. Dickson et al. (43) have reported that pre-incubation of cells in dansylcadaverine results in depletion of the pool of high affinity cell surface receptors for α_2M without altering the low affinity subset. They interpret their findings as suggesting that amines interfere with the processing, possibly mediated by transglutaminase, of low affinity receptors to a high affinity state. Van Leuven and his colleagues (6) reached a similar but slightly different conclusion. These workers observed that methylamine and dansylcadaverine reduced the maximal rates of α_2M endocytosis, suggesting a depletion of the number of receptors. This depletion was independent of the presence of ligand and may be due to the ability of the amines to block receptor recycling. Van Leuven concluded that amines inhibit the tonic ligand-independent recycling of α_2M receptors possibly through a transglutaminase-mediated process thus causing an accumulation of unoccupied receptors inside the cell. Hence both the studies of Dickson and Van Leuven point to an effect of amines on the number of high affinity cell surface receptors for α_2M. Whether transglutaminase plays any part in this process is still unresolved.

Recycling of α_2M receptors includes ligand–receptor accumulation in receptosomes where acidification releases the ligand from the receptor. Because lysosomotropic drugs such as chloroquine and ammonia block the entry of a variety of macromolecules into cells (50–53), it is possible that the inhibitory effects of alkylamines on receptor recycling are simply due to a perturbation of lysosomal pH. Alkylamines are able to concentrate in lysosomes by virtue of their lipophilicity and the presence of an ionizable amine (54, 55). Furthermore, concentrations of methylamine that block α_2M endocytosis also raise intralysosomal pH (30). Nevertheless, several lines of evidence indicate that the effect of alkylamines on receptor recycling is not due solely to effects on lysosomal pH. Neither inhibition of lysosomal function by chloroquine nor inhibition of lysosomal acidification by bicarbonate interferes with α_2M endocytosis in fibroblasts (6, 45). Secondly, the ability of alkylamines to alter lysosomal function shows no structural specificity (55). There is no difference in the potency of either short and long chain amines or straight, as opposed to branched, chain amines (55). Finally, dansylcadaverine, which interferes with α_2M endocytosis in a manner entirely analogous to the alkylamines, inhibits recycling at concentrations which produce no perturbation of lysosomal pH (30). Thus the inhibition of receptor recycling by alkylamines does not seem to be through their ability to perturb intracellular or intralysosomal pH.

The ability of dansylcadaverine to reproduce the inhibitory effects of alkylamines but at much lower concentrations is one of the strongest arguments for a role of transglutaminase in endocytosis. This argument depends on the specificity of dansylcadaverine as an inhibitor of transglutaminase, however the drug is not entirely specific. At the levels at which it blocks clustering of α_2M on cells (10–50 μM) dansylcadaverine appears quite innocuous to cultured cells. Cells can be grown indefinitely in these concentrations without alterations in morphology or viability. At levels above 200 μM however, cells show marked evidence of deleterious effects of the drug (47, 56). Cells detach from the substratum and protein synthesis is depressed (56). Dansylcadaverine is very similar in structure to a group of halogenated naphthylsulfonamides identified by Hidaka as potent inhibitors of calmodulin (57). Dansylcadaverine itself can inhibit calmodulin-dependent enzyme activation at high concentrations (200–500 μM) although the drug is much less active than the most potent of the halogenated compounds, W-7, (N-(6-aminohexyl)-5-chloro-1-naphthylenesulfonamide) (56, 58). High concentrations of dansylcadaverine have also been reported to inhibit the activity of 17-ketosteroid reductase in testicular interstitial cells (59). Thus the effects of dansylcadaverine on endocytosis of α_2M in fibroblasts could be due to inhibition of metabolic events other than transglutaminase-mediated reactions.

Bacitracin is another compound that has received considerable attention as an inhibitor of α_2M endocytosis. Bacitracin is a cyclic peptide containing an ϵ-(β-asparaginyl)-lysine cross-bridge. The compound does not inhibit purified tissue transglutaminase but it does produce partial inhibition of transglutaminase activity in a variety of cell extracts (4, 7, 60). For this reason the ability of bacitracin to inhibit endocytosis has been used as further evidence for a link between transglutaminase and endocytosis. In fact it appears likely that the effects of bacitracin are quite different from those of the alkylamines or dansylcadaverine. Bacitracin appears to behave as a competitive inhibitor of the binding of α_2M to its receptor (49, 61). The effects of the drug do not require preincubation and the inhibition of endocytosis is immediately reversible upon removal of the compound (49). Therefore it does not appear that the inhibitory effects of bacitracin on α_2M endocytosis are related to the activity of tissue transglutaminase.

Alkylating reagents also have been reported to inhibit α_2M endocytosis in fibroblasts (5, 46). These compounds produce an irreversible inhibition of endocytosis. The reagents that have been reported to inhibit irreversibly α_2M endocytosis are blocked diazoxonorvaline (BDONV), α-bromo-4 hydroxynitro-acetophenone (BHNA), β-bromophenacyl bromide, several chloroketone derivatives of phenylalanine and N-ethylmaleimide. These compounds are also potent inhibitors of transglutaminase (5, 46) because the reactive center thiol group of the enzyme is readily inactivated by many electrophilic alkylating agents. However, alkylating agents react with many enzymes and proteins and their effects can not be directly linked to inhibition of transglutaminase. One exception to this is the experiments reported by Levitzki et al. (5) on the action of alkylating analogues of norvaline and norleucine. Both the chloroketone and diazooxo derivative of norvaline (BDONV) were irreversible inhibitors of transglutaminase and endocytosis. The diazooxo derivative of norleucine, with differs only by the presence of an additional methylene group in the side chain, did not inhibit transglutaminase and did not block endocytosis. Therefore the site of inhibition of the alkylating groups is likely an enzyme with marked structural specificity and this chemical specificity is at least comparable to the substrate specificity of transglutaminase.

While studies with inhibitors provide circumstantial evidence for a role of transglutaminase in endocytosis they are at best ambiguous. Therefore we have developed a more direct approach to examining the issue by searching for cells that lack transglutaminase and then examining their capacity to endocytose α_2M (58). Balb 3T3 cells have low but detectable levels of tissue transglutaminase (7 ng per mg cell protein) and they bind and internalize α_2M. SV-40 transformed Balb 3T3 (SV-3T3) cells have no measurable transglutaminase activity and antibodies to tissue transglutaminase detect no enzyme in extracts of these cells (<1 ng per mg cell protein). In spite of a complete lack of tissue transglutaminase, SV-3T3 cells internalize α_2M as well as Balb 3T3 cells (58), therefore the enzyme cannot be required for the internalization of α_2M. If the effects of alkylamines are in fact due to interference with endogenous transglutaminase activity it can only be an effect on an accessory system such as the formation of macroscopic cell surface aggregates. It is relevant to note that the effects of methylamine and dansylcadaverine have been most prominent in cells such as CHO cells or WI-38 cells that have high transglutaminase activity (58). Cells that have low levels of transglutaminase, such as Balb 3T3 cells or SV-3T3 cells show no effects of dansylcadaverine on α_2M endocytosis (58, 62). It is possible that incubation of cells that contain transglutaminase with alkylamines derivatizes proteins that normally are involved with cell surface clustering of ligands such as α_2M.

Transglutaminase and the endocytosis of epidermal growth factor

Epidermal growth factor is a polypeptide hormone that binds to cell surface receptors on a variety of cultured cell lines. The unoccupied receptors are uniformly diffuse in their distribution and only following EGF binding do the receptors and ligand cluster in coated pits (63). Initial studies with fluorescent derivatives of EGF suggested that, like α_2M, the endocytosis of this hormone could be inhibited by methylamine, dansylcadaverine and bacitracin (4, 45, 64). Experiments on the endocytosis of radiolabelled EGF did not confirm these findings. Methylamine and bacitracin did not block EGF endocytosis and dansylcadaverine blocked the process but only at levels much higher than those required to inhibit α_2M endocytosis (62, 65, 66). The discrepancy between the findings with the fluorescent preparations and the radiolabelled preparations was ultimately resolved with the recognition that contaminating peptides in the fluorescent EGF preparation were contributing to the observed inhibition of endocytosis by amines (66, 67).

We have compared the endocytosis of EGF in cells that either have abundant tissue transglutaminase (WI-38 cells) or cells that completely lack the enzyme, such as SV-40 transformed WI-38 cells (VA-13) (58). Both types of cells internalize EGF equivalently. Furthermore, levels of dansylcadaverine that block EGF internalization in the WI-38 cells are equally effective in the VA-13 cells in spite of the complete lack of transglutaminase in these latter cells. Therefore the ability of high levels of dansylcadaverine to block EGF internalization is clearly due to mechanisms other than interference with transglutaminase activity. In summary there is no evidence that links the activity of tissue transglutaminase with any aspect of EGF internalization.

Transglutaminase and the receptor-mediated endocytosis of other ligands

Both alkylamines and dansylcadaverine have been found to be effective inhibitors of a variety of endocytic processes. Among the ligands whose endocytosis can be blocked by either methylamine or dansylcadaverine are neurotransmitters and hormones such as catecholamines (7), acetylcholine (68) and tri-iodothyronine (69); peptide hormones and growth factors such as insulin (9, 70) or interleukin-2; bacterial toxins such as diphtheria (32) and pseudomonas toxin (51); transport proteins such as low density lipoprotein and vitellogenin (72) and several viruses such as Semliki forest virus

(52) or vesicular stomatitis virus (53, 73). In many instances the inhibitory effects of amines are reproduced by chloroquine and other lysosomotropic drugs making it likely that the activity of the amines is related to their ability to alter lysosomal pH. The endocytosis of insulin however is inhibited by dansylcadaverine even in the presence of chloroquine (70). Dansylcadaverine pretreatment of fibroblasts blocks the transfer of cell surface insulin to an intracellular pool and also blocks the effect of insulin on overall protein degradation rates (70). Dansylcadaverine also blocks the ligand-induced down-regulation of insulin receptors in cultured human fibroblasts (74). The inhibitory effects of dansylcadaverine on endocytosis are also observed with methylamine and BDONV, a selective irreversible inhibitor of transglutaminase (9). The inhibition of insulin endocytosis appears to follow a pattern very similar to the inhibition of α_2M endocytosis and to be different from the inhibition of EGF endocytosis.

It has been suggested that the internalization of catecholamines may depend on the presence of an active transglutaminase (7). Chuang has characterized the ligand-induced endocytosis of β-receptors by frog erythrocytes and demonstrated a marked correlation between the ability of compounds to block endocytosis and their ability to interfere with the activity of frog erythrocyte transglutaminase (7). Inhibition of endocytosis of this receptor also blocked ligand-induced desensitization of the frog erythrocyte adenylate cyclase. We have also observed that low concentrations of dansylcadaverine can block ligand-induced desensitization in hamster fibroblasts (Davies, unpublished observation).

Transglutaminase and ligand and receptor processing

Several recent studies have indicated that a fraction of ligands such as insulin (75), EGF (76) and thrombin (77) become covalently coupled to their receptor. The interaction is a time-dependent, saturable reaction that occurs on the surface of several types of cells. The nature of the cross-link has not been established but it has been suggested that transglutaminase-catalyzed cross-linking might be involved. The coupling is not inhibited by exogenous alkylamines but the concentrations used might not be sufficient to compete out cross-links between non-covalently bound proteins such as ligand–receptor aggregates.

A potential link between transglutaminase and receptor processing has come from studies by Tolleschaug et al. on the post-translational processing of the LDL receptor (78). Studies with pulse-labelled LDL receptors indicate that following synthesis, the LDL receptor becomes processed to a mature form that is covalently cross-linked to a second unidentified protein. This processing event is required for generation of an active LDL receptor. While the mechanism of this processing is not yet established, transglutaminases are one of the few enzymes known to catalyze the covalent cross-linking of intracellular proteins. Thus the possibility exists that, as in the case of the Fc receptor (27), transglutaminase might participate in the post-translational processing of receptors.

Acknowledgements

We would like to express our appreciation to Ms Sandra Hobbs for preparation of this manuscript. Support for studies described in this report comes from NIH grant AM27078 and a grant, BC-334A from the American Cancer Society.

References

1. Folk, J. E. and Finlayson, J. S., 1977. Adv. Protein Chem. 31: 1–120.
2. Folk, J. E., 1980. Ann. Rev. Biochem. 49: 517–531.

2a. Lorand, L., Ong, H. H., Lipinski, B., Rule, N. G., Downey, J. and Jacobsen, A., 1966. Biochem. Biophys. Res. Commun. 25: 629–637.

3. Rice, R. H. and Green, H., 1979. Cell 18: 681–694.
4. Davies, P. J. A., Davies, D. R., Levitzki, A., Maxfield, F. R., Milhaud, P., Willingham, M. C. and Pastan, I., 1980. Nature 283: 162–167.
5. Levitzki, A., Willingham, M. and Pastan, I., 1980. Proc. Natl. Acad. Sci. U.S.A. 77: 3453–3459.
6. Van Leuven, F., Cassiman, J. J. and Van Den Berghe, H., 1980. Cell 20: 37–43.
7. Chuang, D. M., 1981. J. Biol. Chem. 256: 8291–8293.
8. Goldstein, J. L., Anderson, R. W. and Brown, M. S., 1980. Nature 279: 679–685.
9. Pastan, I. H. and Willingham, M. C., 1981. Science 214: 504–509.
10. Schlessinger, J., 1980. Trends in Biochemical Sciences 5: 210–214.

11. Roth, T. F. and Woods, J. W., 1982. In: Differentiation and Function of Hematopoietic Cell Surfaces, pp. 163–181. Alan R. Liss Inc., New York.
12. Silverstein, S. C., Steinman, R. M. and Cohn, Z. A., 1977. Ann. Rev. Biochem. 46: 669–722.
13. Griffin, F. M., Griffin, J. A. and Silverstein, S. C., 1975. J. Exp. Med. 144: 788–809.
14. Griffin, F. M., Bianco, C. and Silverstein, S. C., 1975. J. Exp. Med. 141: 1269–1277.
15. Stahl, P. D., Rodman, J. S., Miller, M. J. and Schlesinger, P. H., 1978. Proc. Natl. Acad. Sci. U.S.A. 75: 1399–1403.
16. Munthe-Kaas, A. C., Kaplan, G. and Selzelid, R., 1976. Exp. Cell Res. 103: 201–212.
17. Kaplan, G., 1977. Scand. J. Immunol. 6: 797–807.
18. Aggeler, J. and Werb, Z., 1982. J. Cell Biol. 94: 613–623.
19. Schroff, G., Neumann, C. and Sorg, C., 1981. Eur. J. Immunol. 11: 637–642.
20. Leu, R. W., Herriott, M. J., Moore, P. M., Orr, G. R. and Birchbichler, P. J., 1982. Exp. Cell Res. 141: 191–199.
21. Murtaugh, M. P., Mehta, K., Juliano, R. L. and Davies, P. J. A., 1983. J. Biol. Chem. (in press).
22. Fesus, L., Sandor, M., Horvath, L. I., Bagyinka, C., Erdei, A. and Gergely, J., 1981. Mol. Immunol. 18: 633–638.
23. Ezekowitz, R. A. B., Austyh, J., Stahl, P. D. and Gordon, S., 1981. J. Exp. Med. 154: 60–76.
24. Bianco, C., Griffin, F. M. and Silverstein, S. C., 1975. J. Exp. Med. 141: 1278–1290.
25. Kannagi, R., Teshigawara, K., Noro, J. and Masuda, T., 1982. Biochem. Biophys. Res. Commun. 105: 164–171.
26. Birckbichler, P. J. and Patterson, M. K., 1980. In: Tumor Cell Surfaces and Malignancy, pp. 845–855. Alan R. Liss Inc., New York.
27. Fesus, L., Erdei, A., Sandor, M. and Gergely, J., 1982. Mol. Immunol. 19: 39–43.
28. Novogrodsky, A., Quittner, S., Rubin, A. L. and Stengel, K. H., 1980. Proc. Natl. Acad. Sci. U.S.A. 75: 1157–1161.
29. Willingham, M. C. and Pastan, I. H., 1980. Cell 21: 67–72.
30. Tycko, B. and Maxfield, F. R., 1982. Cell 28: 643–651.
31. Van Leuven, F., Cassiman, J. J., Van Den Berghe, H., 1979. J. Biol. Chem. 254: 5155–5160.
32. Dickson, R. B., Willingham, M. C. and Pastan, I. H., 1981. J. Biol. Chem. 256: 3453–3459.
33. Kaplan, J. and Nielson, M. L., 1979. J. Biol. Chem. 254: 7323–7328.
34. Mosher, D. and Vaheri, A., 1980. Biochim. Biophys. Acta 627: 113–122.
35. Imber, J. M. and Pizzo, S. V., 1981. J. Biol. Chem. 256: 8134–8139.
36. Willingham, M. C., Maxfield, F. R. and Pastan, I. H., 1979. J. Cell Biol. 82: 614–625.
37. Willingham, M. C., Rutherford, A. V., Gallo, M. G., Wehland, J., Dickson, R. B., Schlegel, R. and Pastan, I. H., 1981. J. Histochem. Cytochem. 29: 1003–1013.
38. Howard, J. B., 1980. J. Biol. Chem. 255: 7082–7084.
39. Gorski, J. P. and Howard, J. B., 1980. J. Biol. Chem. 255: 10025–10028.
40. Mosher, D. F., 1976. J. Biol. Chem. 251: 1639–1645.
41. Mortensen, S. B., Sottrup-Jensen, L., Hansen, H. F., Rider, D., Petersen, T. E. and Magnusson, S., 1981. FEBS Lett. 129: 314–317.
42. Dickson, R. B., Willingham, M. C. and Pastan, I., 1981. J. Cell Biol. 89, 29–34.
43. Dickson, R. B., Willingham, M. C. and Pastan, I. H., 1981. J. Biol. Chem. 256: 3453–3459.
44. Willingham, M. C., Yamada, S. S. and Pastan, I., 1978. Proc. Natl. Acad. Sci. U.S.A. 75: 4359–4364.
45. Maxfield, F. R., Willingham, M. C., Davies, P. J. A. and Pastan, I., 1979. Nature 277: 661–663.
46. Nilsson, J. L. G., Stenberg, P., Junggren, C. L., Hoffmann, J., Lunden, R., Eriksson, O. and Lorand, L., 1972. Ann. N. Y. Acad. Sci. 201: 286–296.
47. Dickson, R. B., Schlegel, R., Willingham, M. C. and Pastan, I. H., 1982. Exp. Cell Res. 140: 215–225.
48. Maxfield, F. R., Willingham, M. C., Haigler, H. T., Dragsten, P., Pastan, I. H., 1981. Biòchemistry 20: 5353–5358.
49. Van Leuven, F., Marynen, P., Cassiman, J. J. and Van Den Berghe, H., 1981. FEBS Lett. 134: 83–87.
50. Ohkuma, S. and Poole, B., 1978. Proc. Natl. Acad. Sci. U.S.A. 75: 3327–3331.
51. Seglen, P. O. and Gordon, P. B., 1980. Mol. Pharm. 18: 468–475.
52. Leppla, S. H., Dorland, R. B. and Middlebrook, J. L., 1980. J. Biol. Chem. 255: 2247–2250.
53. Fitzgerald, D., Morris, R. E. and Saelinger, C. B., 1980. Cell 21: 867–873.
54. Helenius, A., Marsh, M. and White, J., 1982. Gen. Virol. 58: 47–61.
55. Miller, D. K. and Lenard, J., 1981. Proc. Natl. Acad. Sci. U.S.A. 78: 3605–3609.
56. Cornwell, M. M., Juliano, R. L. and Davies, P. J. A., 1983. Biochim. Biophys. Acta (in press).
57. Hidaka, H., Asano, M. and Tanaka, T., 1981. Mol. Pharmacol. 20: 571–578.
58. Davies, P. J. A. Submitted for publication.
59. Moger, W. H., 1982. Can. J. Physiol. Pharmacol. 60: 858–861.
60. Slaughter, R. S., Smart, C. E., Wong, D.-S. and Lever, J. E., 1982. J. Cell Physiol. 112: 141–147.
61. Dickson, R. B., Willingham, M. C., Gallo, M. and Pastan, I., 1981. FEBS Lett. 126: 265–268.
62. Yarden, Y., Gabbay, M. and Schlessinger, J., 1981. Biochim. Biophys. Acta 674: 188–203.
63. Schlessinger, J., Schechter, Y., Willingham, M. C. and Pastan, I., 1978. Proc. Natl. Acad. Sci. U.S.A. 75: 2659–2663.
64. Maxfield, F. R., Davies, P. J. A., Klempner, L., Willingham, M. C. and Pastan, I., 1979. Proc. Natl. Acad. Sci. U.S.A. 76: 5731–5735.
65. Haigler, H. T., Maxfield, F. R., Willingham, M. C. and Pastan, I. H., 1980. J. Biol. Chem. 255: 1239–1241.
66. Haigler, H. T., Willingham, M. C. and Pastan, I., 1980. Biochem. Biophys. Res. Commun. 94: 630–637.
67. Willingham, M. C. and Pastan, I. H., 1981. Ann. Rev. Physiol. 43: 239–250.
68. Higuchi, H., Takeyasu, K., Uchida, S. and Yoshida, H., 1982. Eur. J. Pharmacol. 79: 67–77.
69. Cheng, S., Maxfield, F. R., Robbins, J., Willingham, M. C. and Pastan, I. H., 1980. Proc. Natl. Acad. Sci. U.S.A. 77: 3425–3429.
70. Draznin, B. and Trowbridge, M., 1982. J. Biol. Chem. 257: 11988–11993.

71. Farrar, J. J., Benjamin, W. R., Hilfiker, M. L., Howard, M., Farrar, W. L. and Fuller-Farrar, J., 1982. Immunolog. Rev. 63: 129–166.
72. Tucciarone, L. M. and Lanclos, K. D., 1981. Biochem. Biophys. Res. Commun. 99: 221–227.
73. Schlegel, R., Dickson, R. B., Willingham, M. C. and Pastan, I. H., 1982. Proc. Natl. Acad. Sci. U.S.A. 79: 2291–2295.
74. Baldwin, D., Prince, M., Marshall, S., Davies, P. and Olefsky, J. M., 1980. Proc. Natl. Acad. Sci. U.S.A. 77: 5957–5978.
75. Saviolakis, G. A., Harrison, L. C. and Roth, J., 1981. J. Biol. Chem. 256: 4924–4928.
76. Baker, J. B., Simmer, R. L., Glenn, K. C. and Cunningham, D. D., 1979. Nature 278: 743–745.
77. Linsley, P. S., Blifeld, C., Wrann, M. and Fox, C. F., 1979. Nature 278: 745–748.
78. Tolleschaug, H., Goldstein, J. L., Schneider, W. J. and Brown, M. S., 1982. Cell 30: 715–724.

Received 14 March 1983.

Molecular and Cellular Biochemistry 58, 79–89 (1984).

β-adrenergic receptor internalization and processing: role of transglutaminase and lysosomes

De-Maw Chuang
Laboratory of Preclinical Pharmacology, National Institute of Mental Health, Saint Elizabeths Hospital, Washington, D.C. 20032, U.S.A.

Introduction

Membrane-bound β-adrenergic receptors are linked to adenylate cyclase using a guanine nucleotide binding protein as the linkage (1). It is well known that an extensive stimulation of these β-adrenergic receptors by β-receptor agonists attenuates the response of adenylate cyclase to further receptor stimulation – an event variously termed desensitization, subsensitivity, or down-regulation. In isolated frog erythrocytes (1, 2) and other systems (3–7), desensitization of β-adrenergic receptors is associated with a reduction in the number of membrane-bound receptor sites, suggesting that the density of surface receptor binding sites may control the magnitude of the receptor response.

The loss of membrane-bound receptor sites during receptor desensitization may be related to a variety of events, including: a) persistent receptor occupancy by an agonist; b) receptor inactivation resulting from covalent modification; c) increased receptor degradation; d) decreased receptor synthesis; e) changes in the microviscosity of the membrane; f) receptor exocytosis; and g) receptor endocytosis or internalization.

Since it has been demonstrated that receptors for a variety of polypeptide ligands can be internalized following their binding to the surface bound receptors, we have attempted to access whether a similar mechanism is involved in the down-regulation of β-adrenergic receptors. Using frog erythrocytes as a model system we have investigated the effects of β-adrenergic agonists on the distribution of β-receptor binding sites in an intracellular fraction of red blood cells. We have found that the loss of membrane-bound β-adrenergic receptor binding sites during isoproterenol induced desensitization is associated with an increase in the soluble fraction of the number of binding sites for ^{3}H-dihydroalprenolol (^{3}H-DHA), a β-receptor antagonist ligand (8–10). Displacement experiments using various β-receptor agents verify that these soluble ^{3}H-DHA binding sites are indeed β-receptor binding sites (8). However the physical and chemical properties of these soluble receptors differ in several respects from the β-receptors present in the plasma membrane. For instance, the soluble β-receptors are uncoupled from adenylate cyclase (9, 10), resistant to phospholipase digestion (9), and insensitive to guanine nucleotides for agonist binding (9, 10). The soluble β-receptor binding sites remain in the supernatant following centrifugation up to four hr at 100 000 g (9) and appear to devoid of the activity of adenylate cyclase or GTP binding protein (9), indicating that the binding sites do not reside in fragments of the plasma membrane. The soluble β-receptors are more labile than the particulate receptors to heat inactivation (unpublished results). A summary of these differences is presented in Table 1. The requirements for the appearance of the soluble receptor binding sites resemble those reported for the internalization of receptor-ligand complexes in other systems. For example these events are temperature and time dependent, energy required and independent of protein synthesis (8–10). Thus we have suggested that β-adrenergic receptors in frog erythrocytes are redistributed during β-receptor subsensitivity and that this receptor redistribution reflects internalization of the receptor binding sites.

Morphological studies have demonstrated that, for most systems, receptor internalization is initiated by ligand-induced clustering of receptor-ligand complexes in plasma membrane pits coated with clathrin (11). The pits containing the ligand-receptor complexes then invaginate to form receptosomes (12, 13) or endocytic coated vesicles (11) which are eventually delivered to lysosomes. Previous investigations of the molecular mechanisms underlying the receptor clustering indicate that the class of transglutaminases (glutaminyl-peptide γ-glutamyltransferase (EC 2.3.2.13) may be involved in promoting this event. Transglutaminases are Ca^{2+}-dependent enzymes capable of cross-linking proteins to form isopeptide bond between lysine and glutamine residues of two proteins. Various primary amines, which are competitive inhibitors of this enzyme, inhibit the ligand-induced clustering and consequently the internalization of receptors for α_2-macroglubulin (14, 15), insulin (12), T_3 (16), epidermal growth factor (17), low density lipoprotein (12) and vitellogenin (18). There is a good correspondence between the concentration of those compounds to inhibit transglutaminase and their ability to inhibit internalization of α_2-macroglobulin (14, 15) and, possibly, T_3 (12, 16) and vitellogenin (18) but not of epidermal growth factor (17) or low density lipoprotein (12). The present report describes our investigation of β-adrenergic receptors, as well as whether soluble β-receptor binding sites are derived from the processing of the internalized β-receptors in lysosomes.

Table 1. Comparison of the properties of soluble and membrane-bound β-adrenergic receptors.*

	Soluble receptors	Membrane-bound receptors
Receptor subtype	β_2	β_2
Pronase sensitivity	Yes	Yes
Centrifugation at $10^5 \times$ **g**	Soluble	Pelleted
Adenylate cyclase activity	No	Yes
GTP binding protein	No	Yes
Phospholipase sensitivity	No	Yes
Thermostability	Labile	Stable

* These statements were based on data obtained from our previous publications (8–10).

Materials and methods

Isolation of frog erythrocytes

Erythrocytes were isolated from bull frogs (Rana catesbeiana) as described previously (8, 9) except that the centrifugation force used during isolation was reduced from 700 to 300 **g**. Such a modification was found to improve the purity of erythrocyte preparation. Routinely, erythrocytes constituted more than 95% of the cell population.

Assay for transglutaminase activity

Transglutaminase was assayed by Ca^{2+}-dependent incorporation of ^{3}H-putrescine into casein using the filter paper method (19). The reaction mixture contained in 200 μl of 20 mM Tris-HCl, pH 7.4, 2.85 mM, $CaCl_2$, 1 mM dithiothreitol, 5 mM [1,4-^{3}H]putrescine, 2 mg/ml of casein, 0.5–1.0 mg of erythrocyte cytosol protein, and various amounts of this enzyme inhibitor. The incubation time was 5 min at 30 °C and the value obtained at zero time incubation was considered as the blank.

Measurement of the binding of ^{3}H-dihydrolaprenolol (^{3}H-DHA) to the cytosol and particulate fraction in erythrocytes

These procedures have been described previously (8, 9). Briefly, the cells were incubated in the absence or presence of (–)isoproterenol bitartrate (5 μM) in modified Ringer's medium for 2 h at 30 °C. Isoproterenol was removed by repeated centrifugation washes and cells were lysed by a hypotonic buffer containing 25 mM Tris-HCl, pH 7.4 and 1 mM $MgCl_2$. The homogenate was centrifuged at 30 000 **g** and the supernatant was clarified by a further centrifugation. The binding of ^{3}H-DHA to the cytosolic (supernatant) and particulate fractions was determined by chromatography with a Sephadex G-50 column and vacuum filtration over a glass fiber filter, respectively.

Assays of lysosomal enzymes

With the exception of cathepsin D, all the lysosomal enzymes were assayed essentially according to the procedures of Hall et al. (20). Purified erythrocytes were collected by centrifugation, washed three times with 0.9% NaCl and resuspended in saline to give 50% hamatocrit. The cell suspension was then subjected to freezing-thawing followed by grinding in a Teflon-glass homogenizer in the presence of saline. For most of the acid hydrolases, activities were linear at least up to 10 mg/ ml of the erythrocyte homogenate. Reactions were stopped by heating at 90 °C for 5 min followed by centrifugation to remove the protein aggregate.

Cathepsin D was assayed essentially in accordance with the procedure of Barrett (21) using acid-denatured hemoglobulin as the substrate. The incubation was at 37 °C for 1 h. The value obtained at 0 times was considered as the blank. The reaction was stopped by the addition of 5% trichloroacetic acid and the mixture was filtered over a Whatman 3-mm filter. The definition of 1 unit of activity is the quantity that would produce an extinction (E_{280}) of 1.0 in the filtrate during 1 h of incubation.

Percoll gradient centrifugation of erythrocyte homogenate

Erythrocytes were washed 5 times by centrifugation in Ringer's medium followed by three more washes with 0.25 M sucrose (pH 7.0) after various treatments. The cells were then homogenized in 4 ml of 0.25 M sucrose by 40 strokes in a glass homogenizer with a loosely fitted Teflon pestle. Breakage of cells as judged by light microscopy was virtually complete. Aliquots of the homogenate were centrifuged immediately over a layer of Percoll in the following manner: one-ninth volume of 2.5 M sucrose was added to the original Percoll solution. A 30% solution (v/ v) was prepared from this stock using 0.25 M sucrose. To a nitrocellulose tube (13 ml) was added 1.0 ml of 1.8 M sucrose. This was overlayed with 9.0 ml of the 30% Percoll solution, followed with 1.0 ml of the erythrocyte homogenate. Centrifugation was performed in a 50-Ti fixed angle rotor (Beckman) at 28 000 rpm for 20 min. Fractions of 0.4 ml were collected from the top of the tube. The distribution of lysosomal activity was determined by removing 200 μl from the fractions and assaying for α-L-fucosidase activity according to Hall et al. (20) except that 0.1% Triton X-100 was present in the assay mixture. The density of the gradient was determined by weighing 100 μl from fractions under mineral oil to prevent evaporation. The binding of ^{3}H-DHA to the particulate material was determined by removing 100 μl from the fractions and assaying them in 500 μl of buffer containing 25 mM Tris-HCl, pH 7.4 and 1 mM $MgCl_2$ as described previously (8, 9).

Materials

Levo-[propyl-1,2,3-^{3}H]-dihydrolaprenolol was obtained from New England Nuclear and [1,4-^{3}H]-putrescine from Amersham. Percoll was the product of Pharmacia. Blocked diazooxonorvaline (BDONV) and blocked chlorooxonorvaline (BCLONV) were the generous gifts of Dr. J. F. Folk at the National Institute of Health.

Results

Characterization of transglutaminase in frog erythrocyte homogenate

Transglutaminase activity in frog erythrocytes was assayed by measuring Ca^{2+}-dependent incorporation of ^{3}H-putrescine into casein through the formation of γ-glutamyl-lysine bonds. The data in Table 2 show the general properties of this enzyme in frog erythrocyte homogenate. In the complete system the activity was 0.54 nmol of ^{3}H-putrescine incorporated/ mg protein/ minute. In the absence of Ca^{2+}, less than 10% of the activity was detected. However, a considerable incorporation was obtained when casein was deleted from the reaction mixture, indicating the presence of endogenous substrate for transglutaminase. The enzyme requires the presence of dithiothreitol for maximal

Table 2. Activity of transglutaminase in frog erythrocyte lysate.*

Assay conditions	[^{3}H]-Putrescine incorporated (nmol/ mg prot/ min)
Complete system	0.54
-Erythrocyte lysate	0
-$CaCl_2$	0.05
-Casein	0.10
-Dithiothreitol	0.34
+Methylamine	0.22
+n-Propylamine	0.31
+n-Hexylamine	0.36

* Assay conditions are as described in Materials and methods. When used the concentrations of amine inhibitors were 10 mM.

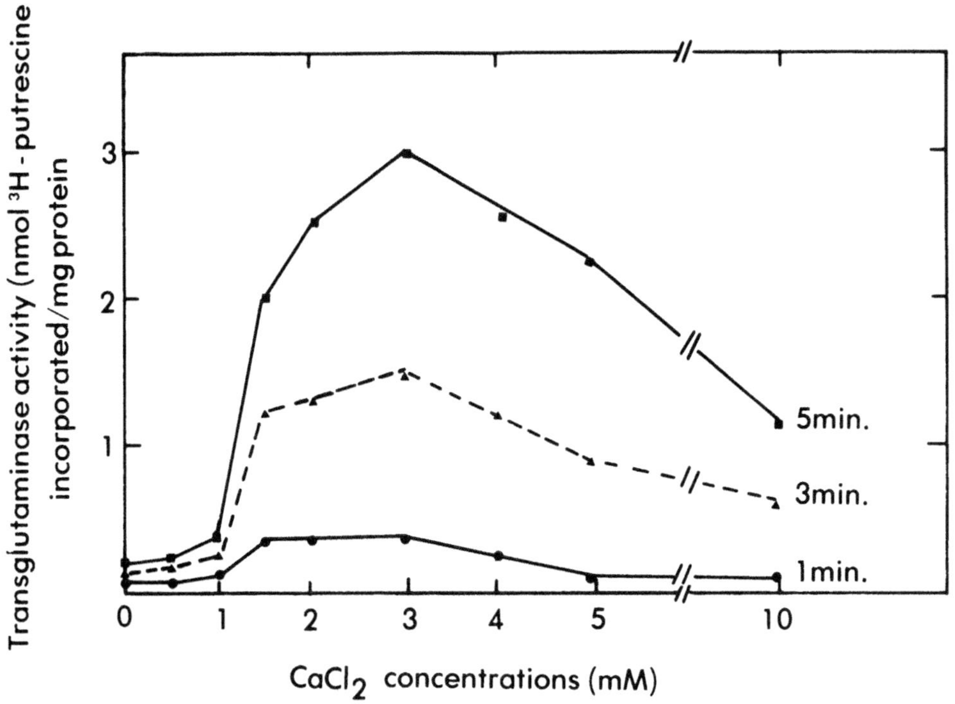

Fig. 1. Effect of Ca^{2+} on the activity of transglutaminase in frog erythrocyte homogenate. Assay of transglutaminase was described in Materials and methods except that the reaction mixture contained 1.25 mM of EGTA and indicated concentrations of Ca^{2+}. The reactions were terminated after incubation at 30 °C for indicated periods of time.

activity. Various aliphatic amines inhibit the enzymatic activity in the millimolar concentration ranges. The Km of ^{3}H-putrescine for the incubation was about 1.10 mM. The time course and the dependence of Ca^{++} for the incorporation of ^{3}H-putrescine is shown in Fig. 1. The incorporation was linear at least up to 5 minute incubation at 30 °C. Maximal incorporation was obtained when about 3 mM of exogenous $CaCl_2$ was added; higher concentrations of this divalent ion were inhibitory for the reaction. Since 1.25 mM of EGTA was present in the reaction mixture, the Ka of $CaCl_2$ required for promoting the incorporation was no more than 200 μM. About 80% the total enzyme activity was found in the 30 000 **g** supernatant of erythrocyte homogenate with the remaining activity present in the pellet fraction. In both preparations the characteristics of the enzyme and its sensitivity to various amine inhibitors are rather similar. Therefore, for convenience, the supernatant was used as the source of the enzyme for the subsequent studies.

Evidence for the involvement of transglutaminase in the internalization of β-adrenergic receptors

To determine if transglutaminase activity is involved in the internalization of β-adrenergic receptors, isolated frog erythrocytes were preincubated with methylamine, one of the potent inhibitors of transglutaminase, prior to the exposure to isoproterenol, a selective β-adrenergic agonist. The results in Figure 2 show that preincubation with methylamine caused a dose-dependent inhibition of the internalization of β-adrenergic receptor binding sites induced by stimulation with isoproterenol. This receptor internalization was accessed by the presence of the binding sites for ^{3}H-DHA in the 30 000 **g** supernatant. The concentration of methylamine required for a 50% inhibition was estimated to be 3.1 mM. The loss of membrane-bound β-adrenergic receptor binding sites was also prevented by methylamine with an equal potency. The methylamine concentration which inhibited by 50% the

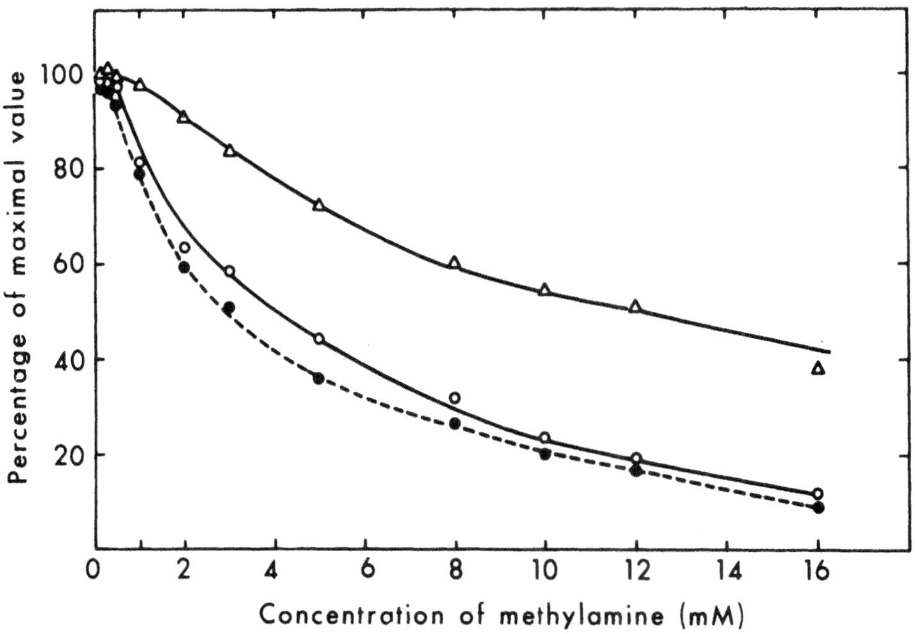

Fig. 2. Effects of methylamine on the internalization (●---●) and the loss of membrane-bound β-adrenergic receptors (○—○) in isoproterenol-treated erythrocytes and the activity of transglutaminase in erythrocyte lysate (△—△). Isolated frog erythrocytes were incubated with indicated concentrations of methylamine for 30 min at 30 °C prior to the addition of isoproterenol. Cells were then washed extensively and the binding of [^{3}H]-DHA to the cytosol and particulate fraction was then determined as described under the Materials and methods. Results were taken from a previous publication (22).

transglutaminase activity was about 13 mM (Fig. 2). Since this compound is a competitive inhibitor of the enzyme, a Ki value of 3.7 mM can be obtained using the formula:

$$Ki = \frac{I_{50}}{1 + (S/Km)}$$

where I_{50} is the concentration of methylamine required to produce a 50% inhibition of the enzymatic activity: S is the concentration of labeled putrescine in the assay system and Km is the Michaelis constant for putrescine. From these data one may conclude that methylamine is about equally potent as an inhibitor for internalization and down-regulation of β-adrenergic receptors and for transglutaminase in frog erythrocytes. Scatchard analyses of binding data verify that the effect of methylamine on the binding of ^{3}H-DHA to the soluble and particulate fractions was due to changes in the Bmax with no apparent effect of the Kd (Table 3). The maximal effect of methylamine on β-receptor number in frog erythrocyte requires the co-presence of the amine and isoproterenol during incubation; removal of methylation from erythrocytes by repeated centrifugations prior to the addition of isoproterenol abolished almost totally the changes in Bmax of ^{3}H-DHA binding induced by the amine. Moreover inhibitory concentrations of methylamine added to isolated plasma membranes affected neither the binding of ^{3}H-DHA nor the activity of isoproterenol-sensitive adenylate cyclase. In the experiment described in Figure 3, the values of Ki, IC_{50} and Ka of several transglutaminase inhibitors were compared. The most potent inhibitor of erythrocyte transglutaminase was dansylcadaverin with a Ki of 0.38 mM. The potency for inhibiting the enzymatic activity decreased in the following order: glycinamide; methylamine; n-propylamine; n-hexylamine; bacitracin; ethylamine, and t-butylamine. When IC_{50} for preventing internalization was compared with Ki for inhibiting transglutaminase, a striking linear correlation (with a $r = 0.94$) was obtained for these compounds (Fig. 3A). A similar result was obtained when Ka for preventing receptor loss was plotted versus Ki for inhibiting the enzyme activity (Fig. 3B). These results strongly suggest that these three events are related. It is of

Table 3. Scatchard analyses of [³H]-DHA binding to the cytosolic membrane fractions in isoproterenol-treated cells: effect of methylamine.*

Treatment	Kd (nM)		Bmax (fmol/ml cells)	
	Cytosol	Membrane	Cytosol	Membrane
-Methylamine	1.16 ± 0.21	1.10 ± 0.13	1 100 ± 173	4 050 ± 264
+Methylamine	1.22 ± 0.26	1.21 ± 0.08	430 ± 64	5 236 ± 259

* Cells were preincubated in the absence of presence of methylamine (15 mM) for 30 min at 30 °C prior to the exposure to isoproterenol (5 μM). The binding of [³H]-DHA to the cytosolic and membrane fractions were described in Materials and methods using, however, various concentrations of [³H]-DHA (0.15–40 nM). The results are mean ± S.E. of 3 separate experiments.

importance to point out that with the exception of bacitracin, none of these compounds added to isolated plasma membrane preparation significantly affected the binding of ³H-DHA to β-adrenergic receptors or the activation of adenylate cyclase by isoproterenol. Bacitracin was found to inhibit isoproterenol-sensitive adenylate cyclase with a IC_{50} of about 5 mM. This side effect may explain why this peptide inhibitor is more potent in preventing receptor internalization and desensitization than in inhibiting transglutaminase activity. Two irreversible inhibitors of transglutaminase, BDONV and BCLONV (23), were also found to be potent inhibitors of isoproterenol-sensitive adenylate cyclase (data not shown), making it difficult to decipher their role in preventing receptor internalization mediated through inhibition of transglutaminase.

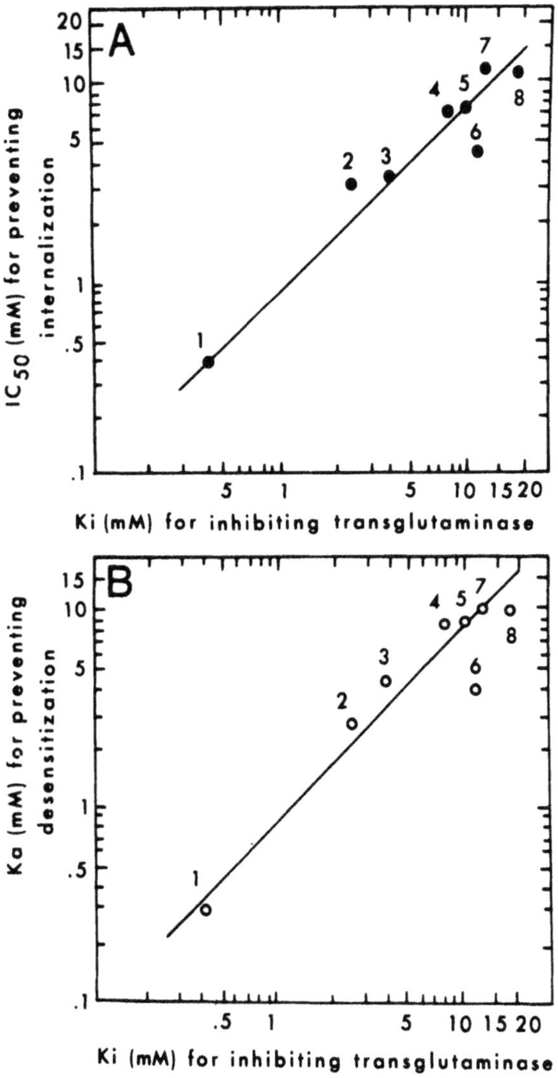

Fig. 3. Comparison of various compounds for their potencies in inhibiting transglutaminase activity and their abilities in preventing the appearance of receptor sites in the cytosol and the receptor loss from the membrane. Compounds tested were: 1, dansylcadavarin; 2, glycinamide; 3, methylamine; 4, n-propylamine ; 5, hexylamine; 6, bacitracin; 7, ethylamine; 8, t-butylamine. The Ki for inhibiting transglutaminase, the IC_{50} for preventing internalization and the Ka for preventing the loss of membrane-bound β-receptors during desensitization were calculated in the experiment as described in Fig. 2. Results were taken from a previous report (22).

Involvement of lysosomes in the processing of internalized β-adrenergic receptors

In light of the findings that many internalized ligand-complexes are delivered by vesicular structures to lysosomes where they are at least partially degraded (11, 12, 24–26), I have attempted to investigate whether soluble β-adrenergic receptors detected during desensitization result from the processing of internalized β-receptors by lysosomal enzymes. In order to document the presence of lysosomes in frog erythrocytes, the activities of various acid hydrolases were measured. The data in

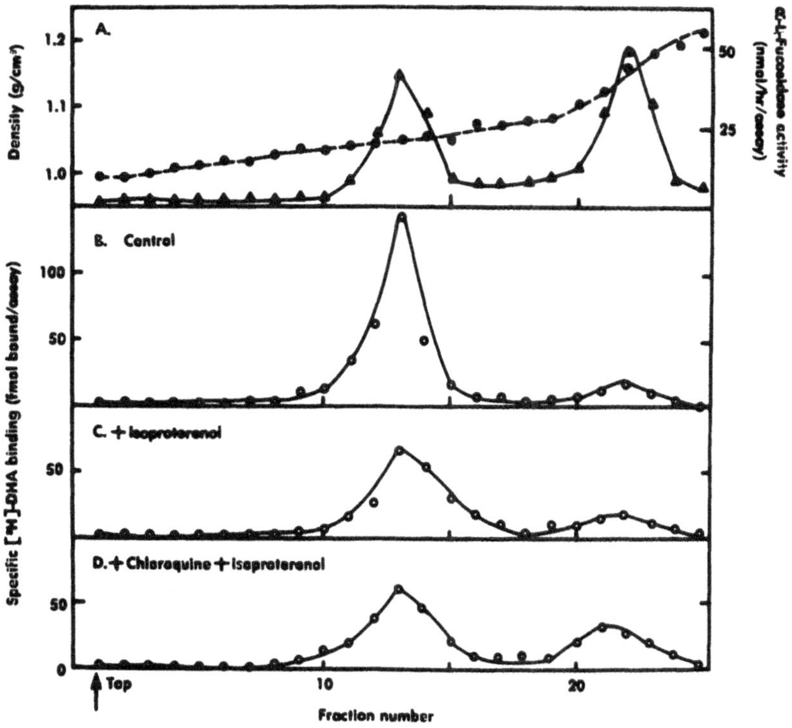

Fig. 4. Patterns of [^{3}H]-DHA binding to frog erythrocyte fractionated by Percoll gradient centrifugation. Erythrocytes were incubated in the absence or presence of chloroquine (500 μM) for 1 h at 30 °C. Buffer or isoproterenol was added for 2 h of incubation. Following the washing of the cells, homogenate was prepared and then subjected to centrifugation in a self-generated gradient of Percoll. Details of the experimental procedures were described in the Materials and methods. Results were taken from a previous paper (28).

Table 4 show that several lysosomal enzymes including cathepsin B and D were detected in frog erythrocyte homogenate. The purity of the cell preparation was greater than 95%. Moreover further removal of most of the residual lymphocytes and platelets by centrifugation through a layer of Ficoll-Paque did not alter the specific activities of these lysosomal enzymes, indicating that these enzymes are indeed present in frog erythrocytes. In collaboration with Dr Pertti Panula in our laboratory, we have also obtained preliminary ultrastructural evidence for the presence of lysosomes in frog erythrocytes (unpublished results).

Various lysosomotropic drugs are known to inhibit acid hydrolases by raising the intralysosomal pH after they are trapped by protonation within this organelle. The data in Table 5 show that pretreatment of red blood cells with chloroquine, cocaine, tetramethylethylenediamine (TEMED), and ammonium chloride prior to the addition of isoproterenol caused a dose-dependent inhibition of the binding of ^{3}H-DHA to the cytosol, but failed to alter the binding of ^{3}H-DHA to the membrane preparation derived from the same cells. Scatchard analyses revealed that these drugs effects on the soluble β-receptor binding sites are due to a de-

Table 4. Activities of lysosomal enzymes in frog erythrocytes.*

	Activities
Experiment A	(nmol/hr/mg protein)
α-N-acetylglucosaminidase	25.9 ± 2.3 (n = 3)
β-galactosidase	9.6 ± 0.8 (n = 5)
α-mannosidase	16.9 ± 3.2 (n = 3)
β-glucuronidase	78.5 ± 5.2 (n = 3)
α-L-fucosidase	30.6 ± 3.9 (n = 5)
β-hexosaminidase	118.0 ± 11.1 (n = 4)
Acid phosphatase	64.4 ± 7.1 (n = 4)
Experiment B	(unit/hr/mg protein)
Cathepsin B	0.10 ± 0.01 (n = 5)
Cathepsin D	0.14 ± 0.02 (n = 5)

* Lysosomal enzymes were assayed according to Hall et al. (20) in Experiment A and Barrett (21) in Experiment B. Details of the enzyme assays were described in Materials and methods.

Table 5. Effects of various lysosomotropic drugs on the binding of [^{3}H]-DHA to the cytosol and particulate fractions of cells preincubated with isoproterenol.*

Preincubation conditions	[^{3}H]-DHA Bound to Cytosol	[^{3}H]-DHA Bound to Membranes
None	100%	100%
Chloroquine		
20 μM	70	92
200 μM	39	100
Cocaine		
0.2 mM	71	98
2.0 mM	45	105
TEMED		
0.2 mM	72	103
2.0 mM	46	106
NH_4Cl		
1 mM	79	98
10 mM	54	99

* Isolated frog erythrocytes were incubated with indicated concentrations of drugs for 30 min at 30 °C. Isoproterenol was added to give a final concentration of 5 μM and cells were further incubated for 2 h. Cells were washed extensively and the binding of [^{3}H]-DHA to the cytosol and membrane fractions was determined. The 100% value of [^{3}H]-DHA binding in cells preincubated without drug were 706 ± 57 fmol/ml cells and 3 396 ± 324 fmol/ml cells for the cytosol and membrane fractions, respectively. Results were taken from a prevous publication (28).

crease in the Bmax with no apparent changes in the Kd of ^{3}H-DHA binding.

The inhibitor effects of lysosomotropic drugs cannot be attributed to the inhibition of isoproterenol-sensitive adenylate cyclase. Moreover these drug effects were time dependent. A minimal of 30-min preincubation before the stimulation with isoproterenol was required to display their effect and maximal inhibition was seen after 1–2 hr of preincubation. Extensive washings of these cells to remove the drugs prior to the addition of isoproterenol diminished the chloroquine-induced decrease of soluble receptor binding by about 50%. Taken together; these results suggest that lysosomotropic drugs do not affect the down-regulation or internalization of β-adrenergic receptors but do inhibit the processing of the internalized β-receptor binding sites within the lysosome.

The above conclusion is further supported by the experiment described in Fig. 4 using Percoll gradient centrifugation to separate lysosomes from the erythrocyte homogenates. When erythrocyte homogenates prepared in isotonic sucrose solution were fractionated by a self-generated gradient of Percoll, two species of particles containing roughly equal amounts of activity of α-L-fucosidase were detected with a density of about 1.07 and 1.15 g/ml respectively (Fig. 4A). Similar results were obtained when α-mannosidase and β-hexosaminidase were used as enzyme markers, suggesting that both particles are lysosomal. The light lysosomes co-migrated with the plasma membrane; but the heavy lysosomes appeared to be free of several markers of the plasma membrane. In the untreated cells, most of the β-adrenergic receptors were located in the fractions containing the plasma membrane (Fig. 4B). However, a small amount of binding activity for ^{3}H-DHA could be detected in the fraction containing the heavy lysosomes. In cells desensitized with isoproterenol, the β-receptor sites associated with the plasma membrane was reduced by about 50%, whereas the binding of ^{3}H-DHA to the fractions containing the heavy lysosomes was unchanged (Fig. 4C). Thus the proportion of the lysosome-associated β-receptor binding sites was significantly increased by treatment with isoproterenol. Such an increase in the proportion was further enhanced by pretreatment of cells with chloroquine which increased the binding of ^{3}H-DHA to the lysosomal fractions by about 2-fold (Fig. 4D). Pretreatment with chloroquine in the untreated cells had little or no effect on the binding of ^{3}H-DHA (data not shown). These results strongly suggest that at least the heavy lysosomes in frog erythrocytes can retain β-adrenergic receptors and such a

Table 6. Activity of soluble β-adrenergic receptors in erythrocyte homogenate fractioned by Percoll gradient centrifugation.*

Treatments	Specific [^{3}H]-DHA Binding (fmol/assay)
Control	15 ± 2
+Isoproterenol	65 ± 10
+Chloroquine + Isoproterenol	21 ± 3

* The top 6 fractions from a Percoll gradient were pooled and centrifuged at 150 000 **g** for 2 hours to pellet the Percoll. Aliquots of 200 μl of the supernatant were used to assay for [^{3}H]-DHA binding assay using Sephadex G-50 column chromatography. The nonspecific binding was determined in the presence of 5 μM of unlabeled alprenolol and was subtracted from the total binding. Values given were mean ± S.E. of three determinations. Details of the experiments were described in Materials and methods.

retention is enhanced during receptor desensitization induced by isoproterenol. It is of interest to mention that Percoll gradients bind ^{3}H-DHA strongly when assayed by gel filtration with a Sephadex G-50 column, thereby interfering with binding assays for soluble receptors. However when the top fractions (1 to 6) of the gradient were combined and centrifuged at higher speed to pellet the Percoll, binding activity for ^{3}H-DHA assayed by gel filtration could be detected in the supernatant obtained from the control cells (Table 5). This binding activity was enhanced by about 4-fold following exposure to isoproterenol, but this isoproterenol-induced effect was markedly inhibited by pretreatment with chloroquine.

Discussion

The results obtained in this investigation have provided strong evidence that soluble β-adrenergic receptor binding sites detected during receptor desensitization induced by isoproterenol are due to the processing of internalized receptors by the lysosomal enzymes in frog erythrocytes. This conclusion is supported by several observations. Pretreatment of cells with lysosomotropic drugs caused a time and dose-dependent decrease in the number of soluble β-receptor binding sites in desensitized cells, without affecting the binding sites located on the plasma membrane in the same cells (Table 5). Percolll gradient centrifugation revealed that a small population of membrane-bound β-receptor sites is associated with the heavy lysosomes (Fig. 4). The proportion of these lysosome-associated β-receptors is increased when desensitized cells are pretreated with chloroquine. Since the light lysosomes are contaminated with the plasma membrane, it is unknown whether this species of lysosomes is also involved in the processing of the internalized β-receptor binding sites. It is of interest to point out that in human fibroblasts both the heavy and light lysosomes were detected and that both species were able to take up lysosomal enzymes by a receptor-mediated process (27). From the present study it is unclear which lysosomal enzyme(s) in frog erythrocytes participates in the solubilization of internalized β-receptors. However, since the level of most species of lysosomal enzymes in frog erythrocytes is lower than in other systems reported, one may surmise that these hydrolytic enzymes digest only the endocytic vesicles such as receptosomes, while sparing the enclosed receptors; these internalized receptors are thereby converteed to a soluble form and released into the cytoplasm.

Various inhibitors of transglutaminase also reduced the number of soluble β-adrenergic receptor binding sites in isoproterenol-treated cells. Unlike the lysosomotropic drugs, they also prevented the loss of membrane-bound β-receptor binding sites in the desensitized cells, indicating that these inhibitors prevented both the down-regulation and internalization of β-adrenergic receptors. As there is a good correlation between the ability of these compounds to inhibit transglutaminase and their potency to block β-receptor internalization, these compounds most like exert their effects through direct inhibition of transglutaminase. This results in a diminished clustering of ligand-receptor complexes on coated pits as was first suggested for the internalization of α_2-macroglobulin in CHO cells (14) by Pastan and coworkers. However, our results differed from those reported in the laboratory of Pastan in some respects. First, dansylcadaverin, methylamine and other alkylamines inhibit transglutaminase activity in CHO extracts biphasically and only the high affinity component appears to be related to the internalization of α_2-macroglobulin-receptor complex (14). Secondly, in contrast to the present report, in CHO extracts 5–10 times more inhibitor was required to produce inhibition of receptor clustering than was required to produce inhibition of high affinity transglutaminase (14). The reason for these discrepancies is unknown but may be related to the extreme abundance of transglutaminase present in CHO cells (which is about 10–20 times higher than in frog erythrocytes and most other cell types). Perhaps transglutaminases in CHO cells are heterogeneous and only a minor fraction is related to the internalization of α_2-macroglobulin. The presence of a subtype of transglutaminase in certain cells may also explain the poor sensitivity to various transglutaminase inhibitors for the internalization of epidermal growth factor (17) and low density lipoprotein (12). Whether transglutaminase actually catalyzes cross-linking of receptor protein with another protein in the coated pit during receptor clustering, has never been verified. In any event clathrin (the structural protein of coated pits) may not be the substrate for transglu-

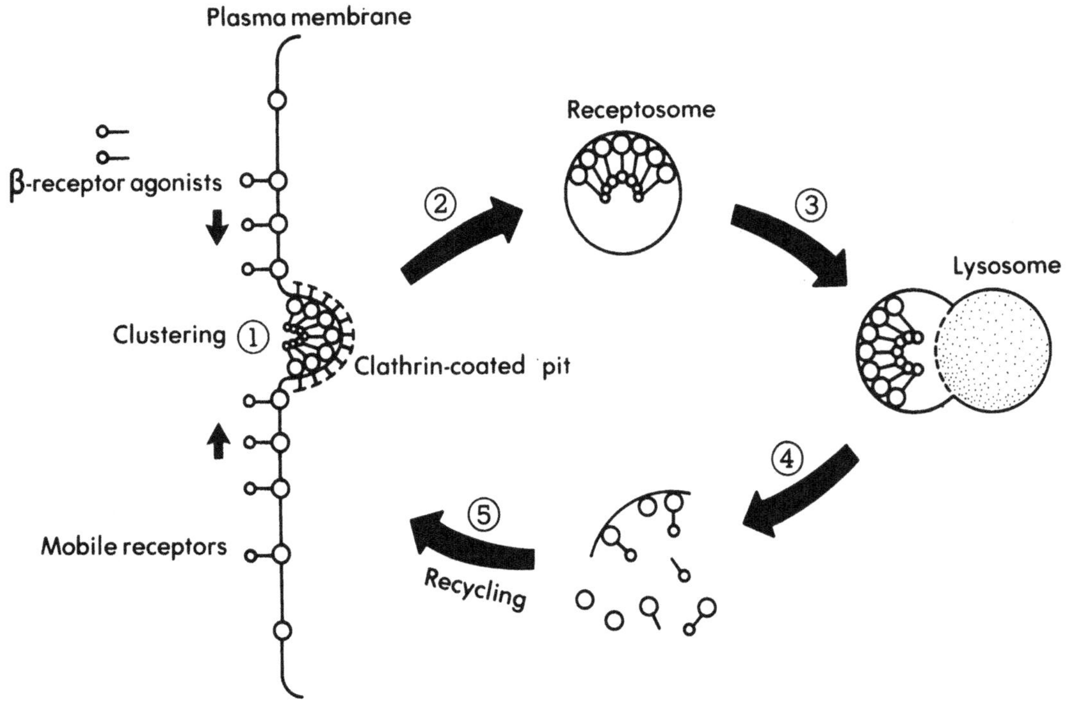

Fig. 5. A proposed model for the internalization of β-adrenergic receptor binding sites in frog erythrocytes.

taminase in frog erythrocyte because ^{3}H-putrescine failed to be incorporated when casein was replaced with clathrin purified from bovine brain in the assay (unpublished date).

Based on the present study a model is proposed for the sequences of events that occur during β-receptor internalization in frog erythrocytes stimulated with isoproterenol (Fig. 5). Initially β-receptors are evenly distributed in the plasma membrane. A clustering of receptors may be triggered by the binding of a β-receptor agonist and this receptor clustering on clathrin-coated pit (step 1) is proposed to be mediated by the enzyme transglutaminase. The coated pit eventually pinches off the plasma membrane to form an endocytic vesicle (step 2), which may be the receptosome as was reported previously (12, 13). The receptosome then carries the β-receptor-isoproterenol complex to the lysosome (step 3) where the receptosome is partially degraded but the β-receptor sites are spared (step 4). Transglutaminase inhibitors are believed to act on receptor clustering, whereas lysosomotropic drugs affect the solubilization of receptors within the lysosome. The dependence of temperature, requirement of energy and effect of concanavalin for β-receptor internalization reported previously (8–10) are most likely related to the invagination of coated pits to form receptosomes and/or the movement of receptosomes to lysosomes. Since the level of soluble β-receptors was returning to normal when β-receptor sites were restored during receptor resensitization (9) and cyclohexamide failed to block these events (unpublished results), it is proposed that internalized β-receptor binding sites can be recycled to the plasma membrane (step 5) to restore the receptor sites and functional sensitivity that are lost during receptor subsensitivity. It should be stressed that this model is tentative and its final proof depends on future morphological studies. The search for a labeled β-adrenergic agonist which has high affinity for the receptor site and is capable of inducing receptor internalization but is not taken up into red blood cells nonspecifically, would be the direction of future investigation.

References

1. Lefkowitz, R. J., Limbird, L. E., Williams, L. T. and Wessels, M., 1978. J. Supramol. Structure 8: 501–510.
2. Mickey, J. V., Tate, R., Millikin, D. and Lefkowitz, R. J., 1976. Mol. Pharmacol. 12: 409–419.

3. Su, Y.-F., Harden, T. K. and Perkins, J. P., 1980. J. Biol. Chem. 255, 7410–7419.
4. Shear, M., Insel, P. A., Melmon, K. L. and Coffino, P., 1976. J. Biol. Chem. 251: 7572–7576.
5. Fishman, P. H., Mallorga, P. and Tallman, J. F., 1981. Mol. Pharmacol. 20: 310–318.
6. Moyer, L. H., Greenberg, L. H., Frazer, A., Brunswick, D. J., Mendels, J. and Weiss, B., 1979. Life Sci. 24: 2237–2244.
7. Sulser, F., 1979. Trends Pharmacol. Sci. 1: 92–94.
8. Chuang, D. M. and Costa, E., 1979. Proc. Natl. Acad. Sci., U.S.A. 76: 3024–3028.
9. Chuang, D.-M., Kinner, W. J., Farber, L. and Costa, E., 1980. Mol. Pharmacol. 18: 348–355.
10. Chuang, D.-M., Farber, L., Kinnier, W. J. and Costa, E., 1980. In: Receptors for Neurotransmitters and Peptide Hormones (Pepeu, G., Kuhar, M. J. and Enna, S. J., eds.) Raven Press, New York, pp. 143–150.
11. Goldstein, J. L., Anderson, R. G. W. and Brown, M. S., 1979. Nature 279: 679–685.
12. Pastan, I. and Willingham, M. C., 1981. Science 214: 504–509.
13. Willingham, M. C. and Pastan, I., 1980. Cell 21: 67–77.
14. Davies, P. J. A., Davies, D. R., Levitzki, A., Maxfield, F. R., Milhaud, P., Willingham, M. C. and Pastan, I., 1980. Nature 283: 162–167.
15. Levitzki, A., Willingham, M. and Pastan, I., 1980. Proc. Natl. Acad. Sci. U.S.A. 77: 2706–2710.
16. Cheng, S., Maxfield, F. R., Robins, J., Willingham, M. C. and Pastan, I. H., 1980. Proc. Natl. Acad. Sci. 77: 3425–3429.
17. Haigler, H. T., Willingham, M. C. and Pastan, I., 1980. Biochem. Biophys. Res. Commun. 94: 630–637.
18. Tucciarone, L. M. and Lanclos, K. D., 1981. Biochem. Biophys. Commun. 99: 221–227.
19. Lorand, L., Campbell-Wilkes, L. K. and Cooperstein, L., 1972. Anal. Biochem. 50: 623–631.
20. Hall, C. W., Liebaers, I., DiNatale, P. and Neufeld, E. F., 1978. Methods in Enzymology, pp. 439–456, Academic Press, New York.
21. Barrett, A. J., 1970. Biochem. J. 117: 601–607.
22. Chuang, D.-M., 1981. J. Biol. Chem. 256: 8291–8293.
23. Folk, J. E. and Gross, M., 1971. J. Biol. Chem. 246: 6683–6691.
24. McKanna, L. A., Haigler, H. T. and Cohen, S., 1979. Proc. Natl. Acad. Sci. U.S.A. 76: 5689–5693.
25. King, A. C., Hernaez-Davis, L. and Cuatrecasas, P., 1980. Proc. Natl. Acad. Sci. U.S.A. 77: 3283–3287.
26. Ashwell, G. and Morell, A. G., 1974. Adv. Enzym. 41: 99–128.
27. Rome, L. H., Garvin, A. J., Allietta, M. M., and Neufeld, E. F., 1979. Cell. 17: 143–153.
28. Chuang, D.-M., 1982. Biochem. Biophys. Res. Commun. 105: 1466–1472.

Received 11 June 1982.

Molecular and Cellular Biochemistry 58, 91–97 (1984).

Transglutaminase catalyzed incorporation of putrescine into surface proteins of mouse neuroblastoma cells

Kuang Yu Chen
Department of Chemistry, Rutgers – The State University of New Jersey, New Brunswick, NJ 08903, U.S.A.

Summary

Transglutaminase, purified from guinea pig liver, was used to catalyze the incorporation of [^{14}C]putrescine into exposed surface proteins of intact mouse neuroblastoma cells. This method specifically labeled two surface proteins ($Mr = 92\,000$ and 76 000) in the N-18 mouse neuroblastoma cells and three surface proteins ($Mr = 92\,000$, 76 000, and 72 000) in the NB-15 mouse neuroblastoma cells. In addition, transglutaminase also catalyzed cross-linking reactions of exposed surface proteins. In both the N-18 and NB-15 cells, differentiation was accompanied by a 2-fold increase of specific radioactivity incorporated into trichloroacetic acid insoluble cellular material, suggesting that the differentiated mouse neuroblastoma cells may possess greater amount of accessible peptide-bound glutaminyl residues on their surface than their malignant counterparts. Sodium dodecyl sulfate-polyacrylamide gel electrophoresis and fluorographic method revealed that while the [^{14}C]putrescine-labeled protein patterns of undifferentiated and differentiated mouse neuroblastoma cells were similar, the intensity of labeling of individual bands was specifically modulated by cell differentiation.

Abbreviations

PMSF, phenylmethylsulfonylfluoride; Bt_2cAMP, $N^6,O^{2'}$-Dibutyryl adenosine 3′:5′-cyclic monophosphate; IBMX, 3-isobutyl-1-methyl xanthine; SDS-PAGE, sodium dodecyl sulfate – polyacrylamide gel electrophoresis; HEPES, N-2-hydroxylethylpiperazine-N′-2-ethanesulfonic acid.

Introduction

Transglutaminases, a class of enzymes widely occurring in animal tissues and body fluids, catalyze a calcium-dependent acyl transfer reaction between the γ-carboxamide group of peptide-bound glutaminyl residue and primary amino group of various acceptor substrates (1). Depending on the acceptor substrates, either amine or peptide-bound lysine residue, transglutaminase can mediate a labeling reaction or an inter-peptide cross-linking reaction (1, 2).

Purified guinea pig liver transglutaminase has an apparent molecular weight of 85 000 and, when added to intact cells, would not penetrate cell membranes (3, 4). Since the conditions of transglutaminase catalyzed reactions are mild, purified transglutaminase has been used to identify exposed surface proteins of intact cells. For example, Dutton and Singer (4) have used guinea pig liver transglutaminase and dansylcadaverine to specifically label the Band 3 protein of mouse erythrocytes. Okumura and Jamieson (5) have shown that transglutaminase catalyzes the incorporation of [^{14}C]glycine ethyl ester specifically into the glycoprotein II of human platelet membrane.

Early investigators have demonstrated that among various primary amines, polyamines (putrescine, spermidine and spermine) serve as good acceptor substrates for transglutaminase (6). Numerous reports have established that polyamines are involved in a wide spectrum of growth regulatory processes (7–9). Other studies have also provided evidence that cellular transglutaminase may play a role in growth regulation (10–13). Due to wide-occurrence of transglutaminase and polyamines in animal tissues and body fluids, it is likely that endogenous polyamines may become covalently bound to some surface proteins via transglutaminase catalyzed reaction. Studies from several laboratories have suggested that indeed polyamines may serve as physiological substrates for transglutaminase in vitro (13, 14) and in vivo (15, 16). In view of these considerations it appears likely that the transglutaminase catalyzed incorporation of polyamines into surface proteins may be used to probe the surface structure and to identify potential surface sites where polyamines can be covalently conjugated.

In this report, the cell surface proteins of the undifferentiated (malignant) and differentiated cultures of two mouse neuroblastoma cell lines, the N-18 and the NB-15 cells, were studied and compared using the transglutaminase/[^{14}C]putrescine labeling method.

Materials and methods

[1,4-^{14}C]putrescine · 2HCl (60 mCi/mmol) and [^{14}C]spermidine · 3HCl (122 mCi/mmol) were purchased from Amersham, Arlington Hgts., Ill. Dulbecco's modified Eagle medium, fetal calf serum and Earle's balanced salt solution were obtained from Gibco, Grand Island, N.Y. Phenylmethylsulfonyl fluoride (PMSF), $N^6,O^{2'}$-dibutyryl adenosine 3':5'-cyclic monophosphate (Bt_2cAMP) and dithiothreitol were purchased from Sigma Chem. Co., St. Louis, Mo. 3-Isobutyl-1-methyl xanthine (IBMX), an inhibitor of phosphodiesterase, was from Aldrich Chem. Co., Milwaukee, WI. N,N-dimethylcasein was obtained from Calbiochem-Behring Corp., San Diego, CA. QAE-Sephadex ion exchanger was purchased from Pharmacia Fine Chemicals, Piscataway, N.J. Hydroxyapatite was purchased from Bio-Rad Lab., Richmond, CA.

Purification of guinea pig liver transglutaminase

Guinea pig liver transglutaminase was purified by a modified version of the method of Connellan et al. (17). Transglutaminase activity was determined by the method of Lorand et al. (18). Briefly, a 100 000 **g** supernatant from guinea pig liver homogenate was chromatographed twice over a QAE-Sephadex column. Fractions containing transglutaminase activity were pooled and further purified by hydroxyapatite column chromatography.

Cell culture and differentiation

Mouse N-18 and NB-15 neuroblastoma cells were grown as monolayer cultures in Dulbecco's modified Eagle medium (with 4 500 mg glucose per liter, without sodium pyruvate) supplemented with 10% fetal calf serum. Cells were plated at a density of 2×10^4 cells/cm^2 and were maintained at 37 °C in a Forma water jacketed CO_2 incubator (95% air, 5% CO_2). Differentiation of neuroblastoma cells was induced by adding Bt_2cAMP (1 mM) and isobutylmethylxanthine (0.5 mM) to sparse culture 15 h after plating (19). The undifferentiated and differentiated neuroblastoma cells at the early stationary phase of growth (approximately 4–5 days after subculture) were used for the labeling experiments.

Transglutaminase/[^{14}C]putrescine labeling of surface proteins

Cells grown in monolayer cultures (100 mm tissue culture dish) at the early stationary phase of growth were rinsed three times with phosphate-buffered saline (0.15 M NaCl, 10 mM sodium phosphate; pH 7.4) and removed from the substratum by gentle flushing with a stream of Earle's balanced salt solution. Cells were then sedimented by centrifugation at 1 000 **g** for 2 min. The cell pellet was resuspended in Earle's balanced salt solution (pH 7.6), containing 20 mM Tris-HCl, 9.5 mM dithiothreitol, and 4.5 mM $CaCl_2$. The cell concentration was kept at 1.5×10^7 cells/ml. [1,4-^{14}C]putrescine (5 ~ 10 μCi/ml) was added to the cell suspension, and the labeling reaction was initiated by the addition of purified transglutaminase (60 μg/ml) to the cell suspension. The mixture was incubated at 35 °C for 1 h with occasional gentle agitation. The cell suspension was then diluted with Earle's balanced salt solution containing 5 mM cold putrescine, and washed three times by centrifugation at 1 000 **g**. The final cell pellet was suspended in a 10 mM Tris-HCl (pH 7.2) buffer containing 1 mM

PMSF. Samples were then prepared for SDS-PAGE according to methods described below. Alternatively, after labeling, the cell pellet was resuspended and homogenized in 10 mM HEPES (pH 7.2), containing 1 mM $MgCl_2$, 2 mM CaCl, and 0.5 mM PMSF. The plasma membrane enriched fraction was prepared by discontinuous sucrose gradient (15%, 30% and 50%) centrifugation at 23 300 g for 30 min as previously described (20). The incorporation of [^{14}C]putrescine into acid-insoluble cellular material was determined by a filter-paper assay method (21). Protein concentrations were measured by the procedure of Lowry et al. (22) using bovine serum albumin as the standard.

SDS-polyacrylamide gel electrophoresis and fluorographic procedure

Protein samples obtained after labeling experiments were mixed with one-fifth volume of SDS-stop solution containing 12% SDS, 0.5 M Tris-HCl (pH 8.0) 10% β-mercaptoethanol, 5 mM EDTA, 25% glycerol and 0.005% pyronin Y and heated at 100 °C for 5 min. The protein samples were then subjected to SDS-PAGE on a linear gradient slab gel (7.5–15% acrylamide) as previously described (20). Gels were stained for protein with Coomassie blue. Fluorographs were prepared according to the procedure of Bonner and Lasky (23). Apparent molecular weights of the radioactive protein bands were estimated by the method of Fairbanks et al. (24). Standard proteins (and their molecular weights) used were cytochrome C (12 000), chymotrypsinogen (25 000), aldolase (40 000), ovalbumin (45 000), catalase (57 000), bovine serum albumin (67 000), phosphorylase a (94 000) and β-galactosidase (130 000). Fluorographs were scanned with a Schoeffel SD-3 000 spectromicrodensitometer, and the peak areas of the optical tracings were used as a quantitative measure of the incorporation of radioactivity.

Results

When intact mouse neuroblastoma cells were incubated with [^{14}C]putrescine in the presence of purified guinea pig liver transglutaminase, several surface proteins were specifically labeled. Figure 1 shows the specific labeling of two protein bands with apparent molecular weight of 92 000 and 76 000 of the N-18 cells and three protein bands with apparent molecular weight of 92 000, 76 000 and 72 000 of the NB-15 cells by [^{14}C]putrescine. In addition to these labeled proteins, a band of radioactivity was also apparent at the top of the gradient gel in both N-18 and NB-15 cells. It seems likely that this very high molecular weight radioactive band may represent cross-linked proteins using [^{14}C]putrescine as a cross-linking agent. This notion is supported by the finding that this band did not penetrate appreciably into the gel after electrophoresis even when acrylamide concentration was reduced to 5% (for example see Fig. 3B).

The amount of [^{14}C]putrescine incorporated into

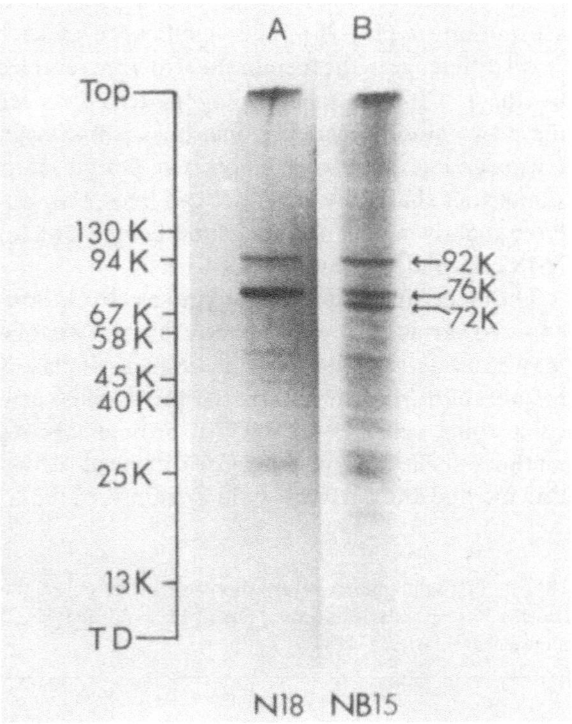

Fig. 1. Fluorograph illustrating the transglutaminase catalyzed incorporation of [1,4-^{14}C]putrescine into surface proteins of the N-18 and NB-15 mouse neuroblastoma cells. Both N-18 and NB-15 cells were labeled under identical conditions and plasma membrane fractions of each cell line were prepared (see Methods). The plasma membrane preparations used in this experiment were 9-fold purified judging by the increase in specific activity of 5′-nucleotidase (see Methods). The protein samples were then solubilized in SDS-Stop solution and subjected to analysis by SDS-PAGE as described in Methods. Each lane contained 50 μg protein. T.D. refers to the dye front of tracking dye, pyronin Y.

the major labeled protein bands in N-18 and NB-15 cells was calculated from the densitometric tracings of the fluorographs and was expressed as percentage of radioactivity incorporated (Table 1). While the three labeled surface proteins ($Mr = 92\,000$, 76 000 and 72 000) were of approximately equal intensity in NB-15 cells, the 76 000-dalton band of N-18 cells alone accounted for about 50% of the total radioactivity incorporated.

The difference in the labeling pattern between N-18 and NB-15 cells is unexpected in view of the relatedness of these two clonal lines. Both the N-18 and NB-15 mouse neuroblastoma cells are clonal lines derived from the original tumor line C-1300 mouse neuroblastoma cells (25). Both cell lines can be induced to undergo differentiation by the addition of cyclic AMP analogs and/or phosphodiesterase inhibitors (19, 26). The significance of such small difference in the membrane structure revealed by the [^{14}C]putrescine labeling pattern between these two closely related clonal lines is not clear. However, it is interesting to note that, despite many similarities shared by these two cell lines, they differed slightly in their doubling time; being ~21 h for N-18 cells and ~25 hrs for NB-15 cells.

The fact that [^{14}C]putrescine can selectively label exposed surface proteins suggests that this method may be used to monitor possible changes of plasma membrane during the differentiation of mouse neuroblastoma cells. Other surface protein labeling methods such as lactoperoxidase catalyzed iodination method and galactose oxidase catalyzed reductive tridiation method have been used to study the changes of plasma membranes in cell differentiation and malignant transformation (20, 27).

In both the N-18 and NB-15 neuroblastoma cells, the amounts of radioactivity incorporated into the trichloroacetic acid-insoluble material of the differentiated cells were approximately 2-fold higher than that of their undifferentiated (malignant) counterparts (Table 2).

The electrophoretic pattern of [^{14}C]putrescine-labeled surface proteins of the undifferentiated and differentiated N-18 neuroblastoma cells is shown in Fig. 2. The intensities of the gel top protein band and the 76 000-dalton band of the differentiated N-18 cells were greater than that of the undifferentiated cells by 2.1-fold and 2.3-fold, respectively. However, the intensity of the 92 000-dalton band of the differentiated N-18 cells was diminished when compared to that of the undifferentiated cells. The reciprocal relationship of the intensities of the gel top band the 92 000-dalton band in the differentiated N-18 cells could be due to a more preferential cross-linking of the 92 000-dalton band in the differentiated N-18 cells than that in the undifferentiated counterparts.

The labeling patterns of the undifferentiated and differentiated NB-15 cells are shown in Fig. 3. In contrast to results obtained with the N-18 cells, the intensity of all the labeled protein bands of NB-15 cells increased during cell differentiation. In addition, differentiation of NB-15 cells, was also accompanied by increases of labeling of some low

Table 1. Transglutaminase catalyzed incorporation of [^{14}C]putrescine into various surface proteins of the N-18 and NB-15 mouse neuroblastoma cells.

	Percentage of the total amount of radioactivity incorporated into individual protein bands	
	N-18	NB-15
gel top band	20 ± 5[a]	17 ± 3
92 000	15 ± 3	20 ± 5
76 000	48 ± 6	16 ± 3
72 000	0	18 ± 4
others	~15	~25

[a] Mean ± S.D. of three separate experiments
The percentage of the total amount of radioactivity incorporated into individual protein bands was estimated from the densitometric tracings of the fluorographs.

Table 2. Transglutaminase catalyzed incorporation of [^{14}C]putrescine into acid-insoluble material of mouse neuroblastoma cells.

Cell line	Morphology	Trichloroacetic acid insoluble radioactivity (cpm per mg protein)
N-18	Undifferentiated	55 000 ± 800[a]
	Differentiated	114 000 ± 2 000
NB-15	Undifferentiated	77 000 ± 900
	Differentiated	153 000 ± 3 000

[a] Mean ± S.D. of three separate experiments
Undifferentiated and differentiated mouse neuroblastoma cells were labeled by the transglutaminase/[^{14}C]putrescine method under identical conditions. The specific radioactivity incorporated into acid insoluble cellular material was determined by filter-disc assay method (21).

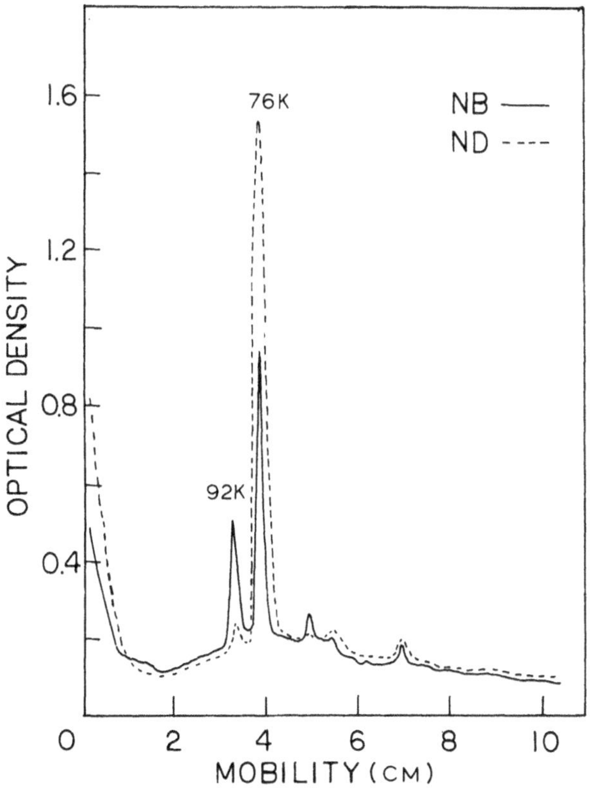

Fig. 2. Densitometric tracing of fluorograph illustrating the pattern of transglutaminase/[^{14}C]putrescine labeled surface proteins of the undifferentiated (NB) and differentiated (ND) N-18 mouse neuroblastoma cells. Undifferentiated and differentiated N-18 cells were labeled under identical conditions. Labeled proteins were separated on a linear gradient (7.5–15%) of SDS-polyacrylamide slab gel. The fluorograph was made on a SB-5 X-ray film (Eastman Kodak). Densitometric tracings were obtained by scanning the fluorographs. For both the NB and ND cells, 50 μg of cellular protein were used for gel electrophoresis.

molecular weight protein bands (Fig. 3A). Fig. 3B emphasizes the fact that the very high molecular weight band in both undifferentiated and differentiated NB-15 cells did not penetrate appreciably into the 5% gel suggesting that this gel top band is a cross-linked band.

The results illustrated in Fig. 2 and Fig. 3 also demonstrated that while the patterns of proteins labeled by the transglutaminase catalyzed incorporation of [^{14}C]putrescine of the undifferentiated and differentiated mouse neuroblastoma cells were similar, the intensity of labeling of the individual bands was modulated by cell differentiation.

Discussion

In this study, the transglutaminase/[^{14}C]putrescine labeling method was used to compare and probe the possible differences in surface architectures of undifferentiated and differentiated neuroblastoma cells.

A significantly greater amount of radioactive putrescine was incorporated into the surface proteins of the differentiated N-18 and NB-15 cells than that of their undifferentiated counter parts (Figs. 2 and 3). These results indicated that the differentiation of mouse neuroblastoma cells was accompanied by an increase of available acyl donor groups (i.e., glutaminyl residues) at cell surface proteins. Such an increase may be due to any one or a combination of the following mechanisms: (a) both the undifferentiated and differentiated neuroblastoma cells have the same number of peptide-bound glutaminyl residues, but in the undifferentiated cells these residues are partially occupied by endogenous polyamines; (b) the conformation of surface proteins of the differentiated cells is such that they are more susceptible to the transglutaminase catalyzed transamidation reaction than their undifferentiated counterparts; (c) there are more copies per cell of those surface proteins which can be labeled by [^{14}C]-putrescine in the differentiated cells than that in the undifferentiated cells.

It has been shown that transglutaminase can cross-link proteins through two mechanisms, one is the direct cross-linking of two proteins via γ-glutaminyl-lysine linkage, (2, 4) the other is using diamine or polyamine as a bridge to cross-link proteins via the γ-glutaminyl-diamine (polyamine)-γ-glutaminyl linkage (28). The incorporation of radioactive putrescine into protein band at gel top of both N-18 and NB-15 cells suggests that [^{14}C]putrescine may be used as a cross-linking agent under the present experimental conditions. If so, the increased cross-linking of surface proteins in the differentiated N-18 and NB-15 cells suggest that membrane structure of the differentiated cells may be different from that of the undifferentiated cells.

Polyamine-conjugated proteins have been described in human plasma (15), human amniotic fluid (16), and possibly human lymphocytes (29). Since transglutaminase and polyamines exist both in tissues and tissue fluids, transglutaminase may catalyze the incorporation of extracellular polyamines

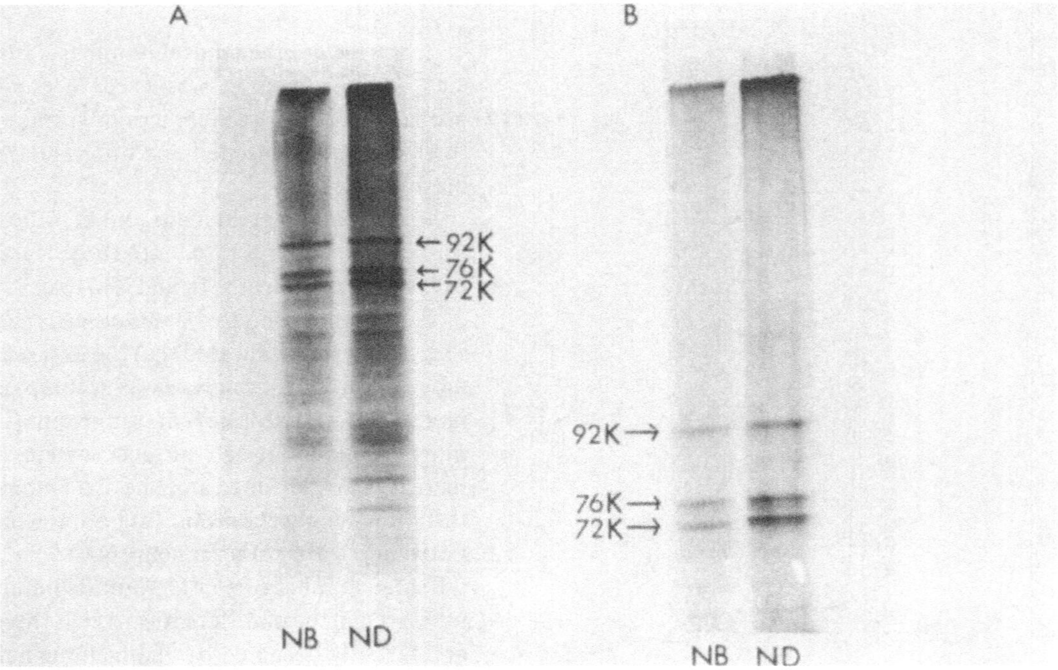

Fig. 3. Fluorograph illustrating the transglutaminase/[^{14}C]putrescine labeled surface proteins of undifferentiated (NB) and differentiated (ND) NB-15 mouse neuroblastoma cells.
Panel A: labeled proteins were separated on a 7.5–15% linear gradient SDS-polyacrylamide slab gel.
Panel B: labeled proteins were separated on a 5% SDS-polyacrylamide gel. Each lane contained 50 μg protein.

into cell surface proteins in vivo. In light of this, it is of interest to note that Quash et al. have shown that anti-polyamine antibody binds tightly to the surface of chick embryo cells (30) and is able to lyse L cells and BHK Py cells (31) in the presence of complement (31). Their results suggest that tightly-bound polyamines do exist at cell surface, however, the chemical properties of such binding remain to be identified. In the present study, I have shown that putrescine can covalently bind to certain specific surface proteins of mouse neuroblastoma cells. If indeed there are tightly bound polyamines at cell surface, these labeled surface proteins (92 000- and 72 000-dalton proteins for N-18 cells and 92 000-, 76 000- and 72 000-dalton proteins for NB-15 cells) may represent potential binding sites for polyamines in vitro and in vivo.

Acknowledgement

I wish to thank Dr. M. Takahashi of the Department of Physiology, Rutgers Medical School for the access to his unpublished method of purifying transglutaminase, Ms. R. Kozlowski, Ortho Pharmaceutical Research, N.J., for providing us with the fresh guinea pig livers, and Dr. A. Y-C Liu of the Department of Pharmacology, Harvard Medical School, for critical reading and discussion of the manuscript. This work was supported by grant CA 24479, NIH, grant PHS RR7058-15, NIH and grant 07-2188, Rutgers Research Council.

References

1. Folk, J. E., 1980. Ann. Rev. Biochem. 49: 517–531.
2. Folk, J. E. and Finlayson, J. S., 1977. Adv. Protein Chem. 31: 1–133.
3. Folk, J. E. and Cole, P. W., 1966. J. Biol. Chem. 241: 5518–5525.
4. Dutton, A. and Singer, S. J., 1975. Proc. Natl. Acad. Sci. USA 72: 2568–2571.
5. Okumura, T. and Jamieson, G. A., 1976. J. Biol. Chem. 251: 5944–5949.
6. Clarke, D. D., Mycek, M. J., Neidle, A. and Waelsch, H., 1959. Arch. Biochem. Biophys. 790: 338–354.

7. Cohen, S. S., 1971. Introduction to the Polyamines, pp. 1–179, Prentice-Hall, N.J.
8. Bachrach, U., 1973. Function of Naturally Occurring Polyamines, pp. 1–211, Academic Press, New York.
9. Russell, D. H. and Durie, B. G. M. Polyamines as Biochemical Markers of Normal and Malignant Growth, pp. 1–178, Raven Press, N.Y.
10. Novogradsky, A., Quittner, S., Rubin, A. L. and Stenzel, K. H., 1978. Proc. Natl. Acad. Sci. USA 75: 1157–1161.
11. Obinata, A. and Endo, H., 1979. J. Biol. Chem. 254: 8487–8490.
12. Davies, P. J. A., Davies, D. R., Levitzki, A., Maxfield, F. R., Milhaud, P., Willingham, M. C. and Pastan, I. H., 1980. Nature 283: 152–157.
13. Williams-Ashman, H. G., Beil, R. E., Wilson, J., Hawkins, M., Grayhack, J., Zunamon, A. and Weinstein, N. K., 1980. Adv. in Enzyme Regulation 18: 239–258.
14. Williams-Ashman, H. G., Wilson, J., Beil, R. E. and Lorand, L., 1977. Biochem. Biophys. Res. Commun. 79: 1192–1198.
15. Seale, T. W., Chan, W. Y., Shulka, J. and Rennert, O. M., 1979. Arch. Biochem. Biophys. 198: 164–174.
16. Seale, T. W., Chan, W., Shukla, J. and Rennert, O., 1979. Clin. Chim. Acta 95: 461–472.
17. Connellan, J. M., Chung, S. I., Whetzel, N. K., Bradley, L. M. and Folk, J. E., 1971. J. Biol. Chem. 246: 1093–1098.
18. Lorand, L., Campbell-Wilkes, L. K. and Cooperstein, L., 1972. Anal. Biochem. 50: 623–631.
19. Chen, K. Y., Presepe, V., Parken, N. and Liu, A. Y-C., 1982. J. Cell. Physiol. 110: 285–290.
20. Chen, K. Y. and Rinehart, C. A. Jr., 1981. Biochim. Biophys. Acta. 685: 61–70.
21. Chen, K. Y., Tsai, C. M. and Canellakis, E. S., 1975. Cancer Res. 35: 2403–2412.
22. Lowry, O. H., Rosebrough, N. J., Farr, A. L. and Randall, R. J., 1951. J. Biol. Chem. 193: 265–275.
23. Bonner, W. M. and Laskay, R. A., 1974. Eur. J. Biochem. 46: 83–88.
24. Fairbanks, G., Steck, T. L. and Wallach, D. F. H., 1971. Biochemistry 10: 2606–2617.
25. Prasad, K. N., 1975. Biol. Rev. 50: 129–265.
26. Prasad, K. N. and Hsie, A. W., 1971. Nature (New Biology) 233: 141–142.
27. Littauer, W. Z., Giovanni, M. Y. and Glick, M. C., 1980. J. Biol. Chem. 255: 5448–5453.
28. Schrode, J. and Folk, J. E., 1978. J. Biol. Chem. 253: 4837–4840.
29. Folk, J. E., Park, M. H., Chung, S. I., Schrode, J., Lester, E. P. and Cooper, H. L., 1980. J. Biol. Chem. 255: 3695–3700.
30. Quash, G. A., Niveleau, A., Aupoix, M. and Greenland, 1976. Exp. Cell Res. 98: 253–261.
31. Quash, G., Gresland, L., Delain, E. and Huppert, J., 1972. Exp. Cell Res. 75: 363–368.

Received 11 June 1982.

Molecular and Cellular Biochemistry 58, 99–111 (1984).

Epidermal and hair follicle transglutaminases and crosslinking in skin

Larry L. Peterson and Kirk D. Wuepper
Dept. of Dermatology, The Oregon Health Sciences University, 3181 S. W. Sam Jackson Park Road, Portland, OR 97201, U.S.A.

Summary

Epidermal and hair follicle transglutaminases crosslink structural proteins in the skin by epsilon-(gamma-glutamyl)-lysine bonds. This crosslinking produces protein polymers that are extremely insoluble and, until recently, difficult to characterize.

Epidermal transglutaminase is localized to the granular layer of the epidermis. It catalyzes the crosslinking of a soluble cytoplasmic precursor to form the cornified envelope that lines the inner membrane of the mature keratinocyte in the stratum corneum.

Hair follicle transglutaminase is localized to the inner root sheath and medulla of the hair follicle. It crosslinks a poorly characterized citrulline-rich protein.

The enzymes and their substrates have been shown to be important markers of normal differentiation. Regulation of these processes is currently under investigation.

I. Introduction

The epsilon-(gamma-glutamyl)-lysine intermolecular crosslink catalyzed by transglutaminases is of major importance in the keratinization of epidermis and hair structures. Many of the characteristics attributed to these tissues like rigidity and impermeability are due to insoluble keratins, proteins crosslinked with epsilon-(gamma-glutamyl)-lysine bonds and intercellular-linking desmosomes. Due to the relative insolubility of many skin proteins, biochemical characterization has been hindered until the last decade. Since keratinization is a complex process, we will present a brief overview of keratinization and the structures it involves. The interested reader is referred to several fine textbooks of dermatology (1, 2, 3).

The first layer of the epidermis is attached to the dermis by half desmosomes (Fig. 1). This basal or germinative layer is one cell layer thick, and most of the mitotic activity of the epidermis is limited to this layer and 1–2 layers above it. In the spinous layer (also called the Malpighian layer), the structural proteins are synthesized, including alpha keratin. This layer received its name because it looks spine-like by conventional microscopy. The granular layer is named for the numerous keratohyalin granules that stain with hematoxylin. Keratohyalin functions to orient and bind the alpha keratin fibrils. Transglutaminase and its substrate are present in this layer and contribute to the cornified envelope. The stratum corneum is the visible layer of the epidermis and is essentially layers of dead cells filled with keratin. The time required for a basal cell to proliferate, undergo terminal differentiation, reach the stratum corneum and be shed is approximately 28 days.

The hair also undergoes keratinization (Fig. 2). The outer root sheath is the outermost layer and is continuous with the epidermis. Next is the inner root sheath which is the first layer to undergo keratinization and acts like a sheath or mold for the developing hair shaft. This layer is shed as it reaches the scalp. The hair shaft itself is composed of the shingle-like cuticle, the cortex and the inner-most layer, the medulla. In human hair the medulla is discontinuous.

The inner root sheath and the medulla keratinize in a manner similar to the epidermis utilizing a keratohyalin equivalent, trichohyalin. These layers also contain the epsilon-(gamma-glutamyl)-lysine isodipeptide bond and a hair follicle transglutaminase. The cortex keratinizes without trichohyalin or transglutaminase. The reader may want to refer back to Figs. 1 and 2 occasionally while reading the remainder of this chapter.

While epidermis and nails grow continuously, hair grows in cycles. Anagen, the growing phase of hair, usually lasts for several years. The hair then undergoes an involutionary period called catagen followed by telogen, the resting phase which lasts a few months. Mitoses are limited to the region of the hair bulb. The substrates of transglutaminase from epidermis, inner root sheath and medulla have been

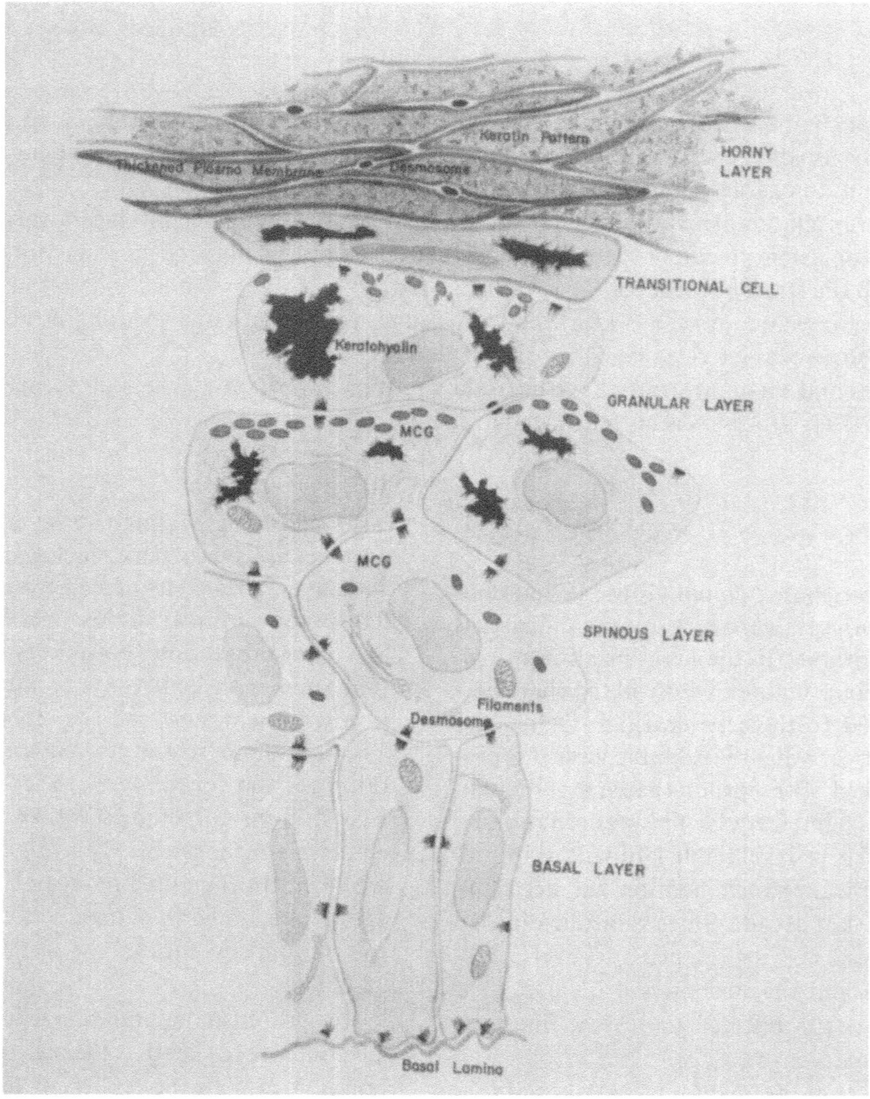

Fig. 1. Schematic diagram showing the morphology of the epidermis (from: The Structure and Function of Skin, 3rd edition, Montagna and Parakkal, 1974, Academic Press, New York. Reprinted by permission).

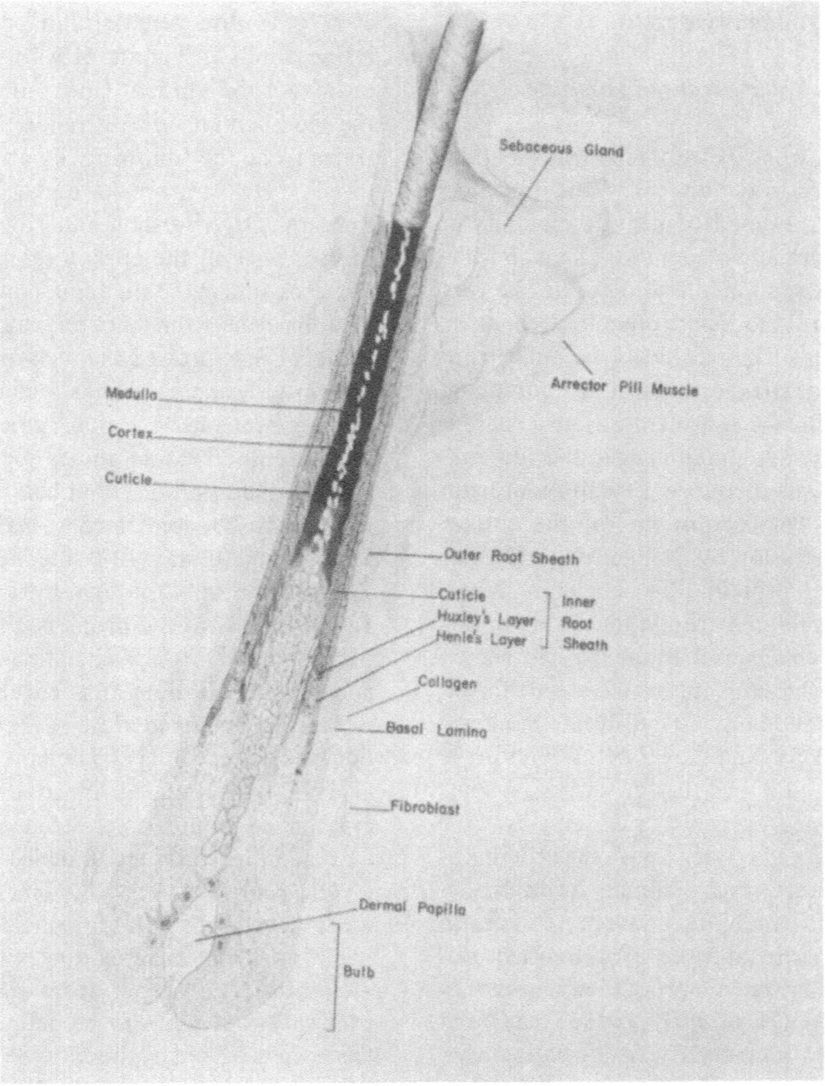

Fig. 2. Diagram of a growing (anagen) terminal hair follicle (from: The Structure and Function of Skin, 3rd edition, Montagna and Parakkal, 1974, Academic Press, New York. Reprinted by permission).

shown to contain citrulline (Fig. 3). This unusual amino acid is an intermediate in arginine synthesis and was previously thought not to be associated with proteins. The substrates have been shown to be distinct from keratin, but a few epsilon-(gamma-glutamyl)-lysine bonds have been detected also in the keratin proteins.

Our review is written from a historical perspective starting with work carried out in the nineteen-fifties and -sixties and concluding with the most recent advances of the early eighties. A better understanding of keratinization and the role of transglutaminase in skin will allow us to approach problems of the skin more intelligently.

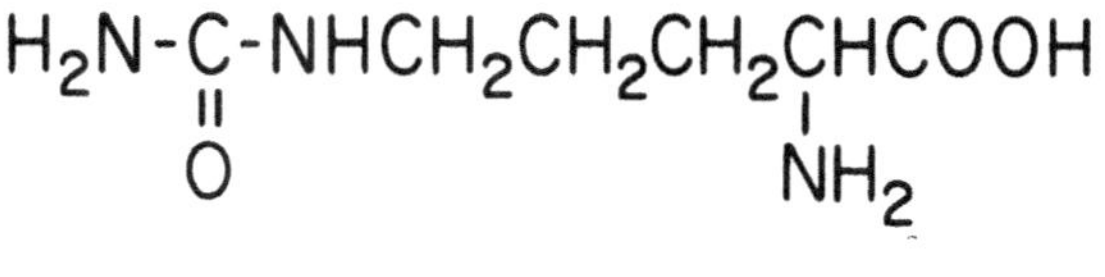

Fig. 3. Structure of citrulline.

II. The epsilon (gamma[4]-glutamyl)-lysine bond and hair follicle transglutaminase

Substrates of hair follicle transglutaminase

The pioneering work by wool biochemists in the nineteen-sixties led to the discovery of the epsilon (gamma-glutamyl)-lysine isodipeptide crosslink in hair follicle proteins and recognition of a hair follicle transglutaminase soon followed. In 1962 Rogers (4) reported a citrulline-rich protein located in the inner root sheath of hair follicles, medulla from hair, and related structures such as porcupine quills. This citrulline-rich protein was also characterized by a very high glutamic acid and low cysteine content. It was therefore very different from the surrounding keratin proteins of the cortex. With the demonstration by Steinert et al. (5) that the citrulline-rich protein was covalently crosslinked, research centered around the nature of this covalent bond. Rogers and Bittar (6) had shown that keratins of the hair cortex were crosslinked mainly by disulfide bonds. The disulfide bond appeared as an unlikely candidate for crosslinks in the inner root sheath and medulla for several reasons. First, as mentioned earlier, the cysteine-cysteine content was known to be very low in the proteins of the inner root sheath and medulla. Additionally, the differences in solubility between the keratin proteins of the cortex and the citrulline-rich proteins of the medulla and inner root sheath were very striking. Matoltsy (7), as early as 1953 had used this fact to separate medullary cells from keratin. Gillespie (8) in 1964 and Maclaren and Kilpatrick (9) in 1969 showed that the medullary cell protein remained insoluble and was resistant to extremes of pH, salt concentration, chaotropic agents, organic solvents and detergents. Cortical keratin proteins under similar conditions underwent dissolution quite easily. Unlike keratin, the citrulline-rich proteins were easily digested by proteolytic agents (10).

Evidence that the citrulline-rich protein was also high in glutamic acid led some investigators to question the possible role of glutamine in covalent bond formatin. Pisano et al. (11) in 1969 had shown that the crosslinking of glutamine and lysine in the fibrinogen-fibrin transformation was catalyzed by coagulation Factor XIII, a transglutaminase, and the bond formed was an epsilon (gamma-glutamyl)-lysine bond. Asquith et al. (12) in 1970, using an enzymatic hydrolysis technique developed by Cole et al. (13), discovered epsilon (gamma-glutamyl)-lysine bonds in digests of wool keratin. Asquith estimated the amount (micromoles/g protein) of the isodipeptide in Australian Merino wool extracts to be the following: keratin-15, alpha keratose-9, beta keratose-30, and gamma keratose-8. Subsequently, Harding and Rogers (14) reported the presence of the epsilon (gamma-glutamyl)-lysine crosslink in the citrulline-containing protein of medulla cells from hair and quills. Some of the values (15), expressed as moles of epsilon (gamma-glutamyl)-lysine crosslinks/1 000 moles of amino acids, were: rat-29.5, porcupine-2.3, guinea pig-28.1, rabbit-25.9, and guinea pig keratin-0.8. This value for keratin was about half the value obtained by Asquith but confirmed a small amount of epsilon (gamma-glutamyl)-lysine bonds in keratins. The vast majority, though, were found in medulla and inner root sheath proteins. Further delineation of the nature of these citrulline-rich proteins was achieved by Harding and Rogers (16) in 1976 by isolating crosslinked proteins that contained citrulline from the medulla and inner root sheath. Amino acid analysis revealed approximately 2/3 of the crosslinked peptides were composed of glutamine and citrulline residues. Labeled antisera to these tryptic peptides stained the inner root sheath and medulla cells in a pattern coincidental with trichohyalin (17). The exact nature of hair follicle transglutaminase substrates from medulla and inner root sheath remains to be defined. Advancement has been hindered by the extreme insolubility of the proteins, once they are polymerized.

Hair follicle transglutaminase

Rogers and Springell in 1959 (18) first established the presence of transamination and transpeptidation in wool. In 1972 Harding and Rogers (15) reported an enzyme in hair follicle homogenates that incorporated (^{14}C) glycine ethyl ester into casein by forming epsilon (gamma-glutamyl) derivatives. The reports of epsilon (gamma-glutamyl)-lysine bonds (12, 15) led Chung and Folk (19) to the first isolation of hair follicle transglutaminase from the guinea pig. Purification of the inner root sheaths of hair follicles yielded two, distinct transglutaminases. The first was similar to the liver transglutaminase they had previously isolated and contained esterase activity. The other enzyme they named hair follicle transglutaminase since it demonstrated properties that were characteristic only to that enzyme including no esterase activity. The enzyme had a molecular weight of 54 000 on gel filtration and dissociated into two apparently identical subunits at 27 000 daltons on polyacrylamide gel electrophoresis containing sodium dodecyl sulfate. Antiserum to the liver enzyme gave no precipitin band upon immunodiffusion with the hair follicle enzyme. The liver and hair follicle enzyme gave no indication of having a zymogen form. Its dependence on calcium was also demonstrated. Kinetic studies of both transglutaminases (20) were consistent with ping-pong mechanism, each having a characteristic K_m and V_{max}.

Buxman and Wuepper (21) found that hair follicle transglutaminase and epidermal transglutaminase were immunochemically distinct. Antiserum to bovine snout epidermal transglutaminase failed to precipitate in agar, or to functionally inhibit transglutaminase activity in hair follicle extracts. These observations were substantiated by the discovery that fluorescein-conjugated antiserum to bovine epidermal transglutaminase localized the enzyme in epidermis but not follicles (22). Ogawa and Goldsmith (23) reported similar results with antiserum to human epidermal transglutaminases.

Differential purification, characterization, and immunochemical localization of epidermal and hair follicle transglutaminase was finally achieved by Peterson and Buxman (24) in 1979 using epidermal extracts from rats 4–5 days old. Epidermal transglutaminase had a Mr of approximately 57 000 by gel filtration and in gels containing sodium dodecyl sulfate while hair follicle transglutaminase had a Mr of 54 000 and was reduced to two apparently identical subunits with a Mr of 27 000 by denaturing media. Purified hair follicle transglutaminase and epidermal transglutaminase had specific activities of 160 and 312 amine-incorporating units/ mg/ h, respectively. Antiserum specific to each enzyme was produced in chickens and no cross reaction was seen between transglutaminases, with human or bovine epidermis or with rat plasma Factor XIII. When conjugated to fluorescein, these

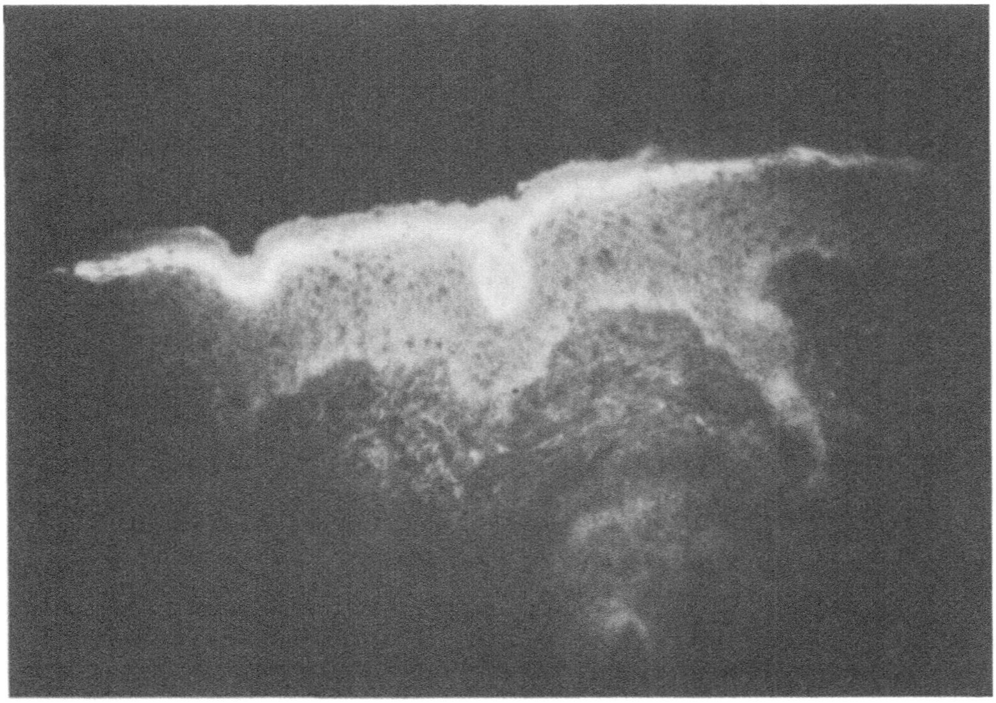

Fig. 4. Direct staining of human epidermis with dansyl cadaverine and calcium (× 62.5). A similar pattern was seen in rat epidermis stained with fluorescein-conjugated antibody to rat epidermal transglutaminase.

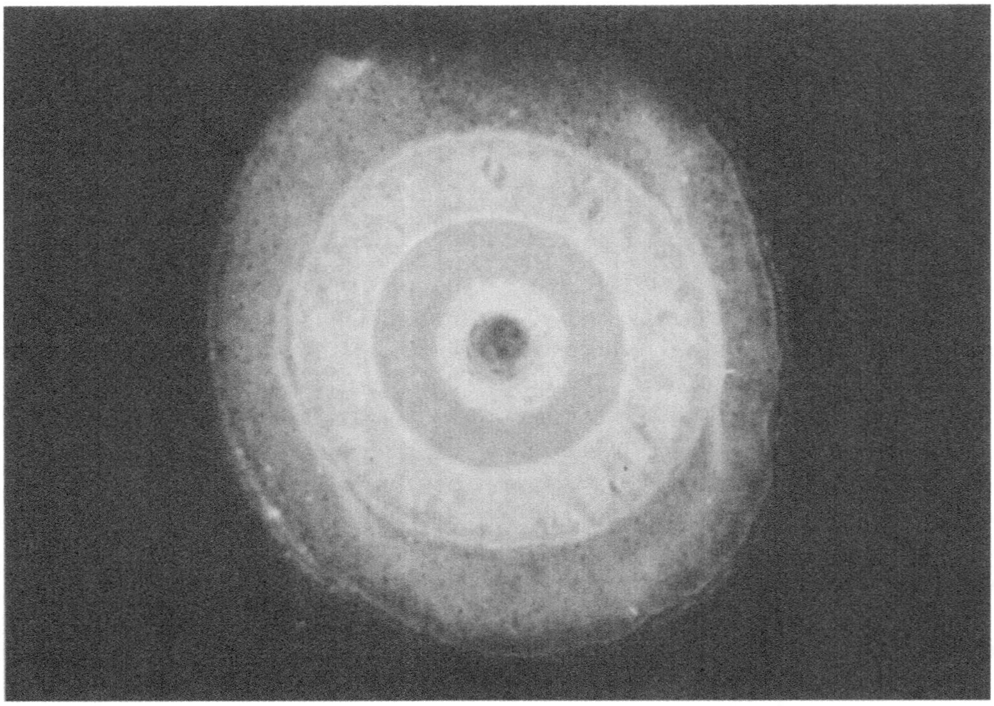

Fig. 5. Direct staining of a cross-section of a human hair with dansyl cadaverine and calcium (× 600). A similar pattern was seen in rat hair stained with fluorescein-conjugated antibody to rat hair follicle transglutaminase. The medulla and inner root sheath are fluorescing.

Table 1.

Author	Chung et al. (19, 55)	Harding and Rogers (56)		Peterson and Buxman (24)
Tissue	Guinea pig	Rat & guinea pig	Sheep	Rat
Metal requirements	EDTA inhibits Ca^{++} activates	EDTA inhibits Ca^{++} activates	EDTA did not inhibit Ca^{++} activates	EDTA inhibits Ca^{++} activates
Substrates tested	^{14}C-Putrescine ^{14}C-Methylamine	^{14}C-Glycine ethyl ester		Dansyl cadaverine
Molecular weight	54 000 (gel filtration) 27 000 (SDS electrophoresis)			54 000 (gel filtration) 27 000 (SDS electrophoresis)
Immunological identity	No inhibition with Ab to liver T-gase	Antihuman Factor XIII inhibits	No cross-reaction with anti-Factor XIII	No cross reaction with rat anti-Factor XIII, rat or bovine epidermal T-gase

antisera localized epidermal transglutaminase and hair follicle transglutaminase to the granular layer of the epidermis (Fig. 4) and the medulla and inner root sheath of hair follicles (Fig. 5), respectively.

The above findings are very interesting since both hair and epidermis develop from ectoderm. Further amino acid analysis and kinetic comparisons need to be done. Table 1 summarizes the hair follicle transglutaminases that have been characterized.

III. Epidermal transglutaminase and its substrates

Epidermal transglutaminase

The discovery of epsilon (gamma-glutamyl)-lysine bonds and hair follicle transglutaminase in hair structures led to the investigation of glabrous epidermis for an epidermal transglutaminase. Buxman and Wuepper (25, 26) found transglutaminase activity in cow snout epidermis, a source of hair-free epidermis, and were the first to purify the enzyme and produce a monospecific antibody to epidermal transglutaminase. The antisera did not cross react with bovine or human Factor XIII. Additionally, localization of epidermal transglutaminase was achieved by direct histochemical staining using dansyl cadaverine. The fluorescence was cytoplasmic over the mid-malpighian layer and increased in intensity in the granular layer. Fluorescein-labeled antiserum to bovine epidermal transglutaminase produced similar results. In 1976 they further characterized their purified enzyme (27). Analytical ultracentrifugaton produced a molecular weight of 55 800 from sedimentation equilibrium data. The purified enzyme had a specific activity of 3 000 amine incorporating units/mg/h.

Goldsmith et al. in 1974 (28) most extensively investigated cow snout epidermis, but also found transglutaminase activity in skin extracts from rat, turtle, toad and frog. The bovine epidermal transglutaminase had a relative molecular weight between 50 000 and 60 000 by gel filtration. It was similar to many of the transglutaminases previously studied demonstrating heat stability, inhibition by EDTA, stimulation by reducing agents and the ability to crosslink fibrin. Enzyme activity was highest in the stratum corneum and granular layer. Detection of the epsilon (gamma-glutamyl)-lysine bond was achieved by using the indirect technique of Pisano et al. (11), but actual isolation of dipeptides containing this bond was unsuccessful.

Goldsmith and Martin (29) next looked at human callus for evidence of transglutaminase activity. Fifty-fold purification of the enzyme was achieved and the protein demonstrated a molecular weight between 50 000 and 55 000 by gel filtration. It shared many of the same properties as guinea pig hair follicle transglutaminase and cow snout transglutaminase already mentioned. The enzyme was active from pH 7.0 to 10.4. Ogawa and Goldsmith (30) then purified human epidermal transglutaminase to homogeneity, and reported its molecular weight at 50 000 by sodium dodecyl sulfate electrophoresis. Their immunologic studies of human epidermal transglutaminase (23) showed that the enzyme was immunologically distinct from other transglutaminases. Crude extracts of skin from frog, rat, and mouse, and of hair follicles did not cross-react with antisera to human epidermal transglutaminase.

Buxman et al. (31) have shown that transglutaminase is an important marker for terminal epidermal differentiation. Fetal rat epidermis and hair follicle inner root sheath undergo terminal keratinization late in gestation. Concomitant with the onset of keratinization is a rise in transglutaminase activity in both structures as shown biochemically and histochemically. Their conclusions were consistent with other studies showing transglutaminase activity to be greatly decreased in poorly differentiated cells such as basal cell epithelioma and psoriatic cells (22). Obinata and Endo (32) have confirmed the importance of the enzyme for normal differentiation. They have observed sharp increases in plasma glucocorticoid levels and epidermal transglutaminase around gestational day 14 in the chick embryo. This rise is followed by developmental changes including synthesis of epidermal structural proteins, insolubilization of proteins and envelope formation. They concluded that glucocorticoid induces transglutaminase activity.

Negi et al. (33) suggested that lysosomal proteinases may also activate epidermal transglutaminase at the autolytic stages in the granular layer of the epidermis. Cathepsin D caused a 9.5-fold time-dependent enhancement of epidermal transglutaminase. We (34) have recently reported a two-fold increase in epidermal transglutaminase activity by purified human epidermal calmodulin. Calmodulin is known to be an activator of many calcium-dependent enzyme systems (35–37). The relative importance of regulators of epidermal transglutaminase activity remain to be defined. A summary of epidermal transglutaminases from various species is presented in Table 2.

The cornified envelope and substrates of epidermal transglutaminase

The chemical and electron microscopic studies by Matoltsy (38) and Balsamo and Brody (39) demonstrated that the plasma membrane undergoes many chemical and structural changes as the epidermal cell reaches maturity. The formation of a cornified envelope lining the inner surface of the mature keratinocyte was evident in the upper granular layers. The thickness of the membrane increased from 80 Å to 150 Å, and the envelope appeared electron dense. It was also noticed that the cornified cells were much more rigid and impervious to keratinolytic agents; e.g., urea, dithiothreitol. This was attributed to the extremely insoluble proteins of the envelope. The Matoltsys in 1966 (40) investigated the amino acid composition of the membrane proteins of human callous. The membrane was especially rich in glutamic acid, proline, 1/2 cystine and lysine. The nature of the covalent bonds and proteins of the envelope were unknown at that time.

Identification in the early seventies of epsilon (gamma-glutamyl)-lysine bonds (12, 41) and transglutaminases (19, 21, 28) from epidermis and hair prompted the investigation of substrates for epidermal transglutaminase. Significant advances were made by Buxman et al. (42) in 1976. Bovine epidermal extracts containing endogenous epidermal transglutaminase were incubated with dansyl cadaverine and calcium. Substrates of epidermal transglutaminase were identified by this method. A soluble substrate was isolated and had a molecular weight of 150 000. A highly insoluble substrate was also identified and solubilized by chemical cleavage with cyanogen bromide. Neither of the substrates precipitated in agar with bovine anti-fibrinogen but were quite strongly bound by a sepharose anti-fibrinogen column, being released by thiocyanate. Antisera to the insoluble substrate produced two separate precipitation lines in agar with either the dialyzed cyanogen bromide extract or purified soluble substrate. The insoluble substrate was determined to be neither an alpha helical, fibrous protein nor a sulfur-rich matrix protein. Preliminary amino acid analysis of the purified substrate showed a very low 1/2 cystine content and a high serine, glutamic acid and lysine content. They proposed that the

Table 2.

Author	Hanigan and Goldsmith (37)	Ogawa and Goldsmith (23, 28) Goldsmith and Martin (27)	Buxman and Wuepper (25, 26)	Peterson and Buxman (24)
Tissue	Cow snout epidermis	Human callus	Cow snout epidermis	Rat epidermis
Metal requirements	Ca^{++} activates EDTA inhibits	Ca^{++} activates EDTA inhibits	Ca^{++} activates EDTA inhibits	Ca^{++} activates EDTA inhibits
Substrates tested	fibrin, dansyl cadaverine ($1,4\text{-}^{14}C_2$) putrescine	putrescine fibrin	dansyl cadaverine	dansyl cadaverine
Molecular weight	50 000–60 000 (gel filtration)	50 000 (gel filtration) 51 000 (SDS electrophoresis)	55 000 (gel filtration & SDS electrophoresis) 55 800 (sedimentation equilibrium)	56 000 (gel filtration & SDS electrophoresis)
Immunological identity	not tested	no cross reaction with human Factor XIII, frog, rat, mouse, chicken or human hair follicle transglutaminase crude extracts	no cross reaction with bovine or human Factor XIII	no cross reaction with bovine or rat hair follicle transglutaminase or rat Factor XIII

natural substrate for epidermal transglutaminase was a soluble precursor protein enzymatically crosslinked by transglutaminase into large polymers that formed the envelope of the keratinocyte.

Epsilon(gamma-glutamyl)-lysine crosslinks were now considered a major characteristic of the envelope like the disulfide bond was to keratin. Even epidermal keratin, like hair keratin (12), was found to contain some epsilon (gamma-glutamyl)-lysine bonds. Abernathy et al. (43) found this bond in the very high molecular weight alpha keratin of human stratum corneum.

In 1978 Hanigan and Goldsmith (44) confirmed the presence of two high molecular weight substrates similar to those reported by Buxman et al. They investigated human stratum corneum and newborn rat epidermis.

The development of methods for human epidermal keratinocyte subculture on irradiated 3T3 cells by Reinwald and Green (45, 46) led to advances in understanding the formation of the cornified envelope. Rice and Green (47), Sugawara (48), and Sun and Green (49) established that cells in this culture system stratify and mature in the upper layers similarly to normal epidermis. DNA synthesis and division are absent, and keratin filaments form disulfide bonds in these upper layers. Also, the inner surface of the plasma membrane undergoes envelope formation. The insoluble envelope protein was found to be crosslinked by epsilon (gamma-glutamyl)-lysine bonds. A sharp decline in protein synthesis was noted just before envelope formation. In fact, protein synthesis inhibitors like cyclohexamide or emitine did not inhibit envelope formation but actually accelerated it. They suggested this showed that all the envelope precursors and transglutaminase were present at the time of envelope formation. Transglutaminase inhibitors prevented, while calcium was needed for, envelope formation. Rice and Green (50) were able to induce envelope formation in cell culture by methods that affected cell permeability or increased transglutaminase activity. Triton-X, 0.7–0.9 M NaCl, and the calcium ionophore, X537A, were all able to initiate crosslinking. A soluble precursor of the crosslinked envelope was identified by its ability to bind dansyl cadaverine and purified to homogeneity. The precursor had a molecular weight on gel filtration of 500 000 and on sodium dodecyl sulfate gel electrophoresis of 92 000. Faint fluorescent bands were also seen at 45 000 and 20 000. The 92 000 molecular weight protein had an isoelectric point at pH 4.5 plus/minus 0.3, and it was biochemically and immunochemically distinct from keratin. Immunofluorescent staining was consistent with the precursor being cytoplasmic in the upper epidermal layers and membrane bound in the envelope in the stratum cornium. A puzzling finding was the amino acid composition of the envelope precursor compared to the crosslinked envelope. The former had a glutamine content of almost 46% while the latter had a content of only 15%. They have named this soluble precursor 'involucrin' (from the Latin, involucrum = an envelope).

In further studies, Watt and Green (51) have been able to correlate involucrin synthesis with cell size in human epidermal cell culture. Almost no cells with a diameter of $\leqslant$14 micrometers contained involucrin while most with a diameter of $\geqslant$17 micrometers did. This is of interest since it is known that cells enlarge after they leave the basal layer in normal epidermis. Banks-Schlegel and Green (52), using antibody to involucrin, found the precursor present in human conjunctival, lingual and corneal epithelium and concluded that involucrin was present in stratified squamous epithelia regardless of whether a stratum corneum is present.

Buxman et al. (53) in 1980 proposed that their previously described relative high molecular weight substrates of epidermal transglutaminase were actually intermediate polymers of a lower molecular weight protein. Until this time, substrates had been identified by their ability to incorporate dansyl cadaverine. The problem with this method is that at the same time the substrate proteins are incorporating dansyl cadaverine, they are also to a limited extent crosslinking to each other and forming polymers. With the antibody to the 150 000 molecular weight substrate described earlier (42), they were able to identify a low molecular weight soluble substrate from bovine snout epidermis that cross-reacted with it. This soluble substrate demonstrated a Mr of 36 000 by gel filtration, and in the presence of calcium and transglutaminase, it was converted to higher molecular weight polymers ranging in molecular weight from 75 000 to greater than 200 000. An insoluble polymer was also formed that was not solubilized by urea, sodium dodecyl sulfate or beta mercaptoethanol. Crosslinking did not occur in the absence of transglutaminase or in the presence of putrescine or EDTA. The 36 000 molecular weight substrate dissociated in sodium dodecyl sulfate gels to considerably smaller subunits. The epsilon (gamma-glutamyl)-lysine isodipeptide was identified in high molecular weight products of crosslinking but not in the 36 kilodalton (Kd) precursor. They proposed that the 36 Kd protein was the precursor protein that crosslinked to form the epidermal envelope. Amino acid analysis of this purified substrate revealed a significant quantity (in residues/ 100 residues) of lysine-5.65, glutamic acid-13.46, an citrulline-2.38 and low levels of methionine and cystine. This analysis is similar to that reported by Green fo the envelope protein (50). Electron microscopy demonstrated the polymer to be a globular, amorphous substance lacking a fibrillar structure. Further studies showed that the high molecular weight polymer was immunologically identical to the 36 Kd protein. By indirect immunofluorescence, substrate was located to the cytoplasm and inner cell membrane of the granular layer and to the inner membrane of the stratum corneum (Fig. 6).

We recently have extended these studies of bovine snout epidermis to include human epidermis (54). The predominant human and bovine substrates purified from epidermis each demonstrated a Mr of 36 000 on gel filtration. By using several low

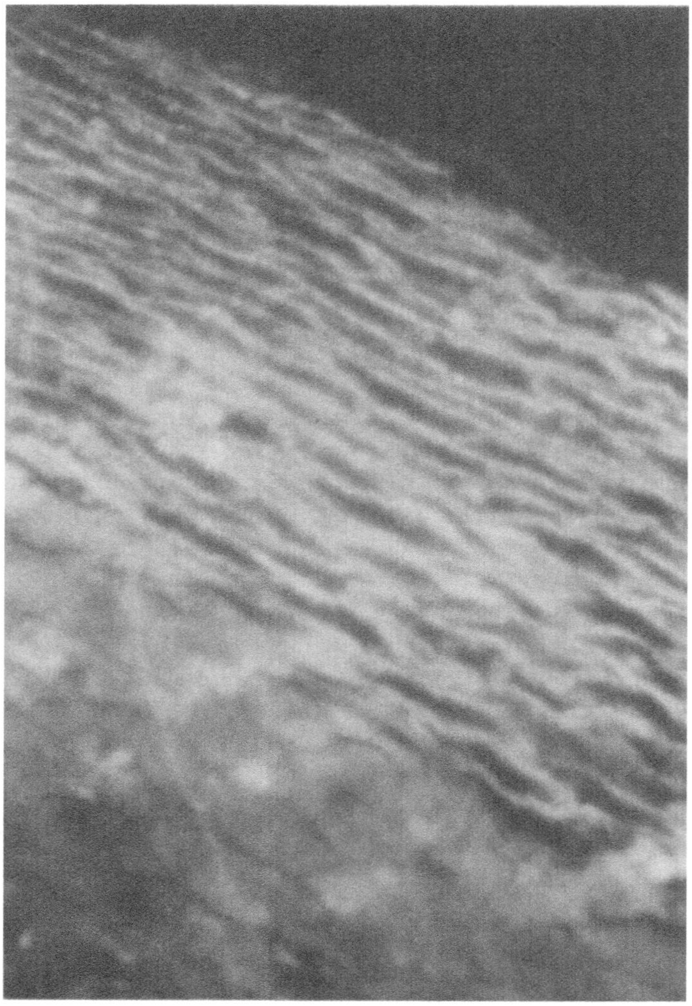

Fig. 6. Direct staining of bovine snout epidermis with fluorescein-conjugated antibody to bovine keratolinin showing fluorescence located on the cell membrane of the stratum corneum (× 200).

Table 3.

Author	Buxman, Buehner and Wuepper (42)	Hanigan and Goldsmith (44)	Rice and Green (50)	Buxman, Lobitz and Wuepper (53); Zettergren, Peterson and Wuepper (54)
Tissue	bovine snout epidermis (early study)	human callus & rat epidermis	human epidermal cell culture	bovine snout & human epidermis
Molecular weight	insoluble - very high molecular weight. soluble form - 150 000 (SDS electrophoresis)	similar to bovine snout in 1st column	500 000 gel filtration 92 000 + traces 40 000 & 20 000 (SDS gel electrophoresis)	36 000 (gel filtration) 6 000 (SDS electrophoresis)
Immunologic identity (to antisera of substrate)	no cross reaction to bovine fibrinogen. cross reacts with insoluble substrate	not tested	cross reacts with conjunctival, corneal & lingual epithelium	soluble 36 000 precursor immunologically identical to polymers ranging from 70 000 to >200 000. bovine & human show no identity
Isoelectric point			4.5	bovine: A 6.3, B 6.0; human: A 5.4, B 5.0

molecular weight standards, we have shown in both species to dissociate in sodium dodecyl sulfate gels to a Mr of 6 000 200. When the substrate were incubated with their respective enzymes, dansyl cadaverine and calcium, fluorescent bands at 6 000, 12 000, 24 000 and 150 000 were apparent. No band has been seen at 92 Kd where involucrin from cultured keratinocytes has been identified. Isoelectric focusing of human or bovine substrate revealed two moieties (human: 5.4/50; bovine: 6.3/6.0). The two proteins were readily resolved by chromatofocusing and each of the moieties of the bovine substrate incorporated dansyl cadaverine and precipitated in a line of identity with antiserum to the soluble bovine substrate. Antiserum to purified human substrate failed to cross react with its bovine counterpart and vice-versa.

We have named the soluble 36 Kd molecular weight envelope precursor *keratolinin* (Gk keratos: horny tissue; Gk linios: to cover the inner surface of). Human and bovine keratolinins appear biochemically similar, but they are immunochemically distinct proteins. The exact relationship of keratolinin from epidermis to involucrin from cell culture remains unclear. Exchange of antiserum and/or substrates between laboratories may clarify the subject. A summary of substrates of epidermal transglutaminase is presented in Table 3.

IV. Conclusion

The epsilon (gamma-glutamyl)-lysine bond has been shown to be an important component of the epidermis and hair. Its presence in the cornified envelope has become an important marker for normal differentiation of the epidermis. Many of the characteristics of the skin including strength, stability and resistance to transepidermal water loss can be attributed in part to the presence of a transglutaminase.

Research recently has been directed almost entirely toward epidermal transglutaminase and its substrate due partly because epidermal cell culture has been so successful. Poorly understood processes such as the assembly of the soluble substrates into the envelope polymer and anchoring mechanisms on the inner membrane surface can now more easily be studied. Evidence that glucocorticoid, calcium, cathepsin, and calmodulin are possible modulators of transglutaminase activity has begun to advance our knowledge of regulatory mechanisms.

Information about envelope abnormalities and transglutaminase changes in tumors and diseases of keratinization is stimulating interest from a wide range of investigators including cancer specialists, cell biologists, and clinical dermatologists.

Disappointingly, new information on the substrate of hair follicle transglutaminase has been almost non-existent over the last five years. The difficulty in obtaining sufficient quantities of starting material, especially human, along with the problems of insolubility discussed earlier is partly to blame. To date, no abnormalities in hair follicle transglutaminase or its substrates have been reported, but little research in this area has been undertaken.

Much has been learned about transglutaminase and intermolecular crosslinks in the skin in the last decade. We are optimistic that this will remain a promising field for future investigation.

References

1. Textbook of Dermatology, 3rd ed. (Rook, A., Wilkinson, D. S. and Ebling, F. J. G., eds.), 1979. Blackwell Scientific Publications, Oxford.
2. Dermatology in General Medicine, 2nd ed. (Fitzpatrick, T. B., Eisen, A. Z., Wolff, K., Freedberg, I. M. and Austen, K. F., eds.), 1979. McGraw-Hill Book Co., New York.
3. Dermatology (Moschella, S. L., Pillsbury, D. M. and Hurley, H. J., eds.), 1975. W. B. Saunders Co., Philadelphia.
4. Rogers, G. E., 1962. Nature 194: 1194–1151.
5. Steinert, P. M., Harding, H. W. J. and Rogers, G. E., 1969. Biochim. Biophys. Acta 175: 1–9.
6. Rogers, G. E., 1969. The Biological Bases of Medicine (Bittar, E. E., ed.), Vol. 6, pp. 21–30, Academic Press, New York.
7. Matoltsy, A. G., 1953. Exp. Cell. Res. 5: 98–103.
8. Gillespie, J. M., 1964. Aust. J. Biol. Sci. 17: 282–286.
9. Maclaren, J. A. and Kilpatrick, D. J., 1969. Aust. J. Biol. Sci. 22: 1081–1087.
10. Rogers, G. E., 1964. The Epidermis (Montagna, W. and Lobitz, W., eds.), pp. 179–236, Academic Press, New York.
11. Pisano, J. J., Finlayson, J. S. and Peyton, M. P., 1969. Biochemistry 8: 871–876.
12. Asquith, R. S., Otterburn, M. S., Buchanan, J. H., Cole, M., Fletcher, J. C. and Gardner, K. L., 1970. Biochim. Biophys. Acta 207: 342–348.
13. Cole, M., Fletcher, J. C., Gardner, K. L. and Corfield, M. C., 1970. 4th International Wool Textile Research Conference, Berkeley.
14. Harding, H. W. J. and Rogers, G. E., 1970. Proceedings of the Australian Biochemical Society 3A: 56.
15. Harding, H. W. J. and Rogers, G. E., 1971. Biochim. Biophys. Acta 257: 37–39.
16. Harding, H. W. J. and Rogers, G. E., 1976. Biochim. Biophys. Acta 427: 315–324.
17. Rogers, G. E., Harding, H. W. J. and Llewellyn-Smith, I. J., 1977. Biochim. Biophys. Acta 495: 159–175.
18. Rogers, G. E. and Springell, P. H., 1959. Nature 183: 993–999.
19. Chung, S. I. and Folk, J. E., 1972. Proc. Nat. Acad. Sci. 69 (2): 303–307.
20. Chung, S. I. and Folk, J. E., 1972. J. Biol. Chem. 247 (9): 2798–2807.
21. Buxman, M. M. and Wuepper, K. D., 1976. Biochem. Biophys. Acta 452: 356–369.
22. Buxman, M. M. and Wuepper, K. D., 1978. J. Histochem. Cytochem. 26 (5): 340–348.
23. Ogawa, H. and Goldsmith, L. A., 1977. J. Invest. Dermatol. 68: 32–35.
24. Peterson, L. L. and Buxman, M. M., 1981. Biochim. Biophys. Acta 657: 268–276.
25. Buxman, M. M. and Wuepper, K. D., 1974. Clin. Res. 22: 157A.
26. Buxman, M. M. and Wuepper, K. D., 1975. J. Invest. Dermatol. 65: 107–112.
27. Buxman, M. M. and Wuepper, K. D., 1976. Biochim. Biophys. Acta 452: 356–369.
28. Goldsmith, L. A., Baden, H. Pl., Roth, S. I., Colman, R., Lee, L. and Fleming, B., 1974. Biochim. Biophys. Acta 351: 113–125.
29. Goldsmith, L. A. and Martin, C. M., 1975. J. Invest. Dermatol. 64: 316–321.
30. Ogawa, H. and Goldsmith, L. A., 1976. J. Biol. Chem. 251 (23): 7281–7288.
31. Buxman, M. M., Kocarnick, M. J. and Holbrook, K. A., 1979. J. Invest. Dermatol. 72: 171–177.
32. Obinata, A. and Endo, H., 1979. J. Biol. Chem. 254 (17): 8487–8490.
33. Negi, M. N., Matsui, T. and Ogawa, H., 1981. J. Invest. Dermatol. 77: 389–392.
34. Peterson, L. L., Zettergren, J. G. and Wuepper, K., 1982. Clin. Res. 30 (1): 159A.
35. Cheung, W. Y., 1970. Biochem. Biophys. Res. Commun. 38: 533–538.
36. Wolff, J., Cook, G. H., Goldhammer, A. R. and Perkowitz, S. A., 1980. Proc. Natl. Acad. Sci. 77: 3841–3844.
37. Means, A. R., Lash, J. S. and Chafouleas, J. G., 1982. Phys. Reviews 62 (1): 1–39.
38. Matoltsy, A. G. and Balsamo, C. A., 1955. J. Biophys. Biochem. Cytol. 1: 339–360.
39. Brody, I., 1959. J. Ultrastruct. Res. 2: 482–486.
40. Matoltsy, A. G. and Matoltsy, M. N., 1966. J. Invest. Dermatol. 46 (1): 127–129.
41. Abernathy, J. and Goldsmith, L. A., 1975. Clin. Res. 23: 226A.
42. Buxman, M. M., Buehner, G. E. and Wuepper, K. D., 1976. Biochem. Biophys. Res. Commun. 73 (2): 470–478.
43. Abernathy, J. L., Hill, R. L. and Goldsmith, 1977. J. Biol. Chem. 252 (6): 1837–1839.
44. Hanigan, H. and Goldsmith, L., 1978. Biochim. Biophys. Acta 522: 589–601.
45. Rheinwald, J. G. and Green, H., 1975. Cell 6: 331–344.

46. Rheinwald, J. G. and Green, H., 1977. Nature 265: 421–424.
47. Rice, R. H. and Green, H., 1977. Cell 11: 417–422.
48. Sugawara, K., 1977. Biochemistry of Cutaneous Epidermal Differentiation. (Seiji, M. and Bernstein, I. A., eds.), pp. 387–397, University of Tokyo Press, Tokyo.
49. Sun, T. T. and Green, H., 1976. Cell 9: 511–521.
50. Rice, R. H. and Green, H., 1979. Cell 18: 681–694.
51. Watt, F. M. and Green, H., 1981. J. Cell. Biol. 90: 738–742.
52. Banks-Schlegel and Green, H., 1981. J. Cell. Biol. 90: 732–737.
53. Buxman, M. M., Lobitz, C. J. and Wuepper, K. D., 1980. J. Biol. Chem. 255 (3): 1200–1203.
54. Zettergren, J. G., Peterson, L. L. and Wuepper, K. D., 1982. Clin. Res. 30 (2): 616A.
55. Chung, S. I., Lewis, M. S. and Folk, J. E., 1974. J. Biol. Chem. 249: 940–950.
56. Harding, H. W. J. and Rogers, G. E., 1972. Biochemistry 11: 2858–2863.

Received 6 May 1982.

Molecular and Cellular Biochemistry 58, 113-119 (1984).

Transglutaminase-mediated cross-linking in mammalian epidermis

Joseph A. Rothnagel and George E. Rogers
Department of Biochemistry, University of Adelaide, Adelaide, South Australia 5001, Australia

Summary

The ε-(γ-glutamyl)lysine isopeptide bond has been identified in certain structural proteins of the hair fibre and the epidermis. The major cross-linked proteins are not keratins and generally have little or no cysteine, but have a high glutamic acid/glutamine residue content. In the hair fibre the cross-link appears in the citrulline-containing proteins of the medulla and the inner root sheath of the follicle. In the epidermis the cross-linked proteins are involved in the formation of the cornified envelope of the stratum corneum cells. In both cases, the cross-linked proteins contribute the characteristic property of chemical resistance to their biological structures.

Introduction

The ε-(γ-glutamyl)lysine cross-link occurs as a posttranslational modification in several structural proteins (reviewed in 1, 2). These cross-links are formed by calcium-requiring transglutaminases which catalyse the formation of a covalent bond between the carboxyl group of the γ-carbon of peptide-bound glutamine residues and a primary amino group. In the epidermis, this group is provided by ε-amino groups of peptide-bound lysine residues. The glutamyl-lysine cross-link is resistant to enzymic and chemical attack and it is presumed that its biological role in the case of the epidermis is a structural one. Transglutaminase activity has been measured in liver and in other tissues, but its physiological function there has yet to be determined. The ε-(γ-glutamyl)lysine isopeptide bond and transglutaminase activity in mammalian skin were first described respectively in the hair fibre and hair follicle. This discovery prompted the search for similar activity in the epidermis and to date most of the research has centered on epidermal tissue.

The γ-cross-link in proteins of the hair and hair follicle

There were several lines of evidence which suggested the possibility of covalent cross-linking in some proteins of the hair fibre and hair follicle, other than that contributed by disulphide bonds. Indeed, the study of these proteins consistently demonstrated their insolubility in the usual protein solvents, including those used to extract keratin (3, 4). Although these proteins were shown to be solubilized by proteolysis in contrast to keratins, the resultant digest was not very amenable to further fractionation, suggesting that the proteins were covalently cross-linked (5). Since the proteins were insoluble in strong reducing and dissociation agents, and contained little or no half-cystine (<1 mole %), disulphide bonds were excluded as being significant and a search for other cross-links was undertaken. The presence of the ε-(γ-glutamyl)lysine cross-link, which accounts for these properties, was first established in enzymic digests of wool fibres (6). In the same year, Harding and Rogers (5) demonstrated the presence of ε-(γ-glutamyl)lysine in the citrulline-containing fraction of medulla cells of hair and quill, and the inner root sheath cells of hair follicles. It was also found to be present extensively in the citrulline-containing medulla proteins of hair from a number of mammalian species (7).

Little data is available on the nature of these proteins, although it is known that they have a very high glutamic acid/glutamine content and a very low half-cystine content. Harding and Rogers (8) analyzed a number of peptides from a proteolytic digest of medulla protein and unequivocally localized the cross-link in the citrulline-containing protein of hair medulla. However, there still is no evidence at present as to whether the cross-link is in fact inter- or intra-chain or both. A putative intracellular precursor to the citrulline-containing protein has been partially purified. The trichohyalin protein, as it has been termed, has a 36 mole % glutamic acid/glutamine content and no half-cystine, and can be cross-linked by hair follicle transglutaminase *in vitro* into a higher molecular weight form as determined by SDS gel electrophoresis (9). The possibility that this fraction may contain a substrate specific for hair follicle transglutaminase has not been followed up.

Hair follicle transglutaminase

The presence of a ϵ-(γ-glutamyl)lysine cross-link in hair medulla protein, prompted an investigation for transamidase activity (transglutaminase) in the hair follicle (10, 11).

Harding and Rogers (10) demonstrated that guinea pig hair follicle homogenates were able to cross-link [^{14}C]glycine ethyl ester, a lysine analogue, with casein, forming a γ-glutamyl glycine compound. Further, they showed that the hair follicle enzyme would cross-link fibrin. These findings, therefore, indicated that the transglutaminase activity in hair follicle homogenates had the capability of forming the ϵ-(γ-glutamyl)lysine bonds found in hair proteins.

Chung and Folk (11) found two transglutaminases in their extracts of guinea pig hair follicles. A minor species was indistinguishable from guinea pig liver transglutaminase and the major species was purified to 85–95% homogeneity. Its molecular weight was determined to be 54 000 and is much lower than that reported for liver transglutaminase and plasma transglutaminase. The hair follicle enzyme was reduced to two apparently identical subunits of 27 000 Mr by SDS gel electrophoresis. This was later confirmed by Peterson and Buxman (12) who isolated the enzyme from the rat hair follicle.

In addition, the hair follicle enzyme could not utilize carbobenzoxy-L-glutamyl glycine as a substrate in contrast to the enzyme found in guinea pig liver homogenates (10). Further, on immunodiffusion, no precipitin bands were observed by follicle homogenates against anti-plasma transglutaminase (10) or antisera against liver transglutaminase (11). Fluorescein-labelled antisera against purified rat hair follicle transglutaminase has localized the enzyme to the medullary and inner root sheath cells. This antiserum did not label epidermal cells (12) nor does the antisera precipitate epidermal transglutaminase. That the hair follicle enzyme is quite different from other tissue transglutaminases was confirmed by preliminary kinetic data (13).

Clearly then, the hair follicle enzyme has unique physico-chemical characteristics, but as is the case for liver and plasma transglutaminases (for a recent review see 14) it requires calcium and is inhibited by sulphydryl-blocking agents such as iodoacetamide (10, 11). The transglutaminase activity in wool follicle homogenates appears to have distinctive properties. Although it was able to form γ-glutamyl glycine derivatives and cross-link fibrin, the enzyme was not inhibited by EDTA and, in fact, was slightly stimulated in the absence of calcium ions (10).

Epidermal transglutaminase

The presence of a transglutaminase activity in hair follicle homogenates led to a search for a similar cross-link activity in the epidermis itself. Goldsmith et al. (15) detected transglutaminase activity in the epidermis of a number of vertebrates. This activity was assayed by measuring the incorporation of dansyl cadaverine into casein and was shown to be dependent on calcium and stimulated by reducing agents. These workers partially purified bovine snout epidermal transglutaminase and assigned it a molecular weight of 40 000–60 000. Since then, epidermal transglutaminases have been purified from a variety of species (12, 16, 17). All have similar molecular weights of around 55 000. There is some evidence that the enzyme may exist as a dimer under certain conditions (16, 17).

Although all epidermal transglutaminases examined so far are calcium-dependent and sulphydryl-activated, there is evidence that other agents and treatments can further enhance enzyme activity (16, 18). Such observations led these authors to propose that enzyme activation may be associated with conformation changes induced by these treatments. Negi et al. (19) found that certain lysosomal proteinases could enhance the activity of human epidermal transglutaminases. These workers proposed that epidermal transglutaminase may be regulated by the activity of lysosomal proteinases, such as Cathepsin D, which are released during keratinocyte development.

The availability of purified epidermal transglutaminase has enabled the immunochemical characterization of these enzymes. Thus the epidermal transglutaminases can be distinguished from other tissue transglutaminases (20, 21) and the species specificity of the enzyme can be determined. Antisera to the human enzyme does not cross-react with frog, rat, mouse or chicken epidermal transglutaminases (21). Hair follicle transglutaminases are not immunochemically related to their epidermal counterpart. Antisera to the human epidermal enzyme did not cross-react with a crude, human hair follicle preparation (21) and antisera against purified epidermal and hair follicle transglutaminases of the rat also did not cross-react (12). Similarly, fluorescein-linked antisera to purified epidermal transglutaminase localized the enzyme in the granular layer but did not stain follicle cells (12).

The γ-cross-link in the epidermis

Evidence for a specific in vivo substrate

In the search for the cellular substrate of epidermal transglutaminases, initial studies were directed towards keratins but the evidence suggested that, although keratins may be so cross-linked, they are not the major substrates of these enzymes. In one of the earliest studies, Goldsmith et al. (15) investigated the α-fibrous proteins (keratins) of the stratum corneum of human and bovine snout epidermis. In this study, posttranslational modifications of these proteins were investigated as a potential function of epidermal transglutaminase. It was determined that 7–8% of the lysine residues in α-fibrous proteins of both species were inaccessible to acrylonitrile and therefore presumed to be involved in an isopeptide bond.

Measurement of acrylonitrile reactivity is a technique developed by Pisano et al. (22), which involves the cyanoethylation of the ϵ-amino groups of lysine residues. Lysines which are cross-linked to glutamine residues are not cyanoethylated and are detected as free lysine after acid hydrolysis. The finding that these lysine residues were cross-linked was verified later when the ϵ-(γ-glutamyl)lysine dipeptide was found to be present in the α-fibrous protein of human stratum corneum to the extent of 9 nmol of the dipeptide/mg protein (23).

In an attempt to study the nature of the presumed substrate, Goldsmith et al. (15) looked at the *in situ* incorporation of [^{14}C]putrescine, a lysine analogue, into the proteins of bovine snout epidermis. The label, [^{14}C]putrescine was detected in all fractions of solubilized epidermis. However, those fractions containing keratin, namely the thiol-urea extract, had low incorporation and showed no significant difference from the urea-soluble prekeratin fraction. Indeed, the highest specific activity was observed in the alkali-soluble fractions, indicating that the major transglutaminase activity is not directed towards the keratins.

In another study, Buxman and Wuepper (20) inferred the existence of an epidermal transglutaminase substrate, using a fluorescent lysine analogue, dansyl cadaverine, to detect an acceptor protein (peptide-bound glutamine). By means of this technique they confirmed the earlier study that the most binding occurred in the alkali-soluble fraction. There was little or no fluorescence in fractions containing the α-fibrous proteins or those containing the sulphur-rich keratins. Interestingly, there was also fluorescence in the fraction soluble in tris-acetate buffer (24). This buffer-soluble fraction was not tested in the [^{14}C]putrescine binding study (15).

Further, Buxman and Wuepper (20) showed *in situ* incorporation of dansyl cadaverine to proteins of the malpighian and granular layers, with an increased graduation in fluorescence towards the granular layer. The nature of this labelling procedure ensures that only those cells with both transglutaminase and a suitable substrate 'acceptor' protein will fluoresce. The same study confirmed the localization of epidermal transglutaminase to those same cell layers using fluorescein-conjugated antibody to label the enzyme *in situ*. These studies suggested that a soluble acceptor substrate (peptide-bound glutamine) is present in the cytoplasm of the living cell layers of the epidermis. It was envisaged that during the course of epidermal differentation, these proteins become increasingly cross-linked because of epidermal transglutaminase activity, resulting in a polymer which could only be solubilized by treatments that cleaved peptide bonds.

Sugawara (25) studied the distribution of the ϵ-(γ-glutamyl)lysine cross-link in various fractions of new-born rat epidermis. Most of the cross-links were localized in a protein fraction insoluble in 0.5 M KOH, confirming that these cross-links confer marked chemical resistance to the epidermis. Sugawara suggested that this fraction was chemically related to membrane protein prepared from the 0.1 M NaOH insoluble fraction of the stratum corneum cell by Matoltsy and Matoltsy (26). This was later confirmed when the cross-link was localized in the cornified cell membranes of the epidermis of rat, bovine snout and human stratum corneum (27), prepared by the method of Matoltsy (26). The ϵ-(γ-glutamyl)lysine cross-link was also found in proteins of the cornified envelope of cultured human epidermal keratinocytes by Rice and Green (28). These chemically-resistant protein envelopes develop beneath the plasma membrane of cells in the stratum corneum during epidermal differentiation. The cross-link was located in the fraction insoluble to thiol-SDS extraction. After enzymic digestion of this insoluble residue, Rice and Green concluded that 18 mole % of the total lysine residues were involved in isopeptide bonds. The fractions containing proteins which could be solubilized in this buffer, including the keratins, had as little as 0.3 mole % of the lysines cross-linked.

Further evidence suggested that the formation of the insoluble cross-linked envelope is associated with epidermal transglutaminase activity, because conditions known to effect enzyme activity also impinged on envelope development. Agents such as EDTA, iodoacetamide and putrescine all inhibited envelope formation in cultured keratinocytes (29). Subsequent studies (30) showed that conditions which increase the cell's permeability to calcium will activate the cross-linking process. Rice and Green concluded that the activation of cross-linked envelope formation is due to the appearance of free calcium in the cytoplasm. That calcium is essential for this process has been confirmed in another study by Hennings et al. (31) using cultured mouse epidermal cells. These authors suggested that calcium may also promote transglutaminase synthesis.

Identification of specific in vivo substrates for transglutaminase

Studies involving the isolation and subsequent characterization of a precursor have been based on the coupling of labelled lysine analogues to presumptive substrate proteins by the action of epidermal transglutaminase, both *in situ* and *in vitro*. Those proteins, whose glutamine residues are more amenable to coupling with the labelled compounds, are presumed to be the favoured substrates of transglutaminase *in vivo*. Hanigan and Goldsmith (32) isolated proteins from human epidermis and from new-born rat epidermis, after identifying those which could be coupled to [^{14}C]putrescine and dansyl cadaverine. Two major species were obtained from each tissue. These proteins from both species had high molecular weights and had similar amino acid compositons. They were low in half-cystine content but had high levels of serine, glycine and glutamic acid/glutamine residues. These proteins could be cross-linked in the presence of epidermal transglutaminase with a progressive decrease in their solubility.

In contrast, evidence of a soluble low molecular weight substrate has come from Buxman et al. (33). These workers used antisera prepared against a high molecular weight epidermal transglutaminase substrate fraction to identify immunochemically small soluble precursor proteins. In this way proteins of 36 000 Mr and 150 000 Mr were identified in bovine snout epidermis (33, 34). Recovery of a buffer-soluble protein with a molecular weight of 150 000, which was immunochemically related to a highly insoluble membrane protein, had been reported in an earlier study where it was identified in crude epidermal extracts by labelling with dansyl cadaverine (24). The purified 36 000 Mr protein was very efficiently cross-linked, by epidermal transglutaminase, into high molecular weight polymers, some of which were only solubilized by conditions that cleave peptide bonds (33). Under certain conditions, *in vitro* cross-linking of the 36 000 Mr protein gave a product resembling the 150 000 Mr protein.

The 36 000 Mr protein itself does not contain ε-(γ-glutamyl)lysine bonds and is dissociated into subunits of approximately 8 000 Mr by 8 M urea and 1% SDS (33). A more recent report quotes a value of 6 000 Mr for these subunits (35). Further, these workers have proposed the name keratolinin for the 36 000 Mr substrate. Amino acid analysis (36) has shown that the 36 000 Mr protein has little half-cystine but has 6 moles % lysine and 14 moles % glutamic acid/glutamine residues. It is also reported to contain citrulline which presumably arose by a posttranslational modification of arginine residues similar to the one that has been described in hair follicle extracts (9).

By employing indirect immunofluorescence with antisera to the purified 36 000 Mr protein, Lobitz and Buxman (36) were able to localize the putative substrate throughout the cytoplasm of the cells of the malpighian layer and cells of the lower granular layer. This corresponds to regions where epidermal transglutaminase has been localized in immunohistological studies (20). In cells of the upper granular layer, fluorescence was seen at the inner layer of the cell membrane as well, but was limited entirely to the inner layer of stratum corneum cells (36). These workers postulated that epidermal transglutaminase substrate is synthesized within the keratinocyte cytoplasm as small subunits which combine to form the initial substrate proteins of 36 000 Mr. Epidermal transglutaminase activity cross-links this substrate into a 150 000 Mr oligomeric intermediate which then associates with the cell membrane and is further cross-linked (33). Although there is no direct evidence for this route, the fate of the precursor clearly involves cross-linking and final localization into the keratinocyte cell membrane where it imparts the property of insolubility.

In a concurrent study, Rice and Green (30) identified a soluble precursor of the cross-linked envelope in extracts of cultured human epidermal cells. Using dansyl cadaverine as the lysine analogue, they detected a single fluorescent species which had a molecular weight of 92 000. The purified 92 000 Mr protein, termed involucrin (37) could be cross-linked into higher molecular weight aggregates by the activity of epidermal transglutaminase *in vitro* (30). Antisera against involucrin was shown to react immunochemically with isolated cornified envelopes and by employing immunofluorescence these workers located this protein in the stratum corneum and the outer living cell layer of human epidermis. Although it was observed in the cytoplasm, it

Table 1. Amino acid composition (moles %) of cell envelopes and their precursors.

	Cornified membranes		Involucrin	Keratolinin
	Human (26)	Bovine (38)	Human (30, 37)	Bovine (35, 36)
Cystine (half)	4.9	3.2	0.3	1.0
Aspartate/asparagine	5.8	5.9	2.8	8.7
Threonine	3.7	4.1	1.6	9.2
Serine	7.4	6.7	1.6	10.4
Glutamate/glutamine	14.0	17.2	45.8	13.5
Proline	13.7	12.5	5.7	4.5
Glycine	14.1	14.1	6.7	8.8
Alanine	5.2	4.3	1.5	8.4
Valine	5.2	5.9	3.7	6.3
Methionine	NS	NS	0.9	0.6
Isoleucine	3.3	2.8	0.4	3.1
Leucine	5.8	5.3	14.6	8.2
Tyrosine	1.1	1.3	0.8	1.6
Phenylalanine	2.3	1.8	0.6	2.7
Histidine	2.1	2.4	4.7	1.1
Lysine	6.6	7.5	7.4	5.7
Tryptophan	NS	NS	0.2	0.7
Arginine	4.6	5.0	0.7	2.5
Citrulline	NS	NS	NS	2.4

NS = not specified by authors.

had a distinct tendency to concentrate at the cell membrane. Amino acid analysis has revealed that involucrin has an unusual amino acid composition with 46 mole % of glutamic acid/glutamine, 15 mole % of leucine and very low levels of half-cystine (0.3 mole %) thereby clearly distinguishing it from keratin. However, the composition of involucrin also differs markedly from that of the whole cornified envelope (Table 1). Rice and Green (30) proposed several explanations for this. These included (i) contaminants in the cornified envelope preparations, (ii) involucrin may undergo processing such as proteolytic cleavage, and (iii) that additional proteins may be involved in the formation of the cornified envelope. In this respect it is of importance to reflect on the possible relationship between involucrin and keratolinin. The amino acid compositions of bovine keratolinin and involucrin differ widely (Table 1). Furthermore, when keratolinin (Mr 36 000) was cross-linked by incubating with epidermal transglutaminase, no 92 000 Mr product characteristic of involucrin was detected (35). These results suggest that both involucrin and keratolinin are distinct proteins which participate together in the formation of the cornified epidermal cell envelope.

Concluding remarks

The investigations of the past decade reviewed here have demonstrated the presence of transglutaminases in mammalian epidermis and in hair follicles. These enzymes are responsible for the establishment, in the differentiated cells, of structures stabilized by γ-cross-links. In the epidermis, the structure is the envelope of the terminal cell of the stratum corneum, whereas in the hair follicle it is the cytoplasmic content of the mature specialized cells of the sheath and medulla layers. It is clear from the survey of studies to the present time that more information on the properties of these tissue-specific transglutaminases is needed in order to establish their relationship, if any, to the transglutaminases found elsewhere, such as in plasma and in the liver. Calcium ions are necessary for transglutaminase activity and it is likely that their intracellular levels regulate enzyme activity during the maturation of epidermal cells. There is some evidence which suggests that calcium may act as an inducer of enzyme synthesis in these cells.

It is known that keratins become cross-linked to a small extent by γ-bonds. However, it is certain from the reported studies that specific precursors, which are the major substrates for transglutaminases, are synthesized in epidermal cells, in the epidermis itself and also in hair follicles. In the epidermis, at least two protein species, involucrin and keratolinin, have been partially defined as precursors of the epidermal cell envelope. Again, further study is needed in order to eliminate species differences and to unequivocally define their individual roles in the molecular events of the cross-linking process and formation of the cell envelope. The putative precursor in the hair follicle, trichohyalin, also requires much further investigation before its biochemical changes during hair growth can be more clearly understood.

Acknowledgments

A postgraduate grant from the Wool Research Trust Fund on the recommendation of the Australian Wool Corporation (to J.A.R.) is gratefully acknowledged.

References

1. Goldsmith, L. A., 1977. In: Biochemistry of Cutaneous Epidermal Differentiation. (Seiji, M. and Bernstein, I. A., eds.), pp. 398–418, Tokyo: University of Tokyo Press.
2. Folk, J. E. and Finlayson, J. S., 1977. Adv. Protein Chem. 31: 1–133.
3. Matoltsy, A. G., 1953. Exp. Cell Res. 5: 98–110.
4. Rogers, G. E., 1964. In: The Epidermis. (Montagna, W. and Lobitz, W., eds.), pp. 179–236.
5. Harding, H. W. J. and Rogers, G. E., 1971. Biochemistry 10: 624–630.
6. Asquith, R. S., Otterburn, M. S., Buchanan, J. H., Cole, M., Fletcher, J. C. and Gardner, K. L., 1970. Biochim. Biophys. Acta 221: 342–348.
7. Harding, H. W. J. and Rogers, G. E., 1972. Biochim. Biophys. Acta 257: 37–39.
8. Harding, H. W. J. and Rogers, G. E., 1976. Biochim. Biophys. Acta 427: 315–324.
9. Rogers, G. E., Harding, H. W. J. and Llewellyn-Smith, I. J., 1977. Biochim. Biophys. Acta 495: 159–175.
10. Harding, H. W. J. and Rogers, G. E., 1972. Biochemistry 11: 2858–2863.
11. Chung, S. I. and Folk, J. E., 1972. Proc. Natl. Acad. Sci. U.S.A. 69: 303–307.
12. Peterson, L. L. and Buxman, M. M., 1981. Biochim. Biophys. Acta 657: 268–276.
13. Chung, S. I. and Folk, J. E., 1972. J. Biol. Chem. 247: 2798–2807.

14. Folk, J. E., 1980. Ann. Rev. Biochem. 49: 517–531.
15. Goldsmith, L. A., Baden, H. P., Roth, S. I., Colman, R., Lee, L. and Fleming, B., 1974. Biochim. Biophys. Acta 351: 113–125.
16. Ogawa, H. and Goldsmith, L. A., 1976. J. Biol. Chem. 251: 7281–7288.
17. Buxman, M. M. and Wuepper, K. D., 1976. Biochim. Biophys. Acta 452: 356–369.
18. Plishker, M. F., Thorpe, J. M. and Goldsmith, L. A., 1978. Arch. Biochem. Biophys. 191: 49–58.
19. Negi, M., Matsui, T. and Ogawa, H., 1981. J. Invest. Dermatol. 77: 389–392.
20. Buxman, M. M. and Wuepper, K. D., 1975. J. Invest. Dermatol. 65: 107–112.
21. Ogawa, H. and Goldsmith, L. A., 1977. J. Invest. Dermatol. 68: 32–35.
22. Pisano, J. J., Finlayson, J. S. and Peyton, M. P., 1969. Biochemistry 8: 871–876.
23. Abernethy, J. L., Hill, R. L. and Goldsmith, L. A., 1977. J. Biol. Chem. 252: 1837–1839.
24. Buxman, M. M., Buehner, G. E. and Wuepper, K. D., 1976. Biochem. Biophys. Res. Commun. 73: 470–478.
25. Sugawara, K., 1977. In: Biochemistry of Cutaneous Epidermal Differentiation. (Seiji, M. and Bernstein, I. A., eds.), pp. 387–397, Tokyo: University of Tokyo Press.
26. Matoltsy, A. G. and Matoltsy, M. M., 1966. J. Invest. Dermatol. 46: 127–129.
27. Sugawara, K., 1979. Agric. Biol. Chem. 43: 2543–2548.
28. Rice, R. H. and Green, H., 1977. Cell 11: 417–422.
29. Rice, R. H. and Green, H., 1978. J. Cell Biol. 76: 705–711.
30. Rice, R. H. and Green, H., 1979. Cell 18: 681–694.
31. Hennings, H., Steinert, P. and Buxman, M. M., 1981. Biochem. Biophys. Res. Commun. 102: 739–745.
32. Hanigan, H. and Goldsmith, L. A., 1978. Biochim. Biophys. Acta 522: 589–601.
33. Buxman, M. M., Lobitz, C. J. and Wuepper, K. D., 1980. J. Biol. Chem. 255: 1200–1203.
34. Buxman, M. M. and Lobitz, C. J., 1980. In: Current Problems in Dermatology. (Mali, J. W. H., ed.), Vol. 10, pp. 245–253, Basel: S. Karger.
35. Zettergren, J. G., Peterson, L. L. and Wuepper, K. D., 1982. J. Invest. Dermatol. Abstract 78: p. 360.
36. Lobitz, C. J. and Buxman, M. M., 1982. J. Invest. Dermatol. 78: 150–154.
37. Watt, F. M. and Green, H., 1981. J. Cell Biol. 90: 738–742.
38. Matoltsy, A. G., 1977. In: Biochemistry of Cutaneous Epidermal Differentiation. (Seiji, M. and Bernstein, I. A., eds.), pp. 93–109, Tokyo: University of Tokyo Press.

Received 22 September 1982.

Molecular and Cellular Biochemistry 58, 121–128 (1984).

Human $\alpha_2$macroglobulin

Primary amines and the molecular mechanisms of endoprotease inhibition and receptor-mediated endocytosis

F. van Leuven
Division of Human Genetics, Department of Human Biology, University of Leuven, Minderbroedersstraat 12, B-3000 Leuven, Belgium

Summary

Human $\alpha_2$macroglobulin combines two unique features: the non-active site directed inhibition of virtually all endoproteases and the selective clearance of α_2M-endoprotease complexes by receptor-mediated endocytosis. To study the molecular details of the mechanisms involved, primary amines were found to be worthwhile probes at three specific levels: the inactivation of native α_2M, the derivatization by factor XIII and the cellular process of receptor-recycling. In this paper published data are supplemented with recently obtained evidence to discuss and speculate on the possible action or involvement of transglutaminase activities, indicated by the effects of the primary amines.

Introduction

α_2Macroglobulin (α_2M) is a major high-molecular weight glycoprotein in the circulation of mammals. In man, the plasma concentration of α_2M varies widely in the entire population (from 1 to 5 mg/ml), with a mean of about 2 mg/ml. α_2M is a general inhibitor of endoproteinases, as virtually all endoproteases are bound by α_2M. Two special features are noteworthy in this regard: the endoproteinases are bound by α_2M in a catalytically active form and the α_2M-protease complexes are efficiently and specifically cleared from the extracellular space through receptor-mediated endocytosis. The combination of these two remarkable features in a protease inhibitor continues to attract the attention of many investigators in biochemistry and cell biology and stands central in the present discussion. Three aspects of the function of human α_2M, are affected by primary amines:

i) the reaction of primary amines with native α_2M, a probe for the molecular mechanism of endoprotease inhibition;
ii) the derivatization of α_2M by primary amines, catalyzed by factor XIII of the blood clotting system;
iii) the effect of primary amines on receptor-mediated endocytosis of α_2M-complexes by cells in culture.

The incentive to discuss these different aspects together stems from a remarkable common denominator: the effect of the primary amines points to the involvement of a transglutaminase-like activity in each of the phenomena. It is the aim of this paper to review published data supplemented with recently obtained evidence and to discuss or speculate on the actions of the transglutaminases.

The mechanism of α_2M-protease complex formation

The fundamental difference between α_2M and other protease inhibitors is that only enzymatically active endoproteases are bound in a non-active site directed fashion. The structural features of α_2M to accomplish this explain the quasi non-specificity of α_2M as an endoprotease inhibitor. These features are: the large molecular weight of α_2M, the characteristics of the bait region and the conformational change.

Human α_2M is a tetramer of identical subunits of M_r 185 000. These are linked two by two by disulfide bridges and two dimers associate non-covalently to form the native molecule of M_r 725 000. Complexes of α_2M with most endoproteases are characterized by a subunit M_r of between 80 000 and 100 000 (1) and by an increased electrophoretic mobility (2). These observations indicate that complex formation with a protease implicates that proteolysis of the α_2M-polypeptide chain occurs, that the α_2M-molecule undergoes a conformational change and that the former triggers the latter. These are the major features of the 'trap hypothesis' (3): proteolysis of the bait region triggers the change in α_2M conformation to trap the protease.

The bait region is a relatively short stretch of residues, located near the middle of the α_2M subunit polypeptide chains (4). To account for the broad specificity (or, in fact, absence of specificity) of α_2M for the endoproteases, the bait region contains a sequence of amino acid residues, reflecting the substrate specificity of all endoproteases.

Whereas these static structural characteristics of α_2M are well studied, the dynamic conformational change is more difficult to tackle. The rapid reaction of endoproteases with native α_2M limits one to comparing the properties of the native molecules to the α_2M-endoprotease complex. The situation is different when primary amines react with native α_2M, although the final α_2M derivatives obtained with either protease or methylamine compare very well with respect to i) the electrophoretic mobility in rate electrophoresis, ii) the complex pattern of bands resolved by isoelectric focusing and iii), most importantly, the kinetics of receptor-mediated endocytosis by fibroblasts in culture (5). The difference is primarily a kinetical one: the reaction of α_2M with methylamine, the best studied primary amine, proceeds much more slowly than complex formation with proteases, and is therefore amenable to kinetic analysis (5, 6, 7, 8). Methylamine becomes covalently incorporated into native α_2M (and in complement components C_3 and C_4) as a γ-glutamyl methylamide in all four subunits (9). As an equivalent number of cysteine thiol groups are detected (while none are present on native α_2M) it was postulated that native α_2M (and C_3 and C_4) contains an internal thioester (10–12).

The cysteinyl and glutamyl residues involved are located only three residues apart, to form a relatively small internal thiolacetone ring. A model peptide, containing this configuration, was recently synthesized (13). As pointed out by these workers, 'the macrocyclic thiolacetone hydrolyzes about 10^3 times faster than the latent binding site of C_3 but about 10^7 times slower than the metastable binding site of nascent C_{3b}'. Clearly, although the cyclic configuration might be necessary, it does in itself not explain the observations that native α_2M (and C_3 and C_4) is stable in an aqueous environment but is rapidly inactivated by nucleophilic reagents like methylamine. Both can only be explained by the assumption that the internal thioester is located in the hydrophobic interior of the α_2M polypeptide chains. This effectively prevents hydrolysis but not aminolysis: methylamine is well known to penetrate hydrophobic sites (e.g., cell membranes) by diffusion of the uncharged molecular species. The pH dependence of methylamine incorporation into α_2M favors this concept (14).

From this basic assumption we can postulate that the internal thioester is formed during the synthesis of the α_2M subunit polypeptide chains on the ribosomes, probably in concert with the folding of the chains. This would effectively protect the thioester from hydrolysis by burial in the hydrophobic interior of the protein. To form the bond, we anticipate that α_2M contains an enzyme active site, transglutaminase-like, containing essentially the cysteine residue. A thio-acyl intermediate is formed with a suitable oriented glutamine residue, analogous to the γ-glutamyl acyl-enzyme intermediate of transglutaminases. The intermediate remains stable until deacylation by nucleophiles, either directly by treatment of α_2M with methylamine or after exposure to external factors resulting from proteolysis of the bait region. This configuration would explain the stability of α_2M in aqueous medium and the inactivation by primary amines. Another reason to invoke an enzyme-like action of the internal thioester in α_2M are our findings on the involvement of the cysteine thiol in the conformational change after reaction with methylamine.

We have shown that the incorporation of methylamine into native α_2M follows biphasic kinetics and involved a stable intermediate in slow form α_2M (5, 7). Since binding of a protease to α_2M destroys the internal thioester in a stoichiometric fashion (6), we examined the effect of methylamine

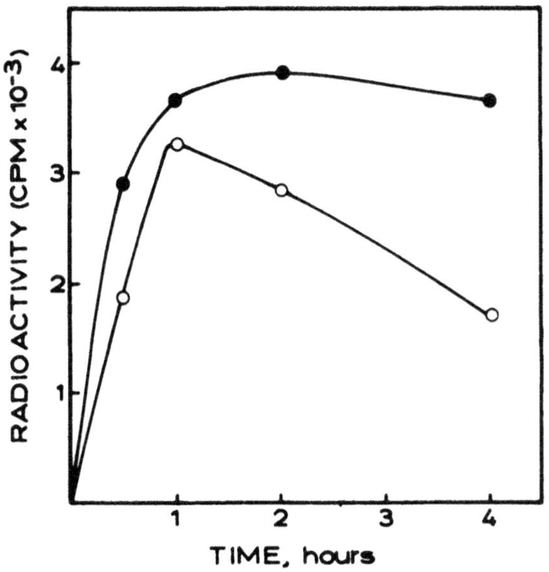

Fig. 1. Effect of EDTA and Ca^{++} on [^{14}C]methylamine incorporation in α_2M. Native plasma α_2M was dialyzed overnight in the cold against either 2 mM EDTA or 10 mM $CaCl_2$ in Tris-HCl (50 mM, pH 8.0). After dilution to 2 mg/ml, α_2M was reacted with 25 mM [^{14}C]methylamine in the same buffers at 25 °C. At the time points indicated, 12.5 μf of protein were separated on 5% polyacrylamide gels. Radioactivity present in slow form α_2M was quantitated by excision of the Coomassie-stained bands and dissolution in concentrated H_2O_2 as described (5), for EDTA-treated (.) and Ca^{++}-treated α_2M (o).

incorporation into α_2M on subsequent complex formation with trypsin. Complex kinetics were observed, which were interpreted as being indicative for the cooperation between subunits in α_2M in the process of conformational change (7). This cooperation was shown to be an absolute requisite for the formation of fully inhibited protease complexes (7). Furthermore, we observed that cyanylation of the thiol group during its liberation by methylamine incorporation prevented the change in conformation of α_2M completely (8). Other sulphydryl specific reagents were less effective in this regard which might be due to steric factors since the effectiveness decreased with bulkier substituents (8). Nevertheless, these results indicate the involvement of the liberated thiol groups in the conformational change of α_2M.

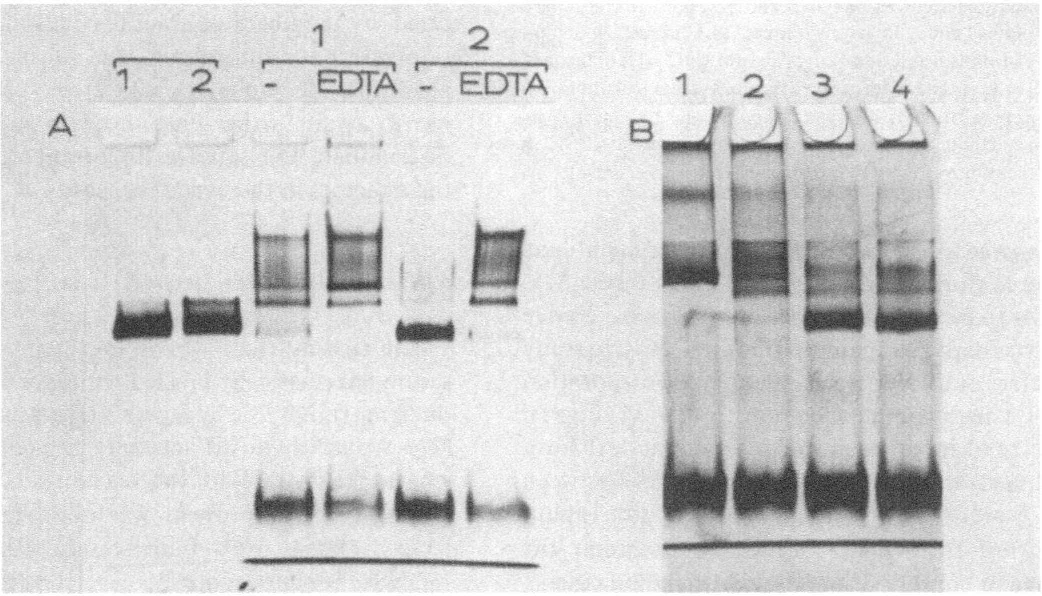

Fig. 2. Tryptic loss of F2B2-determinant on α_2M.MA. Effect of EDTA, Ca^{++} en Zn^{++}.

A) α_2M.MA was prepared from α_2M isolated from plasma of two different donors. Rate electrophoresis on 5% polyacrylamide gel showed both preparations to be largely fast form α_2M (lanes 1 and 2). Part of these preparations were dialyzed against EDTA (1 mM in Tris.HCl 50 mM pH 8.0). Untreated and EDTA-treated samples of the two preparations of α_2M.MA were digested with trypsin (10% w/w, relative to α_2M.MA) for 20 hr at 37 °C, pH 8.0 (Tris. HCl 50 mM).

Monoclonal antibody (F2B2) was added and after 10 min, samples (corresponding to 10 μg α_2M) were examined by rate electrophoresis. Note the large differences in loss of F2B2-determinant in the untreated samples and the insignificant loss after EDTA-dialysis.

B) α_2M.MA (preparation 1 from panel A), dialyzed against EDTA, was incubated without trypsin (lane 1), with trypsin (20%, relative to α_2M.MA) without additions (lane 2) or with 20 mM $CaCl_2$ (lane 3) or with 0.5 mM $ZnCl_2$ (lane 4). After 20 hr at 37 °C, samples were treated with the monoclonal antibody F2B2 and examined by rate electrophoresis for antibody-binding.

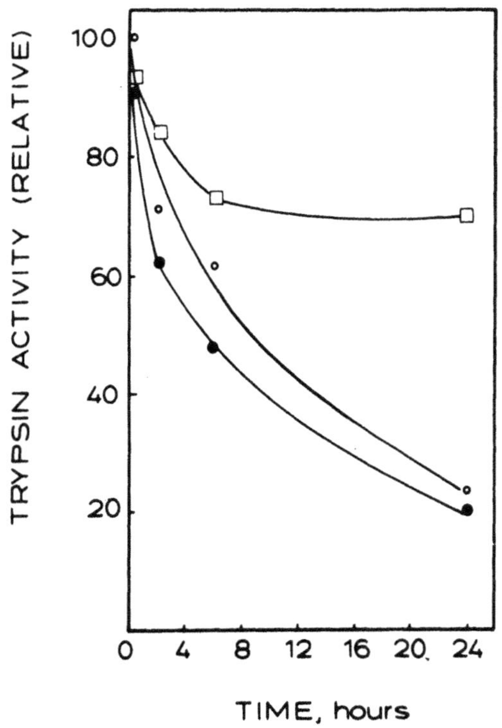

Fig. 3: Time course of autodigestion of trypsin. Trypsin was incubated under conditions used and described in the experiments of Fig. 1 and 2, however without α_2M. Control (○) contained no additions, while addition of 20 mM $CaCl_2$ (□) or 0.5 mM $ZNCl_2$ (●) was examined in parallel. Portions of the incubation mixtures were taken after the time intervals indicated and trypsin amidase activity was measured immediately, with BAPNA as substrate. Enzyme activity is expressed relative to control.

We have also examined the effect of metal ions on the conformational change of α_2M, since α_2M is known to be a major carrier of metal ions. Native α_2M was dialyzed against EDTA or $CaCl_2$, to study the effect of Ca^{++} on the kinetics of incorporation of [^{14}C] methylamine into slow form α_2M (Fig. 1). Since total incorporation (in slow plus fast form combined) did not differ more than 5% between EDTA and Ca^{++}-treated α_2M at any time point, these results indicate that Ca^{++} promotes the change in conformation after methylamine incorporation into α_2M.

A second indication that metal ions influence the conformation of α_2M was found in the proteolysis of the neo-antigenic site defined by the monoclonal antibody F_2B_2 (15). This neo-antigenic site is part of the receptor-recognition site on α_2M-complexes (15, 16). The rate of proteolytic destruction of the antigenic site by trypsin was appreciably increased when Ca^{++} was present and this effect was attributed to the effect of Ca^{++} on autodigestion of trypsin (16). Since α_2M is, however, a major carrier of Zn^{++} and other metal ions in the circulation (17–19) we examined whether Zn^{++} ions would also enhance the proteolysis of the receptor-recognition site. This was indeed observed (Fig. 2) at a concentration of Zn^{++} (0.5 mM) not affecting the trypsin activity or the rate of autodigestion of trypsin (Fig. 3). Moreover, different batches of α_2M, isolated from plasma of different donors, showed large differences in the rate of tryptic proteolysis of the receptor-recognition site. This was invariably reduced to a uniform slow rate after dialysis against EDTA. These results suggest that metal ions bound to α_2M affect the rate of proteolytic degradation, likely by a direct effect on the conformation of α_2M-complexes.

Taken together, our present and published results point to the necessity of cooperative action of the subunits and to the involvement of the liberated sulphydryl groups and metal ions in the conformational change of α_2M. It is tempting to speculate that in fact this conformational change is 'catalyzed' by the liberated thiol residues, in combination with bound metal ions and other amino acid residues on the same or on different subunits. One has to await further data in order to be able to discriminate the relative importance of each of these factors in the overall process.

α_2M, a substrate for factor XIII

The finding that α_2M is the major protein in serum enzymatically labeled with mono-dansylcadaverine (mDNSC) by factor XIII was unexpected (20). Apparently α_2M acts only as γ-glutamyl donor (first substrate) for this reaction since no crosslinking to other proteins was observed (20). We have isolated α_2M fluorescently labeled with mDNSC by endogenous factor XIII activity of clotting human plasma. We observed that modification by factor XIII did not influence the characteristics of α_2M as a protease inhibitor, nor did it decrease the incorporation of [^{14}C]methylamine. Surprisingly, however, it abolished the endocytosis of the labeled complexes by cells in culture (5).

To examine this in more detail we have isolated factor XIII from human plasma. The isolated enzyme was effective in derivatization of α_2M with both mDNSC the reaction within one hour, as indicated by the nearly constant amount of label incorporated in α_2M with prolonged incubation.

In different experiments incorporation of [^{3}H]-putrescine in native α_2M by factor XIII varied from 3.8 to 4.15 mole per mole α_2M, with a mean of 3.95. When factor XIII was omitted from the incubation mixture incorporation was always less than 0.1 mole [^{3}H]putrescine per mole α_2M. This shows the involvement of the enzyme and the absence of [^{3}H]putrescine incorporation in the thioester in native α_2M. Surprisingly, in similar experiments with α_2M-trypsin complexes and with α_2M-methylamine complexes we observed no incorporation of [^{3}H]putrescine with factor XIII. In a typical experiment the incorporation ratios measured for native α_2M, α_2M-trypsin and α_2M-methylamine were respectively 4.04, 0.4 and 0.2 mole [^{3}H]putrescine per mole α_2M. With longer reaction times (up to 4 hours) we observed a minor increase in the incorporation ratios for native α_2M (up to 4.85 mole per mole) but the ratios for α_2M-trypsin and α_2M-methylamine remained low (less than 0.5 mole [^{3}H]putrescine per mole α_2M).

A large batch (1 mg) of native α_2M was derivatized with [^{3}H]putrescine by factor XIII. The labeled derivative was isolated by gel filtration. Derivatization was measured and found to be 3.3 mol [^{3}H]putrescine per mole α_2M. After formation of complexes with trypsin and methylamine (5) the derivatives were examined for their potency to inhibit receptor-mediated endocytosis of ^{125}I-labeled α_2M-trypsin by fibroblasts in culture (5, 21). Inhibition was of the same extent as with similar preparations not treated with factor XIII. Thus derivatization with [^{3}H]putrescine did not abolish effective recognition of the complexes by the cellular receptor. An analogous experiment with mDNSC gave essentially the same results, although in this case we could not determine the amount of incorporation of the fluorescent label in α_2M exactly.

These results indicate that the primary factor XIII catalyzed derivatization of a γ-glutamyl residue by mDNSC or [^{3}H]putrescine is not responsible for the effects observed with the derivative produced in clotting plasma (5). Two possibilities remain: either factor XIII is not involved and another (enzymatic?) process is at work in clotting plasma or factor XIII is still operative but not by the reaction described above. The first possibility can hardly be discussed in the absence of any relevant data. The second option is interesting. Indeed, the reaction catalyzed by factor XIII requires two substrates: a γ-glutamyl donor and an acyl-acceptor, typically a lysine residue. As already mentioned above α_2M does not seem to provide any lysine residue in the role of second substrate for the transglutaminase reaction. It is important to note, however, that the only evidence in this regard is the absence of cross-linking of α_2M to other proteins (20). Evidently, this can be forbidden by steric reasons and it does not preclude that one or more lysine residues on α_2M react with a low-molecular weight γ-glutamyl donor, catalyzed by factor XIII. Since we have shown that one or more lysine residues on α_2M-complexes are essential to effective recognition by the cellular receptor (16), this opens new possibilities to examine this problem.

At least as intriguing is our present observation that not α_2M-trypsin or α_2M-methylamine complexes but only native α_2M is an effective γ-glutamyl donor in the transglutaminase reaction. Recently, a putative transglutaminase cross-linking site in α_2M was located close to the bait region (22). In fast form α_2M this site is clearly not available for enzymatic reaction, independent of whether the bait region was proteolytically nicked (in α_2M-trypsin) or not (in α_2M-methylamine). We conclude therefore that the conformational change has sufficiently altered the configuration of the glutamine residue to preclude access by factor XIII. Perhaps this can be used as an extra probe in the study of the conformation change.

Receptor-recycling and primary amines

The kinetic parameters of the internalization of α_2M-complexes have many features in common with the LDL receptor system: the limited number and apparent affinity of the receptors, the extremely high rate of internalization at saturating ligand concentrations and the rapid intracellular degradation of the proteins ingested (21, 23). α_2M bound to the cell surface receptors is located in coated pits and its internalization via the formation of coated vesicles has been documented (24). These findings provided further evidence for the hypothesis that receptor clustering, and coated pit formation are essential features of receptor-mediated endocytosis (23).

An interesting aspect of receptor-mediated endocytosis is the high rate of internalization of ligand by the limited number of receptors expressed at the plasma membrane. Three explanations for this phenomenon are proposed: i) the receptor mediates only translocation of the ligand and the receptor never leaves the plasma membrane; ii) the receptor is internalized with the ligand and the original number at the cell surface is restored either by synthesis *de novo* or from a large intracellular pool of receptors; or iii) the receptor delivers the ligand intracellularly, segregates and returns to the plasma membrane (recycling of receptors).

The first possibility can be readily eliminated; it is not consistent with the ultrastructural data on the internalization of α_2M, LDL and other ligands by coated pits and coated vesicles.

The other two hypothetical schemes are hard to distinguish and may be complementary. There is evidence however that receptor-recycling is the prominent component. Indeed, without ligand and in the absence of protein synthesis (cycloheximide-treated cells) the number of surface receptors remains fairly constant (23, 25). So either they are not internalized in the absence of ligand or they are replenished. In the presence of ligand (saturating concentration of α_2M-trypsin complexes), thus with ongoing endocytosis, the number of exposed receptors is still not appreciably decreased by simultaneous cycloheximide treatment. Therefore, the receptors must be continuously replaced at the cell surface. One can calculate that in the absence of *de novo* synthesis or recycling of the receptor, the internal pool of receptors should contain at least two orders of magnitude more receptors than the number expressed at the cell surface. This has not been found, neither for the LDL-receptor (23) nor for the α_2M-receptor (26).

Further indications for receptor recycling have come from the effects of primary amines on receptor-mediated endocytosis (25). Methylamine or mono-dansylcadaverine treatment decreases the number of exposed receptors. This effect is time and concentration dependent and fully reversible. The amines, however, do not affect the receptor-ligand interaction at 4 °C, while the accurate correlation between the number of receptors measured at 4 °C (binding only and no internalization) and the rate of endocytosis at 37 °C is maintained. These data indicate that the primary amines interfere with replenishment of the cell surface with receptors.

A quite analogous effect of primary amines and of other compounds was described in a number of receptor systems, confirming our interpretation. Receptor recycling seems established in the following systems: LDL-receptor in fibroblasts (27), asialoglycoprotein-receptor in hepatocytes (28, 29), receptor for lysosomal enzymes (mannose-6-phosphate mediated) in fibroblasts (30), receptor for mannose-conjugated proteins in macrophages (31), α_2M-receptor in macrophages (26) and fibroblasts (25).

Although the phenomenon of receptor recycling is thereby well documented, the intracellular route followed during recycling remains unknown. Receptor-recycling must be related to membrane recycling, which is extensive in a living cell (up to 50% of the original membrane can be internalized per hour, by all types of endocytosis combined). It is indeed likely that receptors, embedded in the lipid bilayer, never leave this hydrophobic environment, hence the two phenomena must be intimately associated.

The plasma membrane is believed to originate from vesicles that bud off from the Golgi apparatus (32). Farquhar (33) demonstrated that elements of the plasma membrane can move retrograde to the Golgi apparatus. This would indicate a connection between endocytosis and exocytosis: receptors might shuttle between plasma membrane and Golgi apparatus. At what point on the intracellular route the ligand segregates from the receptor to end up in the lysosome is unknown.

With this in mind several interesting features of the effect of primary amines and other drugs on receptor-recycling need further consideration. Originally, it was claimed that primary amines would interfere with receptor-clustering at the cell surface and thereby reduce receptor-mediated endocytosis (34). This cannot be retained because our results show that a decreased number of receptors is expressed at the cell surface after treatment with the primary amines (25), which was also found by others (35–39). The observation of Maxfield et al. (34) could be explained by (i) the decreased number of accessible receptors at the cell surface and (ii) the fact that only fluorescence microscopy was employed, which can only yield qualitative data.

The same group extended their original observations on primary amines to a series of drugs in an attempt to find a correlation between receptor-clustering on the one hand and transglutaminase activity on the other and proposed that this enzyme was responsible for covalent linkage of receptors to each other or to unidentified cellular components. This would form the clusters of receptors, the first step in receptor-mediated endocytosis (40). Although a fairly good correlation between inhibition of receptor-clustering and inhibition of transglutaminase activity by the different drugs was observed, the physiological link remains questionable since the same qualitative fluorescence microscopic technique was used. Moreover, bacitracin, one of the drugs used, does in fact inhibit binding of α_2M-complexes to the receptor, without interfering with other aspects of receptor-mediated endocytosis (41).

We interpret these findings differently. We believe that a transglutaminase activity is not involved in receptor-clustering, but might be involved in intracellular receptor-recycling (25, 41). Effects of the primary amines via other mechanisms must however also be considered.

The possibility that the primary amines act on receptor-recycling by affecting intracellular pH is unlikely. These amines are known to accumulate massively intracellularly by passive diffusion of the uncharged molecules through the plasma membrane. The resulting alkalinization of the cytoplasm and of the lysosomal compartment (by proton trapping) decreases the rate of degradation of the ingested α_2M-complexes. No correlation was found, however, between decreased degradation and decreased endocytosis. Moreover, blocking degradation with chloroquine or by alkalinization of the medium (pH 8.0) did not affect receptor-recycling (25).

Another possibility is that, certain amines, such as mono-dansylcadaverine, might affect the function of calmodulin. Recent evidence does indeed indicate that the dansyl derivative W7 is a very potent calmodulin-antagonist (42) while calmodulin was reported to be located in the direct vicinity of coated pits and vesicles (43, 44).

Conclusion

Three, apparently unrelated mechanisms are affected by or involve primary amines: the molecular mechanism of α_2M-protease complex formation, the derivatization of α_2M by factor XIII and receptor-recycling in fibroblasts. In the formation of α_2M-protease complexes, the internal thioester of α_2M is essential: it stabilizes α_2M in the native conformation, it engages eventually in covalent linkage of the protease and circumstantial evidence, presented above, indicate its direct involvement in the change in conformation, which is essential to the function of α_2M as a general endoprotease inhibitor. The characteristics of these processes are reminescent of a transglutaminase reaction, which might form the working hypothesis for further work on the elucidation of the molecular mechanism of α_2M-protease complexation.

The action of the transglutaminase activity of factor XIII on α_2M remains puzzling in several ways. The present data show that the primary derivatization is not directly responsible for abolishing effective receptor-recognition. Moreover, we have provided evidence that only native and not fast form α_2M is a primary substrate for factor XIII, which opens new ways to examine the mechanism of α_2M.

The effect of primary amines on receptor-mediated endocytosis is now clear to be due to a disturbance of intracellular receptor-recycling. This might be exerted by inhibition of a cellular transglutaminase activity but other intervening processes or molecules are equally good or better candidates. Much more work is needed to understand all aspects of these specific points in the function of α_2M. We hope to have provided here some speculative thoughts to stimulate this search.

Acknowledgements

Our own work was supported by Grant 3.0043.79 (National Fund for Scientific Research Belgium) and by a grant 'Geconcerteerde Acties' from the Belgian Government.

References

1. Harpel, P. C., 1973. J. Exp. Med. 138: 508–521.
2. Barrett, A. J., Brown, M. A. and Sayers, C. A., 1979. Biochem. J. 181: 401–418.
3. Starkey, P. M. and Barrett, A. J., 1977. In: Proteinases in mammalian cells and tissues. (Barrett, A. J., ed.) pp. 663–696. North-Holland, Amsterdam.

4. Sottrup-Jensen, L., Lønblad, P. B., Stepanik, T. M., Petersen, T. E., Magnusson, S. and Jörnvall, M., 1981. FEBS Letters 127: 167–173.
5. Van Leuven, F., Cassiman, J. J. and Van den Berghe, H., 1981. J. Biol. Chem. 256: 9016–9022.
6. Van Leuven, F., Cassiman, J. J. and Van den Berghe, H., 1981. J. Biol. Chem. 256: 9023–9027.
7. Van Leuven, F., Cassiman, J. J. and Van den Berghe, H., 1982. Biochem. J. 201: 119–128.
8. Van Leuven, F., Marynen, P., Cassiman, J. J. and Van den Berghe, H., Biochem. J. 203: 405–411.
9. Swenson, R. P. and Howard, J. B., 1979. Proc. Natl. Acad. Sci. 76: 4313–4316.
10. Tack, B. F., Harrison, R. A., Janatova, J., Thomas, M. L. and Prahl, J. W., 1980. Proc. Natl. Acad. Sci. 77: 5764–5768.
11. Sottrup-Jensen, L., Petersen, T. E. and Magnussens, S., 1980. FEBS Letters 121: 275–279.
12. Howard, J. B., 1981. Proc. Natl. Acad. Sci. 78: 2235–2239.
13. Khan, S. A. and Erickson, B. W., 1981. J. Am. Chem. Soc. 103: 7374–7376.
14. Salvesen, G. S., Sayers, C. A. and Barrett, A. J., 1981. Biochem. J. 195: 453–461.
15. Marynen, P., Van Leuven, F., Cassiman, J. J. and Van den Berghe, H., 1981. J. Immunol. 127: 1782–1786.
16. Marynen, P., Van Leuven, F., Cassiman, J. J. and Van den Berghe, H., 1982. FEBS Letters 137: 241–244.
17. Parisi, A. F. and Vallee, B. L., 1970. Biochemistry 9: 2421–2426.
18. Adham, N. F., Song, M. K. and Rinderknecht, H., 1977. Biochim. Biophys. Acta 495: 212–219.
19. Dumitrescu, A., Ganea, D., Olinescu, A., Ghitescu, L. and Szegi, G., 1977. Rev. Roum. Biochemie 14: 167–174.
20. Mosher, D. F., 1976. J. Biol. Chem. 251: 1639–1645.
21. Van Leuven, F., Cassiman, J. J,. and Van den Berghe, H., 1979. J. Biol. Chem. 254: 5155–5160.
22. Mortensen, S. B., Sottrup-Jensen, L., Hansen, H. F., Rider, D., Petersen, T. E. and Magnusson, S., 1981. FEBS Letters 129: 314–317.
23. Goldstein, J. L., Anderson, G. W. and Brown, M. S., 1979. Nature 279: 679–685.
24. Willingham, M. C., Maxfield, F. R. and Pastan, I. H., 1979. J. Cell Biol. 82: 614–625.
25. Van Leuven, F., Cassiman, J. J. and Van den Berghe, H., 1980. Cell 20: 37–43.
26. Kaplan, J., 1980. Cell 19: 197–205.
27. Basu, S. K., Goldstein, J. L., Anderson, R. G. W. and Brown, M. S., 1981. Cell 24: 493–502.
28. Tolleshaug, H. and Berg, T., 1979. Biochem. Pharma. 28: 2919–2922.
29. Steer, C. J. and Ashwell, G., 1980. J. Biol. Chem. 255: 3008–3013.
30. Gonzalez-Noriega, A., Grubb, J. H., Talkad, V. and Sly, N. S., 1980. J. Cell Biol. 85: 839–852.
31. Stahl, P., Schlesinger, P. M., Sigardson, E., Rodman, J. S. and Lee, Y. C., 1980. Cell 19: 207–215.
32. Palade, G. E., 1975. Science 189: 347–358.
33. Farquhar, M. G., 1978. J. Cell Biol. 77: R35–R42.
34. Maxfield, F. R., Willingham, M. C., Davies, P. J. A. and Pastan, I., 1979. Nature 277: 661–663.
35. Hazum, E., Kwen-Jen, C. and Cuatrecasas, P., 1979. Nature 282: 626–628.
36. McKenna, J. A., Haigler, H. T. and Cohen, S., 1979. Proc. Natl. Acad. Sci. 76: 5689–5693.
37. King, A. C., Hernaez-Davis, L. and Cuatrecases, P., 1980. Proc. Natl. Acad. Sci. 77: 3283–3287.
38. King, A. C., Hernaez-Davis, L. and Cuatrecasas, P., 1981. Proc. Natl. Acad. Sci. 78: 717–721.
40. Davies, P. J. A., Davies, D. R., Levitzki, A., Maxfield, F. R., Milhaud, P., Willingham, M. C. and Pastan, I. H., 1980. Nature 283: 762–767.
41. Van Leuven, F., Marynen, P., Cassiman, J. J. and Van den Berghe, H., 1981. FEBS Letters 134: 83–87.
42. Hidaka, H., Sasaki, Y., Tanaka, I., Endo, T., Ohno, S., Fujii, Y. and Nagata, T., 1981. Proc. Natl. Acad. Sci. 78: 4354–4357.
43. Salisbury, J. L., Condeelis, J. S. and Satir, P., 1980. J. Cell Biol. 87: 132–141.
44. Salisbury. J. L., Condeelis, J. S., Maihle, N. J. and Satir, P., 1981. Nature 294: 163–166.

Received 14 April 1982.

Molecular and Cellular Biochemistry 58, 129–137 (1984).

Isopeptide bond formation in epidermis

Joseph Kubilus and Howard P. Baden
Harvard Medical School, Department of Dermatology, Massachusetts General Hospital, Boston, MA 02114, U.S.A.

Summary

Proteins which are major substrates of epidermal transglutaminases can be identified in cultured keratinocytes of human, cow, and new-born rat.

Cow and human keratinocytes both contain substrate proteins which are 30000 to 50000 daltons in size but dissociable in SDS to 12000 daltons or less. In both species these proteins correspond to *in vivo* synthesized proteins which are probable precursors of cornified envelope. Human keratinocytes synthesize a 125000 dalton protein which is also a precursor of cornified envelope both in cells and tissue. By SDS electrophoresis two 100000 dalton substrate proteins are seen in cow keratinocyte extracts and a 23000 dalton substrate protein is seen in rat keratinocyte extracts. Minor substrates of transglutaminase are seen in human keratinocytes, and one has been isolated by preparative electrophoresis. Major structural proteins of epidermis which are *in vitro* substrates of epidermal transglutaminase include the keratins and the stratum corneum basic protein.

Introduction

A major function of skin is to serve as a barrier between the organism and its environment. It is the first line of resistance against environmental insult and must often resist conditions which if directed against other bodily organs and tissue would cause such damage as to result in death to the organism. Skin is aptly suited to this function since it forms a relatively light weight, waterproof, elastic, and toughened structure which can withstand high levels of physical and chemical stress.

Skin is composed of two layers. The innermost layer or dermis is similar to connective tissue. Its main cell type is the fibroblast and one of its major functions is to support the outer skin layer, the epidermis. In this respect it provides a surface upon which the structural cell of the epidermis, the keratinocyte, may proliferate and differentiate, and the synthesis of a basement membrane may proceed. The dermis also appears to direct, to some extent, the course of differentiation followed by the keratinocyte.

Differentiation, as it occurs in the epidermis is called keratinization and is the process by which a basal cell, cubical, relatively undifferentiated, and positioned at the dermal–epidermal junctions loses its capacity to multiply, and becomes part of each successive epidermal layer. As the cell passes through each layer its gross and fine morphologies are altered, this alteration being the consequence of or at least occurring coincident with major biochemical changes.

In the basal layer the keratinocyte is about 10 microns in diameter and has a resilient shape which if isolated by trypsinization, becomes round. When this cell reaches the stratum corneum or outermost layer it has enlarged and become flattened with an effective diameter of nearly 30 microns (1). This large cell is called a corneocyte. It retains its flattened shape even if removed from epidermis by trypsinization. A characteristic of the corneocyte is that it contains a chemically resistant structure referred to as marginal band or cornified envelope.

The marginal band as a morphological entity has been known for many years. Its location is directly beneath the cell membrane and it occurs in cells of the stratum corneum. It is roughly twice the thickness of plasma membrane of non-keratinized cells. As a chemical entity it may be isolated by treatment of the epidermis with sodium hydroxide (2, 3) or by sodium dodecyl sulfate (SDS) and reducing agents (4, 5) or by multiple treatment with 6 or 8 M urea and reducing agents (6). In each case cornified envelope is the residue remaining after the somewhat drastic methods of extraction. Bovine snout cornified envelope has been prepared by NaOH (3) and urea mercaptoethanol (6) extraction methods and the amino acid compositions of the preparations differ only slightly. Cornified envelope has also been prepared from cultured human keratinocytes by extraction with reducing agents and SDS (2, 3) or urea (6). As with bovine snout the different isolation methods produce preparations which are similar in amino acid content. Cornified envelope owes its insolubility to the presence of high levels of the ϵ-(γ-glutamyl) lysine isopeptide bond (5, 6, 8). In human cornified envelope approximately 15 to 20% of the lysine are not free and participate in the isopeptide bond as determined by exhaustive enzymatic digestion and recovery of the γ-glutamyl-ϵ-lysine isodipeptide (5).

Formation of the cornified envelope

The synthesis of the cornified envelope is one of the major biochemical events which occurs during keratinization and has become a popular area for investigation by students of epidermal biology. Although the complete process of envelope synthesis is far from understood, it seems clear that this insoluble structure results from covalent bond formation between pre-existing soluble precursors, an idea originally expressed by Obinata and Endo (8).

The approach that has been most fruitful in elucidating some aspects of cornified envelope synthesis has been to identify protein substrates of epidermal transglutaminase. This is a logical approach since transglutaminase and at least some of its substrates are clearly involved in this process.

The original work utilizing this approach is presented in a series of papers describing the isolation of transglutaminase from bovine snout epidermis and its use to identify protein substrates from that tissue (9–13). Briefly the enzyme was incubated with the tissue extract under transglutaminase active conditions. An incubated extract was fractionated by ACA-34 chromatography and the elution pattern compared to that of an unincubated extract. From this comparison it was concluded that a 150 000 dalton component was arising as a result of incubation of smaller proteins with transglutaminase. After further purification, this cross-link containing derivative was used to raise an antibody. This antibody was found to react with proteins in the unincubated extracts. A low molecular weight protein was isolated on the basis of its cross-reactivity with the antibody to the 150 000 component, and found to be a component of about 36 000 daltons which dissociated to subunits of less than 10 000 daltons when treated with SDS. Antibodies raised to this component reacted both with the cytoplasm and the peripheries of cornified cells (13). When this isolated protein was reacted with isolated transglutaminase, material of high molecular weight was produced. This along with the indirect immunofluorescence data was higly suggestive that the 36K component was not only a transglutaminase substrate but also a cornified envelope precursor.

A second group of workers identified a transglutaminase substrate in cultured human keratinocytes (14). The basic approach to grow human keratinocytes to confluence and to activate the endogenous transglutaminase in keratinocyte extracts while providing ^{14}C-ethanolamine or dansyl cadaverine. Under these conditions, a protein whose SDS dissociated molecular weight was 92K in Weber and Osborn electrophoresis was labeled. This material was isolated by chromatographic procedures and was shown to be a protein of unusually high glutamic acid content. Indirect immunofluorescence with a rabbit antibody to this protein gave a reaction with the periphery of cells in the granular and cornified layer. Absorption of this antiserum with purified cornified envelope abolished its reactivity to the 92K component (14). Later it was shown that this antibody reacted by the indirect immunofluorescence technique to many types of human squamous epithelia and epithelial cells (15).

Recently our own group became interested in cornified envelope assembly and began studying this process as it occurs both *in vitro* and *in vivo*. We looked at human keratinocytes and human epidermis, bovine keratinogcytes, and bovine epidermis and new born rat epidermis and cultured keratinocytes.

Materials and methods

Tissue and cells

Human skin was obtained as foreskin from new-born infants, and as leg skin or seborrheic keratoses from adults. New-born rat skin was obtained from the dorsal area of rats less than 24 hr old and slices of bovine skin were prepared from calf snout. In all cases epidermis was obtained by trypsinization of whole skin overnight at 4 °C and keratinocytes were disaggregated by rapid pipetting. Human keratinocytes were cultured as described previously (6) except that cholera toxin and EGF were used as recommended by Green (16). Bovine keratinocytes were grown as previous described (6) and rat keratinocytes were from a long-lived line which has been recently described in detail (17).

Incubation and dansyl cadaverine labeling

Epidermis and cultured keratinocytes were homogenized in 50 mm Tris 2 mM EDTA, pH 7.5 or at pH 8.3 with 1% Triton X-100. In some experiments epidermis or cultured keratinocytes was homogenized in one of the above buffers which contained 2 mm DTT, 10 mM C_aCl_2 and 1 mg/ml dansyl cadaverine. After 2–3 hr of incubation at 37 °C the extracts were centrifuged and either used immediately or frozen.

Fractionation of extracts

A 1.5 × 90 cm Sephadex G-75 column was used to fractionate both incubated and unincubated crude extracts of both epidermis and cultured kerationcytes. Samples were made 0.15 M in NaCl to inhibit protein aggregation and applied to the column. Samples labeled with dansyl cadaverine were fractionated on the basis of fluorescence, and compared to corresponding fractions produced from unlabeled extracts. Individual fractions were examined by electrophoresis in cylindrical tubes (18) or in 0.75 or 1.5 mm slabs (19).

Isolation of substrates

Electrophoretic runs of 50 to 100 cylindrical gels were done using either 7% or 15% polyacrylamide. Fluorescent bands were removed by slicing and protein was eluted electrophoretically into dialysis bags. SDS was removed by dialysis against BIO Rad Ag 1-X8. Amino acid analysis was done using a Beckman Model 116 analyzer. Samples free of SDS were made 6 N in HCl and hydrolyzed at 100 °C for 24 hr.

Antibodies

Antibodies were raised to human proteins by injecting 200 μg of isolated transglutaminase substrate into rabbits. After 6 weeks animals received booster injections on a bi-weekly basis of 100 μg of isolated protein.

Experimental and discussion

Keratinocytes may be cultured in petri dishes and at confluence consist of a multilayered squamous epithelium. This coherent structure is made up of keratinocytes at various stages of differentiation, with the least differentiated cells in the bottom layer and the most differentiated cells in the topmost layer. Healthy confluent cultures may be rinsed free of serum proteins, removed from dishes by scraping, trypsinization, or other means and homogenized in buffer in a matter of seconds. This leads to extracts which are consistent with respect to the amount and kinds of proteins which are present. If these extracts are reacted with dansyl cadaverine in the presence of calcium ion and a reducing agent, then at least some of the proteins which are substrates of transglutaminase will become labeled with dansyl cadaverine. If dansyl cadaverine is omitted from the incubation mixture than transglutaminase substrate proteins may become cross-linked forming high molecular weight products.

One can look at the proteins of cultured keratinocytes using SDS electrophoresis. In Fig. 1, one can see soluble proteins of human keratinocytes separated on an SDS electrophoretigram. This figure

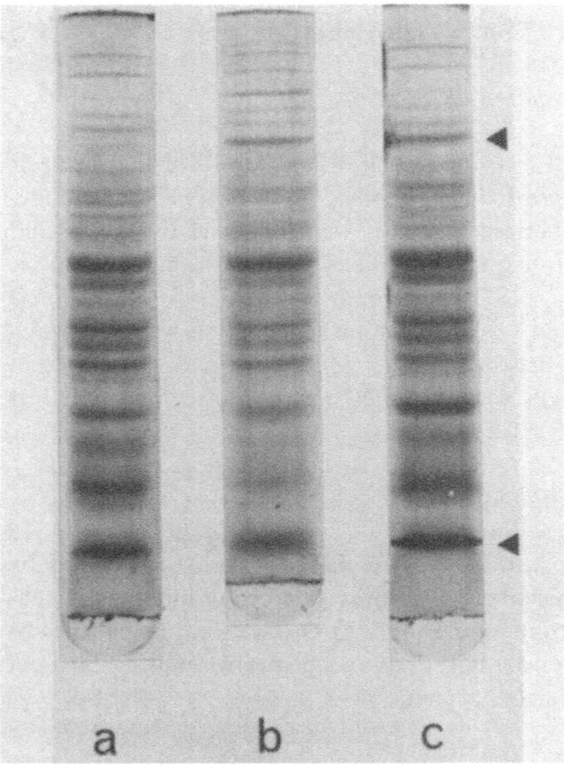

Fig. 1. Electrophoresis of human keratinocyte extracts in 7% SDS gels according to Neville (18); a) extract has been incubated with Ca^{++} ion and a reducing agent; b) extract is unincubated; c) extract incubated with Ca^{++} ion, a reducing agent, and dansyl cadaverine. Arrows denote bands which were fluorescent (Fig. 1 appeared in Ref. 20).

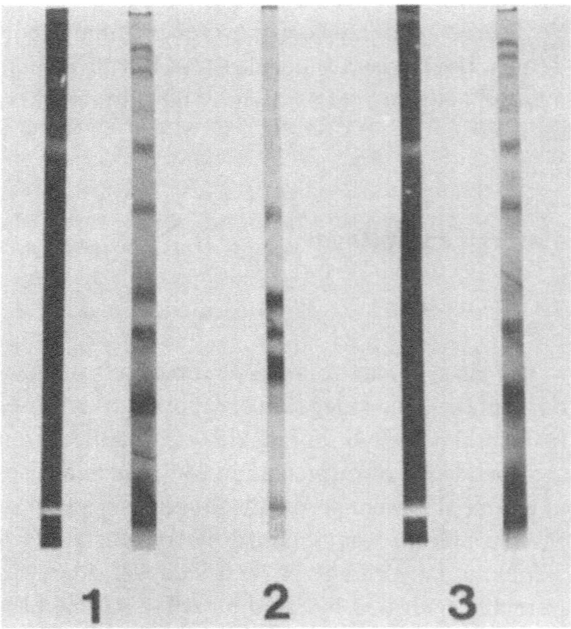

Fig. 2 Electrophoresis of human keratinocyte extracts in 7% SDS gels according to Laemmli (19); in 1 and 3 samples incubated with Ca^{++} ion, DTT, and dansyl cadaverine are shown unstained and viewed in ultraviolet light (left) and stained and viewed in ordinary light (right); 2 is a marker consisting of cow prekeratin (45 000 dalton to 70 000 dalton range) and phosphorylase A (92 000 daltons) (Fig. 2 appeared in Ref. 24).

compares the proteins of freshly extracted cells with those of a fresh extract that has been incubated at 37 °C under transglutaminase active conditions. As can be seen there is no difference in band positions or intensities between the incubated and unincubated gels when dansyl cadaverine is incorporated into the incubation mixture. This is in contrast to the change in protein pattern seen when the incubation is carried out in the absence of dansyl cadaverine. When the lysine analog is present a band of 125 000 becomes strongly fluorescent. Likewise when the analog is absent there is a significant decrease in the amount of the 125 000 dalton component. It is clear that this component has become labeled with dansyl cadaverine and is a major substrate of epidermal transglutaminase. A second band of fluorescence is seen at the electrophoretic front.

When extracts are examined by the more widely used system of Laemmli essentially the same picture is seen (Fig. 2). Fluorescence associated with the 125 000 dalton component is clearly visible as is additional fluorescence in a compact band at the front. These are the major substrates of epidermal transglutaminase present in human keratinocyte extracts. Additional fluorescence is also seen and is associated with polypeptides of about 40 000 and 20 000 daltons and appear to be minor substrates of epidermal transglutaminase under conditions described. One of these has been purified by preparative SDS electrophoresis as shown in Fig. 3.

When extracts labeled with dansyl cadaverine are examined by SDS electrophoresis in denser gels, the band which moves at the electrophoretic front is slowed down and moves about halfway down the gel at a position which corresponds to cytochrome *c* (Fig. 4). Using 7% and 15% polyacrylamide gels we were able to isolate both major substrates of transglutaminase which are present in cultured cells.

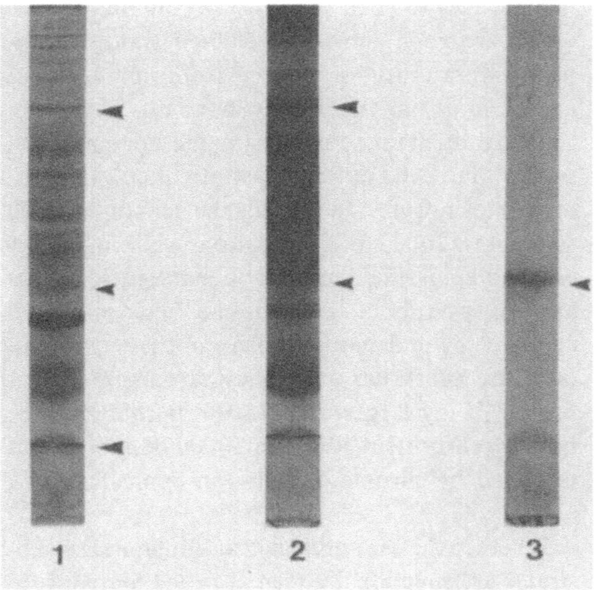

Fig. 3. Electrophoresis of human keratinocyte extracts showing detection and isolation of a minor transglutaminase substrate; 1) is a crude extract incubated with dansyl cadaverine, Ca^{++} ion, and a reducing agent; 2) is the same extract after fractionation on Sephadex G-75; 3) is the sample after isolation by preparative electrophoresis (Fig. 3 appeared in Ref. 24).

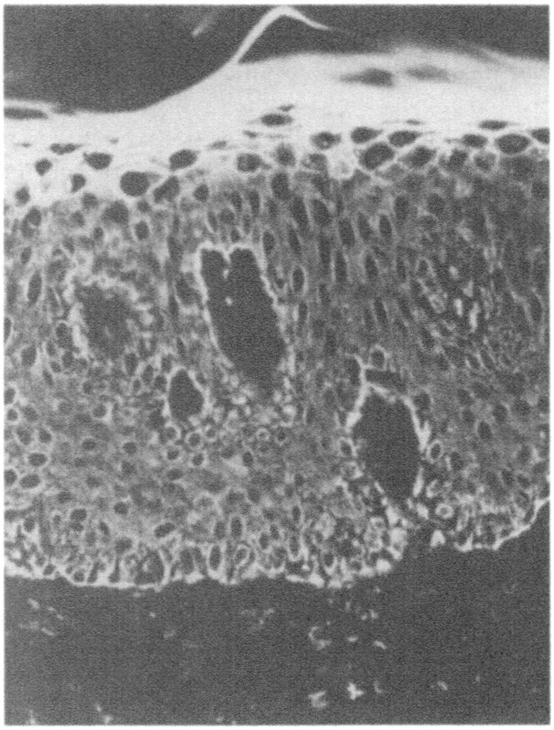

Fig. 5. Indirect immunofluorescence showing the reaction between human epidermis and the antibody raised to the 125000 dalton transglutaminase substrate. Sections were stained with goat FITC conjugated anti-rabbit Igg (Fig. 5 appeared in Ref. 20).

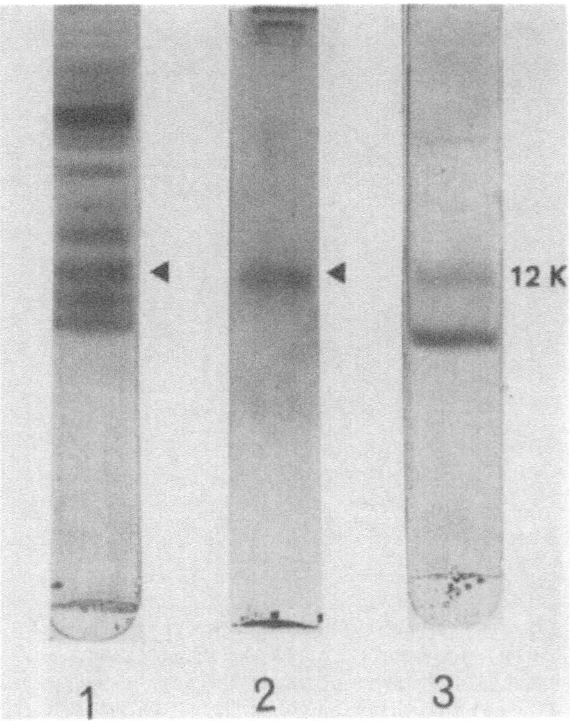

Fig. 4. Electrophoresis of human keratinocyte extracts showing isolation of a major transglutaminase substrate. On the 15% gels are 1) substrate protein partially purified Sephadex G-75; 2) the substrate after preparative electrophoresis; 3) a molecular weight standard showing the 12000 dalton position. Arrows mark fluorescent bands. (Fig. 4 appeared in Ref. 20).

Amino acid analysis of the two components was similar to that of cornified envelope, however, differences between the soluble proteins and the insoluble cornified envelope demonstrated that the cornified envelope was not assembled solely from one soluble protein or the other (20). Antibodies to each of the proteins raised in rabbits, reacted weakly with the isolated SDS treated proteins, but more strongly with semipurified undenatured preparations. Indirect immunofluorescence showed that the large substrate (A) was absent in the lower layers of epidermis but was apparently synthesized in the granular layer (Fig. 5). Similar studies with antibody to the small protein (B) showed that it was a cytoplasmic component of the living layers and was also present in cornified cells (20). A weak cross-reactivity between the small substrate antibody and the large substrate as well as between the large substrate antibody and the small substrate was seen (20). This cross-reactivity suggests that some structural similarities may exist between A and B. This structural similarity probably does not result because one substrate protein is derived from the other by either proteolytic breakdown or isopeptide formation. It may result as a consequence of both proteins being good substrates of epidermal transglutaminase.

It is known that liver transglutaminase and serum factor XIIIa require in their substrates certain amino acid sequences around the glutamine residues which become cross-linked. (21). Although not studied in this regard it seems reasonable that similar requirements for substrate amino acid sequence also exist for the epidermal enzyme. Because of this it would be expected that substrates of the epidermal enzyme would show structural similarities which could explain the weak cross-reactivity between anti A and B and anti B and A (20). The structural features which make each protein a substrate could also constitute an antigenic determinant. In this case, the antigenic site would be determined by primary structure or amino acid sequence. Since antibodies were made to SDS treated proteins this is very likely. Thus it seems to us that the large and small substrates contain regions where the amino acid sequences are very similar. It seems likely that similar amino acid sequences would also occur in other substrates of epidermal transglutaminase.

The large transglutaminase substrate which we have called A seems to be quite similar to the protein, involucrin (14, 15, 22, 23), which also was isolated from cultured human keratinocytes. Although there is some disagreement on the amino acid analysis of the protein, its ability to label with dansyl cadaverine, its electrophoretic mobility in Laemmli SDS electrophoresis (22, 23, Fig. 2), and its position in the granular layer of epidermis as shown by indirect immunofluorescent studies strongly suggest that they are the same. Further studies on involucrin have shown that this protein is widely distributed in various human epithelia (15).

When the keratinocytes of other animals are cultivated, they too can be shown to contain proteins which are good substrates of their transglutaminases. Keratinocytes cultured from bovine snout and from dorsal skin of new-born rats have been shown to contain substrates which are similar in size to the 12000 dalton substrate of cultured human keratinocytes. In rat cells the denatured small substrate is 20000 to 25000 daltons while in the cow cell and epidermis the major denatured substrate appears to be about 9000 daltons (Figs. 6 and 7). In contrast the undenatured forms of these transglutaminase substrates are somewhat larger, the rat protein being larger than 75000 daltons and the bovine cell protein 30000 to 50000 daltons as determined by chromatography on Sephadex G-75 (24).

Whether the rat and cow transglutaminase substrates are functionally related to the human substrate protein B has not yet been demonstrated. However, the 9000 dalton substrate from cultured bovine keratinocytes is nearly identical in size to the transglutaminase substrate protein isolated from bovine snout epidermis (11–13, Fig. 6). This protein has a distribution in bovine epidermis which is sim-

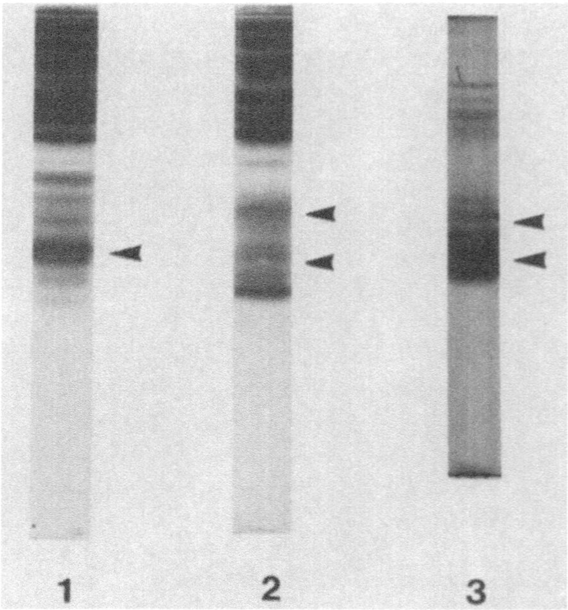

Fig. 6. Electrophoresis of human and bovine extracts labeled with dansyl cadaverine; 1) is a human keratinocyte extract; 2) is a cultured bovine keratinocyte extract; 3) is an extract of bovine snout epidermis. Gels are 15% polyacrylamide and arrows mark positions of fluorescent bands (Fig. 6 appeared in Ref. 24).

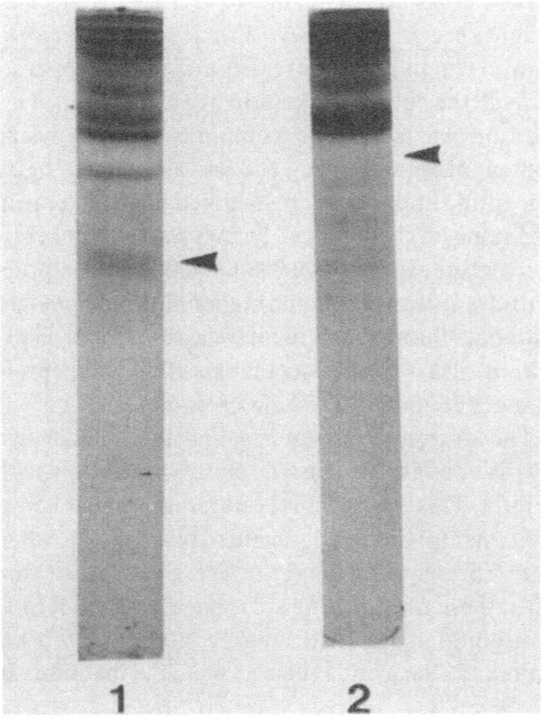

Fig. 7. Electrophoresis of cultured human and rat keratinocytes. Dansyl cadaverine labeled extracts were electrophoresed on 15% polyacrylamide gels. Arrows show the fluorescent bands (Fig. 7 appeared in Ref. 24).

ilar to the distribution of protein B in human epidermis and has been shown to be cross-linkable to high molecular weight forms. We have shown a similar ability of human substrate B to be converted to high molecular weight forms by transglutaminase action (Fig. 8). This evidence along with additional studies strongly suggest that in epidermis the bovine 36000 dalton protein is a precursor of cornified envelope. The amino acid analysis of this protein is similar to our protein B except for the inclusion of relatively high levels of citrulline in the tissue protein. It should be noted, however, that the citrulline content of proteins is due to the action of an enzyme which is active in epidermis but is present in only trace amounts in cultured keratinocytes (Kubilus and Baden, unpublished work).

In the bovine keratinocyte two additional components are visible which label with dansyl cadaverine under transglutaminase active conditions. These can be seen in Fig. 9. The degree of fluorescent labeling which has been achieved with these components is not as great as has been seen with the 9000 component or with the two major transglutaminase substrates of human keratinocytes. If these components are incorporated into the envelope, it is not known whether they correspond to our substrate A or to a cross-linked form of our

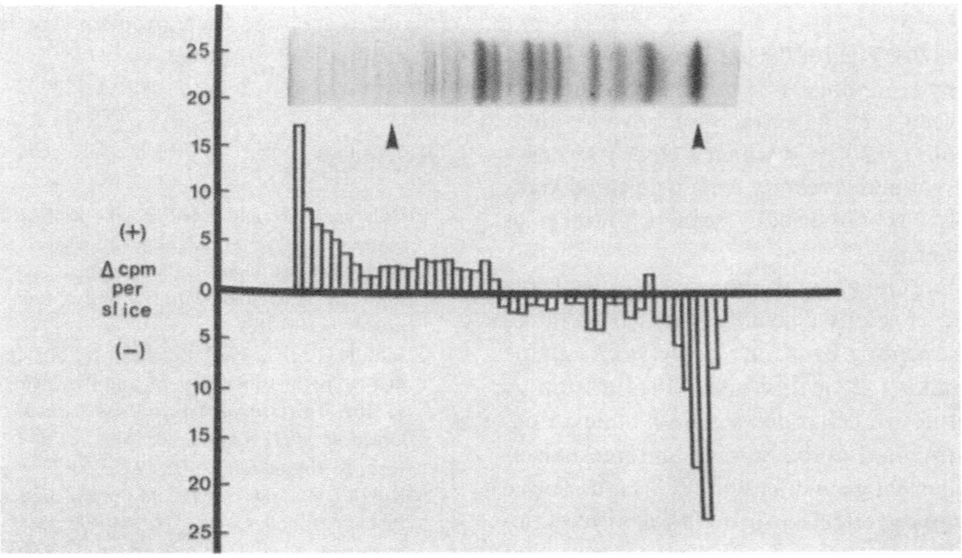

Fig. 8. Conversion to high molecular weight aggregates by transglutaminase. Low molecular weight substrate from [^{14}C-guanido]-L-arginine-labeled human keratinocyte cultures was partially purified on Sephadex G-75. To this was added unlabeled crude extract. Transglutaminase was activated and after incubation the sample was electrophoresed on 7% gels. Gels were stained and sliced. Individual slices were counted and compared to slices from an unincubated gel. Cpms lost after incubation are plotted below the line (−). Cpms gained after incubation are plotted above the line (+) (Fig. 8 appeared in Ref. 20).

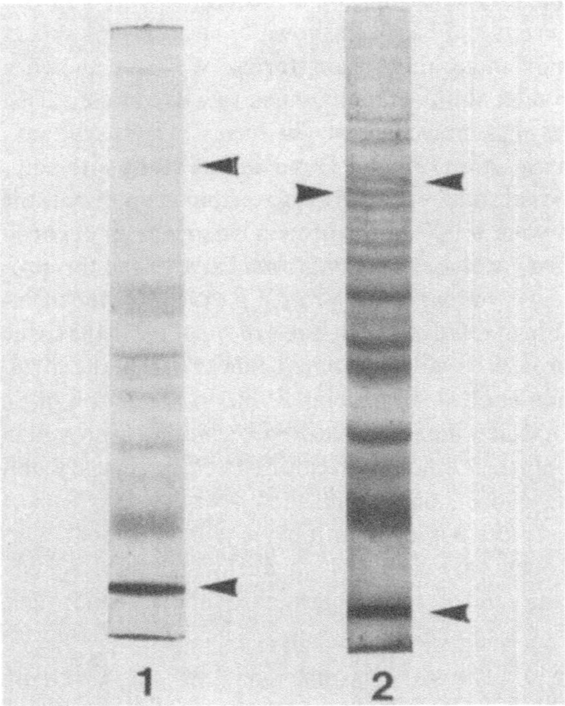

Fig. 9. Electrophoresis of keratinocyte extracts according to Neville (18) on 7% gels. Extracts were labeled with dansyl cadaverine by the action of endogenous transglutaminase; 1) is a human extract; 2) is a bovine extract. Arrows denote fluorescent bands (Fig. 9 appeared in Ref. 24).

substrate B (20). The 100000 dalton dansyl cadaverine labeling components of bovine keratinocytes have not been seen in extracts of bovine snout. However, our substrate A which is clearly an envelope precursor and is readily seen in cultured keratinocytes is also not usually seen in extracts of human epidermis.

Epidermal transglutaminase plays a crucial role in cornified envelope assembly which is to introduce the isopeptide band into or between soluble proteins making them insoluble. Its function in other keratinocyte cell processes is not understood. It may be involved in the binding and internalization of epidermal growth factor (EGF) a hormone with pronounced effects on growth and differentiation of epidermal cells (25). We have shown that proteins important in the keratinization process are *in vitro* substrates of epidermal transglutaminase (26). These include keratin, and the stratum corneum basic protein described by Dale (27). Keratin may be cross-linked *in vitro* to very high molecular weight species, and *in vivo* a small amount of anti-keratin reactive material exists and may be separated from the bulk of the keratin by gel filtration. This fraction was shown to contain a blocked lysine content of 9 nmoles / mg and was assumed to be an isopeptide inter-chain cross-linked form of keratin. The same level of blocked lysines was seen in keratin which consisted of 40 000 to 70 000 dalton polypeptides as was seen in the higher molecular weight material. Because of this it was considered likely that at least some of the keratin polypeptide changes contain intra-chain cross-links.

The stratum corneum basic protein or filaggrin (27) is thought to exist in close association with keratin. This is based on the observation that keratin filaments *in vitro* aggregate to macrofibrils when in the presence of filaggrin. We have shown that when these macrofibrils are treated with epidermal transglutaminase, a structure containing both keratin and filaggrin is formed which is insoluble in 1% SDS.

Thus epidermal transglutaminase plays a major role in cornified envelope synthesis and may also be important in the stabilization of other structural proteins present in skin. The presence of additional soluble proteins in keratinocytes which are weakly labeled with dansyl cadaverine when transglutaminase is activated suggest that transglutaminase-mediated isopeptide formation may be important to additional cell processes.

References

1. Odland, G. F. and Reed, T. H., 1967. Ultrastructure of Normal and Abnormal Skin (Zelickson, A. S., ed.), pp. 54–75. Lea and Febiger, Philadelphia.
2. Matoltsy, A. G. and Matoltsy, M. N., 1966. J. Invest. Dermatol. 46: 127–129.
3. Matoltsy, A. G., 1977. Biochemistry of Cutaneous Epidermal Differentiation (Seiji, M. and Bernstein, I. A., eds.), pp. 93–109. University of Tokyo Press, Tokyo.
4. Green, H., 1977. Cell 11: 405–416.
5. Rice, R. H. and Green, H., 1977. Cell 11: 417–422.
6. Kubilus, J., MacDonald, M. J. and Baden, H. P., 1979. Biochim. Biophys. Acta 578: 484–492.
7. Sugawara, K., 1977. Biochemistry of Cutaneous Epidermal Differentiation (Seiji, M. and Bernstein, I. A., eds.), pp. 387–397. University of Tokyo Press, Tokyo.
8. Obinata, A. and Endo, H., 1977. Nature 270: 440–441.
9. Buxman, M. M. and Wuepper, K. D., 1976. Biochim. Biophys. Acta 452: 356–369.

10. Buxman, M. M., Buehner, G. E. and Wuepper, K. D., 1976. Biochem. Biophys. Res. Commun. 73: 470–478.
11. Buxman, M. M. and Lobitz, C. J., 1980. Biochemistry of Normal and Abnormal Epidermal Differentiation (Bernstein, I. A. and Seiji, M., eds.), pp. 245–254. University of Tokyo Press, Tokyo.
12. Buxman, M. M., Lobitz, C. J. and Wuepper, K. D., 1980. J. Biol. Chem. 255: 1200–1203.
13. Lobitz, C. J. and Buxman, M. M., 1982. J. Invest. Dermatol. 78: 150–154.
14. Rice, R. H. and Green, H., 1979. Cell 18: 681–694.
15. Banks-Schlegel, S. and Green, H., 1981. J. Cell Biol. 90: 732–737.
16. Green, H., 1978. Cell 15: 801–811.
17. Baden, H. P. and Kubilus, J., 1983. J. Invest. Dermatol. 80: 124–130.
18. Neville, D. M., Jr., 1971. J. Biol. Chem. 246: 6328–6334.
19. Laemmli, U. K., 1970. Nature 227: 680–685.
20. Kubilus, J. and Baden, H. P., 1982. In Vitro 18: 447–455.
21. Gorman, J. J. and Folk, J. E., 1980. J. Biol. Chem. 255: 419–427.
22. Watt, F. M. and Green, H., 1981. J. Cell Biol. 90: 738–742.
23. Watt, F. M. and Green, H., 1982. Nature 295: 434–436.
24. Kubilus, J. and Baden, H. P., 1983. Normal and Abnormal Epidermal Differentiation (Bernstein, I. A. and Seiji, M., eds.), in press.
25. Rheinwald, J. G. and Green, H., 1977. Nature 285: 421–424.
26. Baden, H. P. and Kubilus, J., 1982. Stratum Corneum (Marks, R. and Plewig, G., eds.), pp. 19–24. Springer-Verlag, Berlin, Heidelberg.
27. Dale, B. A., 1977. Biochim. Biophys. Acta 491: 193–204.

Received 6 June 1983.

Molecular and Cellular Biochemistry 58, 139–146 (1984).

The role of the insulin receptor in mediating the insulin-stimulated growth response in Reuber H-35 cells

John W. Koontz
Dept. of Biochemistry, Walters Life Sciences Building-F435, University of Tennessee, Knoxville, TN 37996-0840, U.S.A.

Summary

Insulin is able to stimulate a growth response in a variety of different cell types. However, the role of the insulin receptor in mediating this response is not clear. Indeed, it has been reported that the ability of insulin to stimulate a growth response is a result of its interaction with other growth factor receptors rather than the insulin receptor.

We have previously reported that the H-35 hepatoma cell line responded to physiological concentrations of insulin as a growth factor and that the relative potency of proinsulin suggested that this response was mediated by the insulin receptor. In this report, two experimental approaches are used to demonstrate the involvement of the insulin receptor in mediating the growth response. Two different preparations of antibody to the insulin receptor are found to be capable of stimulating this response. In addition, the human insulin-like growth factors (IGF-I and II) show very low cross-reactivity with the insulin receptor and are significantly less potent than insulin in stimulating the growth response.

Abbrevations

IGF = insulin-like growth factor; MSA = multiplication stimulating activity; Hepes = 4-(2-hydroxyethyl)-1-piperazineethanesulfonic acid.

Introduction

A recent report suggests that the ability of insulin to stimulate a growth response in cultured human fibroblasts is dependent upon its interaction with a separate growth factor receptor other than the insulin receptor (1). In addition, the ability of insulin to stimulate the differentiation of cultured L6 myoblasts into myotubes has been attributed to insulin interacting with the receptor for IGF-I or MSA (2, 3). In contrast, the ability of MSA (the rat equivalent of human IGF-II (4)) to stimulate glucose oxidation in adipocytes (1) or IGF-I to stimulate glucose oxidation or antilipolysis in adipocytes (5) and to activate glycogen synthase in skeletal muscle (6, 7) is reported to be dependent upon the ability of these growth factors to bind the insulin receptor.

The stimulation of a growth response in a clone of the H-35 hepatoma cell line by physiological levels of insulin has been reported (8). The relative potency of proinsulin suggests that in this particular cell line the growth response might be mediated through the insulin receptor rather than IFG receptors. The present report describes the results of experiments characterizing the effects of anti-insulin receptor antibody from two separate sources and human IGF-I and IGF-II on both the growth response and the induction of tyrosine aminotransferase in H-35 cells. The results obtained provide compelling evidence that these effects of insulin are mediated through a receptor specific for this hormone.

Material and methods

Cell culture materials were obtained from Grand Island Biologicals (Grand Island, New York) and

M. A. Bioproducts (Moline, Illinois). Protein A and protein A-Sepharose and G-50 fine were obtained from Pharmacia (Uppsala, Sweden). Carrier-free Na ^{125}I and methyl-[^{3}H]-thymidine (20 Ci/mmol) were obtained from New England Nuclear (Boston, Massachusetts). MSA and ^{125}I-MSA were obtained from Collaborative Research (Waltham, Massachusetts). Bovine serum albumin (fraction V) was obtained from Reheis Chemical Co. (Phoenix, Arizona).

The author is indebted to the following individuals for their generosity in providing these materials: porcine insulin (lot 615-07J-50) and porcine proinsulin (lot 615-076-150-1) Dr Ronald Chance, Lilly Research Laboratories, Indianapolis, Indiana; human anti-insulin receptor serum, Dr Jerrold Olefsky, Department of Medicine, University of Colorado Health Science Center, Denver, and Dr David Baldwin, Department of Medicine, University of Chicago Medical School, Chicago, Illinois; rabbit anti-insulin receptor antibody, Dr Steven Jacobs, Wellcome Research Laboratories, Research Triangle Park, North Carolina; human IGF-I (lot 16 SP II) and IGF-II (lot 9 SE IV), Dr René Humbel, Biochemisches Institut der Universität Zürich, Zürich, Switzerland.

The procedures for growth and maintenance of the cells were as described previously (8). In brief, the cells are maintained in monolayer culture in Dulbecco's modified Eagle's medium supplemented with 5% fetal calf serum and 5% calf serum in a humidified atmosphere of 10% CO_2, 90% air. The procedures for measuring the growth response were identical to those described previously (8). Cells are starved of serum for three days and then the media is changed to fresh medium plus or minus the peptides to be assayed. Sixteen hours later, when the cells are in S phase (8), [^{3}H]-thymidine is added (final concentration of 0.5 μM, 0.5 μCi/ml) and after thirty minutes the cells are washed free of media and the acid-precipitable counts incorporated into DNA determined by the method of Bollum (9). The enzyme induction studies were done on confluent monolayers which had been incubated overnight in serum-free medium. The medium was changed to include the inducing agents and incubation continued for varying periods. Cells were washed in phosphate-buffered saline and scraped from the dish in a solution of 0.05 M potassium phosphate, pH 7.4, 0.15 M KCl, 5 mM EDTA. NP-40 is added to the solution to achieve 0.1% final concentration, the solution is mixed and centrifuged at 15 000 × **g** for 15 min. The supernatant is used for assaying tyrosine aminotransferase according to a modification (10) of the method of Diamondstone (11). Protein was determined by the method of Lowry (12). DNA was determined by the method of Burton (13).

Iodination of insulin was carried out by a minor modification of the method of Freychet et al. (14) as described by Terris and Steiner (15). A G-50 (fine) column was used for separation of the iodinated hormone from free iodine. Elution was carried out using 0.1 M acetic acid, 0.1% bovine serum albumin. The fractions containing insulin were assayed for acid precipitability of the iodine and those fractions exhibiting greater than 97% precipitability were pooled, lyophilized and stored frozen at -20 °C. For use in assays, the lyophilized powder was dissolved in Ca^{2+}- and Mg^{2+}-free phosphate-buffered saline containing 1% bovine serum albumin. The biological integrity of the iodinated hormone was determined by its ability to stimulate tyrosine aminotransferase induction in the H-35 cells. Specific activity immediately following iodination varied from 75–125 μCi/g.

The binding of insulin to specific receptors was assayed with cells in monolayer culture. The cells were grown in cluster dishes of 24 × 15 mm wells or 6 × 35 mm wells. Cells were washed twice with ice-cold phosphate-buffered saline and then a binding assay mixture containing serum-free Dulbecco's modified Eagle's medium, 1% BSA, 10 mM Hepes, pH 7.4 and ^{125}I-insulin, 0.5 nM was added to the wells. Non-specific binding was determined by including 2.5 μg/ml of unlabeled insulin in the binding assay mixture. After 3 hr, the reaction was terminated by aspirating the mixture from the well, washing the cells four times with ice-cold phosphate-buffered saline containing 0.1% bovine serum albumin and then dissolving the cell monolayer in 0.1 N NaOH. This solution was then transferred to a counting vial and radioactivity was determined.

The binding experiments were usually conducted at 15 °C to minimize degradation of iodinated ligand. In all the experiments described, the acid precipitability of ^{125}I-insulin was greater than 90% at the end of the incubation period.

The IgG fraction of the sera was purified by

chromatography over protein A-Sepharose as described by Hjelm et al. (16).

In those experiments where applicable, the values for EC_{50} were determined by probit analysis of the concentration-effect curves.

Results

Insulin stimulation of enzyme induction and the growth response

The effect of insulin on tyrosine aminotransferase activity in this particular clone of H-35 cells is essentially identical to other reports on the effect of insulin in the H-35 cells (17, 18). The increase in activity reaches a maximum in approximately 2.5 hr and half-maximal induction occurs with an EC_{50} of about 30 pM (data not shown).

Human anti-insulin receptor antibody

The IgG fractions were affinity purified using protein A-Sepharose chromatography. This fraction has previously been shown to be the immunoglobulin population which interacted with the insulin receptor (19).

The IgG fraction of this serum does not compete with insulin for binding to the insulin receptor in these cells (data not presented). This is in contrast to a report on the action of this antibody in isolated hepatocytes in which insulin binding is inhibited approximately 75% in the presence of 500 μg/ml IgG (19).

However, this same IgG fraction is able to stimulate enzyme induction with an EC_{50} of 2 μg/ml (Fig. 1). This is in good agreement with the value obtained in isolated hepatocytes for the stimulation of the incorporation of labeled glucose into glycogen (20). The time course of the induction (not shown) obtained with the anti-receptor IgG and the magnitude of the increase in aminotransferase activity are indistinguishable from that obtained using insulin. Normal human IgG at 1 mg/ml had no effect on enzyme activity.

The human anti-insulin receptor IgG also stimulates the incorporation of [^{3}H]-thymidine into these cells (Fig. 2). The magnitude of the effect observed with the anti-receptor antibody is essentially the same as that found with optimal concentrations of

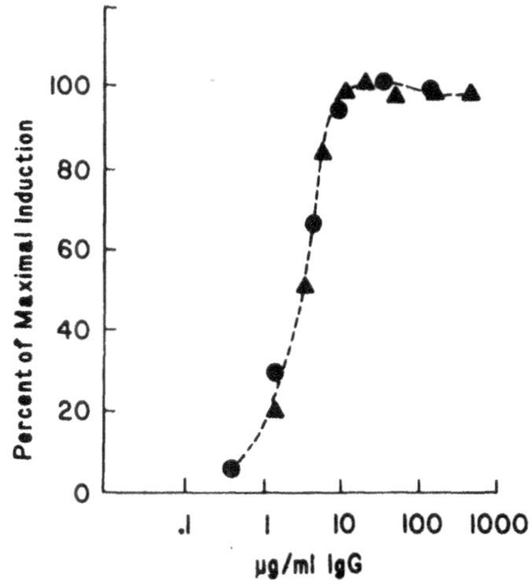

Fig. 1. Concentration effect relationship for induction of tyrosine aminotransferase by the IgG fraction from human serum containing anti-insulin receptor antibodies. Cells which had been on serum-free medium overnight were incubated with the indicated concentrations of IgG for three hours. Cells were harvested and assayed for enzyme-specific activity as described in Materials and methods. The two sets of symbols represent data from two different preparations of IgG from this serum obtained at different times. The experiment was performed with each preparation at least twice and the results represent the average of triplicate determinations made in each experiment.

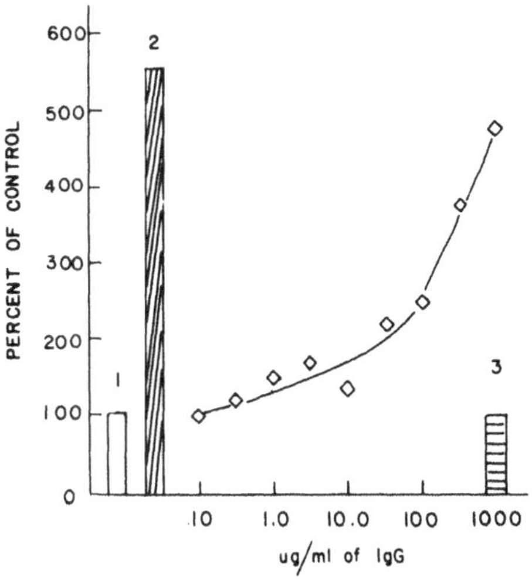

Fig. 2. Stimulation of a growth response by human anti-receptor IgG. Cells which had been starved for serum for 3 days were exposed to the indicated concentrations of IgG for 16 hr. [^{3}H]-thymidine was added to the medium and after 30 min the cells were harvested and the incorporation of thymidine determined as described in Materials and methods. The data are expressed as a percent of the [^{3}H]-thymidine cpm/μg DNA in the cultures treated with serum-free media. The bars represent: 1, serum-free media; 2, 1 nM insulin in serum-free media; 3, 1000 μg/ml normal human IgG. This is representative of 3 separate experiments.

serum or insulin. Logit plot analysis of this data gives an EC_{50} for the effect with the human anti-receptor antibody of approximately 70 μg/ml whereas normal human IgG at 1 mg/ml has no effect on the thymidine incorporation in these cells.

Rabbit anti-insulin receptor antibody

As a means of independently testing the role of the insulin receptor in mediating the growth response in these cells, rabbit anti-receptor antibody was also analyzed for its ability to stimulate enzyme induction and the growth response. Once again the IgG fraction obtained from the serum was used.

The rabbit anti-receptor IgG is unable to competitively displace iodinated insulin from the insulin receptor. Similar results have been previously reported for this antibody preparation (20).

The rabbit anti-receptor IgG stimulates tyrosine aminotransferase induction to the same extent as optimal concentrations of insulin or of human anti-receptor IgG (Fig. 3). In contrast, the IgG fraction from unimmunized rabbits proved incapable of inducing the enzyme (Fig. 3). The EC_{50} for induction is about 5 μg/ml IgG which is very similar to that obtained with the human anti-receptor IgG.

The same rabbit anti-receptor IgG preparation also stimulates [^{3}H]-thymidine incorporation into these cells (Fig. 3). The EC_{50} value for the response is 5 μg/ml and the maximal effect observed is the same as that obtained with optimal concentrations of either insulin or serum. At low concentrations, the IgG fraction from unimmunized rabbit serum had no effect on thymidine incorporation. However, at the highest concentration tested (200 μg/ml), there was a modest increae in [^{3}H]-thymidine incorporation. This was observed only at a concentration that is 10 times greater than the concentration of the anti-receptor IgG necessary for a maximal response.

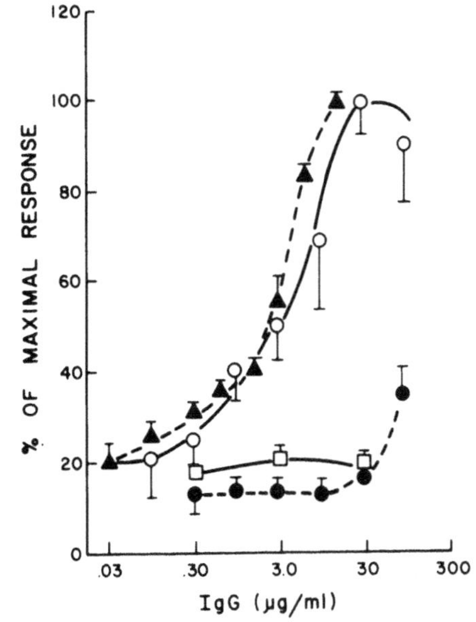

Fig. 3. Stimulation of enzyme induction or a growth response in H-35 cells by the IgG fraction from rabbit anti-insulin receptor antiserum or from non-immune rabbit serum. Cells were prepared for measuring the enzyme induction or growth response as described in the legends to Figs. 1 and 2, respectively. Each of these curves represents data from two separate experiments, each done with triplicate determinations at each concentration. The closed symbols represent determination of the growth response for: ▲, rabbit anti-receptor IgG; and, ●, non-immune rabbit IgG. The open symbols represent determination of enzyme induction in the presence of: ○, rabbit anti-receptor IgG; or, □, non-immune rabbit IgG. The basal activities represent 10% and 20% of the maximal activities for the growth response and enzyme induction, respectively.

Human IGF-I and IGF-II

If insulin does not act through the insulin receptor to stimulate the growth response of H-35 cells, then the most likely receptors it would interact with would be those of the insulin-like growth factors IGF-I and IGF-II. These peptides exhibit significant sequence and structural homology with insulin and proinsulin (21, 22, 23). Furthermore, although specific receptors for these individual peptides have been demonstrated, depending upon the cell type, cross-specificity of binding to the insulin receptor has also been shown. Consequently to determine the possibility of whether insulin might be acting through the IGF receptors, the responses of the H-35 cells of IGF-I and IGF-II were compared to insulin and proinsulin measuring three different parameters: ability to compete with iodinated insulin for binding to the cell surface; induction of tyrosine aminotransferase; induction of the growth response.

Analysis of the competition with iodinated insulin for binding to the cell surface demonstrates that both IGF-I and IGF-II are significantly less potent than insulin (Fig. 4). The EC_{50} values for insulin or proinsulin competing with iodinated insulin for binding to the cell surface are 7.5 nM and 150 nM,

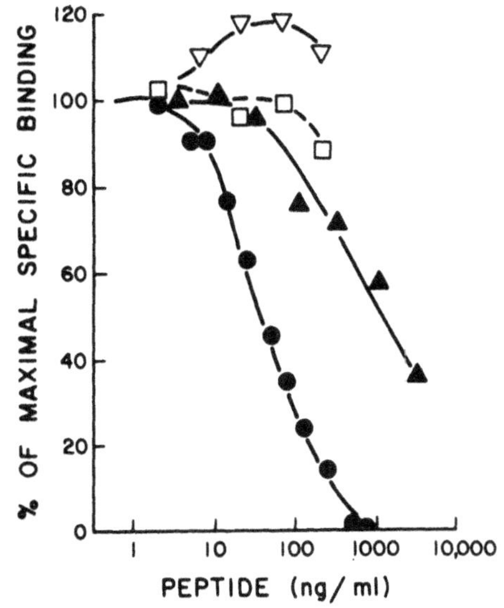

Fig. 4. Binding of ^{125}I-insulin to the H-35 cells in the presence of different unlabeled peptides. Binding to monolayer cultures was performed as described in Materials and methods. The results are the average of triplicate determinations from two separate experiments. The peptides are: ●, insulin; ▲, proinsulin; □, IGF-I; and ▽, IGF-II.

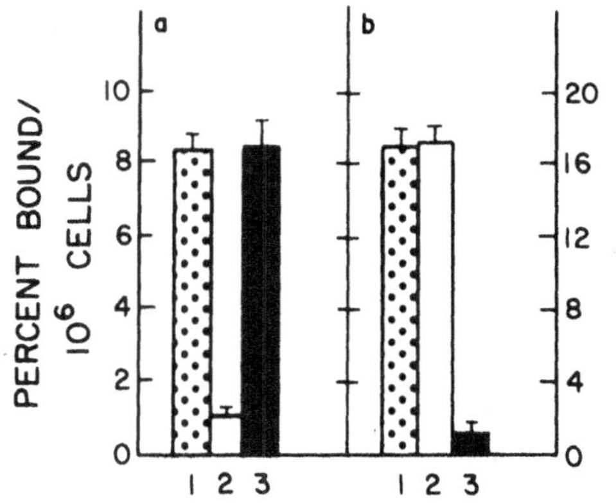

Fig. 5. Binding of ^{125}I-insulin or ^{125}I-MSA to H-35 cells. The cells were incubated with: (a) ^{125}I-insulin, 0.5 nM; or (b) ^{125}I-MSA, 0.3 nM (both represented by bar number 1).

Either 400 nM unlabeled insulin, 2, or 70 nM unlabeled MSA, 3, was included in the incubation mixture. The binding procedure is described in Materials and methods.

respectively. The apparent stimulation of insulin binding by IGF-II was observed in only one of the two experiments of this type.

Because of the limited quantities of human IGF-I and II, displacement curves of these iodinated peptides were not performed. However, it was possible to use a commercially available ^{125}I-MSA preparation for such studies. MSA at 70 nM inhibited the binding of ^{125}I-MSA by greater than 90% (Fig. 5b), but had no effect on the binding of ^{125}I-insulin (Fig. 5a). Conversely, insulin at 400 nM had no effect on ^{125}I-MSA binding (Fig. 5b) and inhibited the binding of ^{125}I-insulin by greater than 90% (Fig. 5a).

Although IGF-I and the IGF-II would not compete effectively with insulin for binding to the H-35 cell surface, both of these peptides were able to induce tyrosine aminotransferase (Fig. 6). They stimulate an increase in the specific activity of the enzyme to the same extent as that observed with insulin or proinsulin. However, the relative potency of both is at least 1 000-fold less than that of insulin and about 10-fold less than proinsulin. Each gives 50% of maximal induction at 5 nM.

Neither of these growth factors is able to provoke a significant increase in [^{3}H]-thymidine incorporation at concentrations below 7 nM (Fig. 7). Because of the limited quantities of these peptides, 70 nM was the highest concentration tested. At this concentration, neither peptide stimulated as great an increase as that obtained with 1 nM insulin in serum-free medium or 10% serum-containing medium. However, it appears that they are not equipotent: IGF-II gives a greater response than IGF-I. Assuming a maximal effect equal to that of insulin, the EC_{50} for IGF-II is 20 nM. A value for IGF-I cannot be extrapolated because of the limited response to this peptide at the concentrations tested.

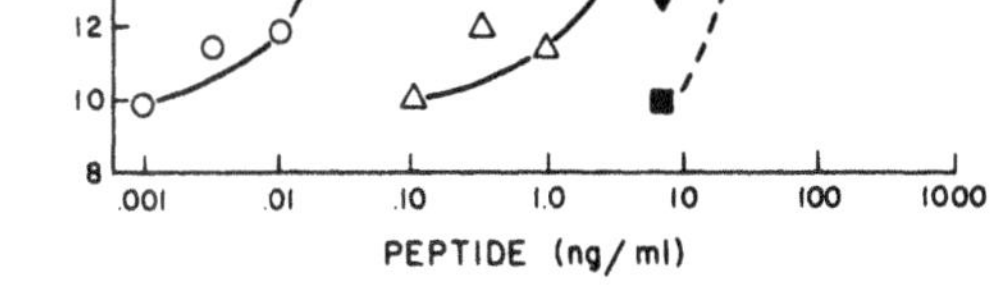

Fig. 6. Concentration effect curves for the induction of tyrosine aminotransferase. This experiment was conducted essentially as described in the legend to Fig. 1. The results are the average of two experiments each performed in triplicate. The peptides are: ○, insulin; △, proinsulin; ■, IGF-I; and, ▼, IGF-II.

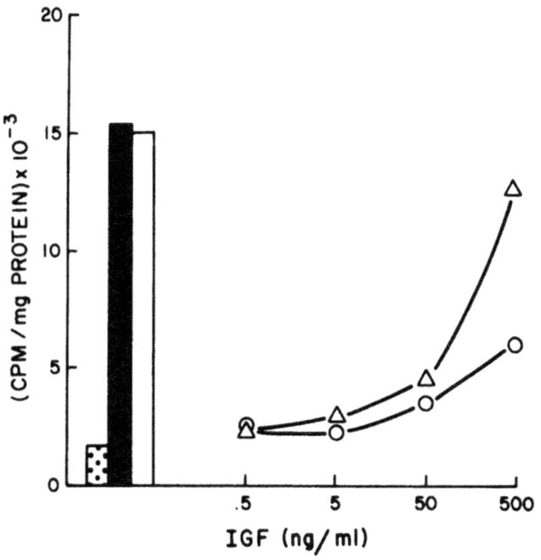

Fig. 7. Ability of IGF-I and IGF-II to stimulate a growth response in H-35 cells. The experiment was conducted as described in the legend to Fig. 2. The bars represent: stipled bar, serum-free medium; closed bar, serum-containing medium; open bar, 1 nM insulin in serum-free medium. The symbols are: △, IGF-II; ○, IGF-I. The data with the IGF peptides are representative of three separate experiments.

Discussion

The antibodies to the insulin receptor have been used to verify the hypothesis that the insulin receptor mediates the growth response in the cultured H-35 hepatoma cell line. Both a human anti-receptor IgG and a rabbit anti-receptor IgG are able to stimulate thymidine incorporation into the H-35 cells. The observation that two separate preparations of human and rabbit antibody to the insulin receptors are both able to mimic the response to insulin strongly suggests that the insulin receptor does mediate this response. This is similar to other reports of anti-insulin receptor antibodies producing insulin-like effects, including glucose oxidation (24), glycogen synthase activation (8), and induction of lipoprotein lipase in 3T3-LI differentiated adipocytes (25). However, this is the only such report demonstrating an effect on thymidine incorporation.

Because these antibody preparations do not compete with iodinated insulin for binding to the insulin receptor of H-35 cells it is not possible to use the monovalent Fab fragments of the antibody to inhibit the insulin-stimulated response. This type of experiment was the basis for the conclusion that insulin did not act through the insulin receptor in stimulating thymidine incorporation in cultured human fibroblasts (1). Unlike the H-35 cells the human fibroblasts did not respond to the anti-receptor antibody IgG preparations (1). The most likely explanation for the differences in the two studies is that the two different cell types respond to insulin action as a growth factor through different receptors. Although the H-35 cells are derived from a hepatoma (26) their ability to respond to insulin as a growth factor through the insulin receptor is probably a reflection of a normal liver function since there is considerable evidence that insulin is an important growth factor for the liver (27–30).

The absolute specificity of these antibody preparations is not established. This was the primary reason for using the two separate antibody preparations. While it is possible that one of the preparations would cross-react with another growth peptide receptor to initiate this growth response it is unlikely that this would occur with both preparations.

Because of the sequence and structural homology of the peptides, insulin would be most likely to interact with the receptors for IGF-I and IGF-II. Furthermore, recent reports suggest at least a subunit homology between the receptors for insulin and IGF-I (31). Consequently, if the anti-insulin receptor antibody preparations were to interact with other growth factor receptors, those for IGF-II and particularly, IGF-I would be most likely. However, the H-35 cells are reported to lack an IGF-I receptor (32, 33). Therefore, although there is a report that this human anti-receptor antibody cross-reacts with the IGF-I receptor, this would not be a factor with the H-35 cells (34).

Comparative studies with IGF-I and IGF-II were performed to establish the relative potencies of these peptides with insulin as an independent means of assessing the possibility that insulin stimulates the growth response through an IGF receptor.

The human IGF-I and IGF-II peptides demonstrated little cross-specificity for the insulin receptor as measured by competition with ^{125}I-insulin for binding to the cell surface. Furthermore, insulin does not compete significantly with ^{125}I-MSA for binding to the cells. This agrees with similar studies using other liver cells or purified liver membranes (35, 36). Examination of biological responses shows that both IGF-I and IGF-II induce tyrosine aminotransferase to the same extent as insulin or proinsulin. The order of potency is insulin $\gg$ proinsulin $>$ IGF-I = IGF-II. Both IGF peptides gave an EC_{50} of approximately 7 nM. In another rat liver cell line (HTC), MSA induces tyrosine aminotransferase with an EC_{50} of approximately 15 nM (37). These authors concluded that MSA was acting through the MSA receptor to mediate this response. They did not examine IGF-I (35). The effects of human IGF-I and IGF-II on tyrosine aminotransferase induction have not been previously reported. The values obtained for IGF stimulation of aminotransferase induction are similar to the EC_{50} values reported for IGF stimulation of [^{14}C]-glucose incorporation into isolated rat adipocytes (5) and MSA stimulation of [^{14}C]-glucose incorporation into human diploid fibroblasts (38).

Both the IGF-I and IGF-II peptides stimulate a growth response in the H-35 cells. However, although the potency of these factors was not accurately determined, they are approximately 1000 times less potent than insulin. IGF-II gave an estimated EC_{50} of 20 nM and IGF-I did not demonstrate a maximal response at the concentrations tested. For comparison, MSA stimulates a growth response in human fibroblasts with an EC_{50} of about 40 nM (1).

Based on the lack of significant cross-reactivity of the receptors with respect to binding and the considerably greater potency of insulin in stimulating the growth response it is concluded that insulin is acting through the insulin receptor to stimulate the growth response.

Although the IGF peptides affect both enzyme induction and growth it is not possible to attribute their action to a specific receptor. Clearly, it must not be through the IGF-I receptor because these cells reportedly lack this receptor (32, 33). However, although they possess the IGF-II receptors, the relative potencies of the peptides leaves open the possibility that they function through the insulin receptor.

Acknowledgement

This work was supported by NIH grant AM 30512, to John W. Koontz, and institutional support grants ASC-IN89L, and NIH Biomedical support grant 7082.

References

1. King, G. L., Kahn, C. R., Rechler, M. M. and Nissely, S. P., 1980. J. Clin. Invest. 66: 130–140.
2. Ewton, D. Z. and Florini, J. R., 1980. Endocrinology 106: 577–583.
3. Ewton, D. Z. and Florini, J. R., 1981. Dev. Biol. 86: 31–39.
4. Marquandt, H., Todaro, G. J., Henderson, L. E. and Oroszlan, S., 1981. J. Biol. Chem. 256: 6859–6865.
5. Zapf, J., Schoenle, E., Waldvogel, M., Sand, I. and Froesch, E. R., 1981. FEBS Lett. 37: 605–609.
6. Poggi, C., Le Marchand-Brustel, Y., Zapf, J., Froesch, E. R. and Freychet, P., 1979. Endocrinology 105: 723–730.
7. Koontz, J. W. and Iwahashi, M., 1981. Science 211: 947–949.
8. Lawrence, J. D., Larner, J., Kahn, D. R. and Roth, J., 1978. Mol. Cell. Biochem. 22: 153–157.
9. Bollum, F. J., 1968. Meth. Enzymol. 12: 169–173.
10. Wicks, W. D., 1969. J. Biol. Chem. 244: 3941–3950.
11. Diamondstone, T. I., 1966. Anal. Biochem. 16: 395–401.
12. Lowry, D. H., Rosebrough, N. J., Farr, A. L. and Randall, R. J., 1951. J. Biol. Chem. 193: 265–275.
13. Burton, K., 1956. Biochem. J. 62: 315–326.
14. Freychet, P., Roth, J. and Neville, D. M., Jr., 1971. Biochem. Biophys. Res. Commun. 43: 400–407.
15. Terris, S. and Steiner, D. F., 1975. J. Biol. Chem. 250: 8389–8398.
16. Hjelm, H., Hjelm, K. and Sjoquist, J., 1972. FEBS Lett. 28: 73–76.
17. Iwamoto, Y., Wong, K. Y. and Goldfine, I. D., 1981. Endocrinology 108: 44–51.
18. Wicks, W. D., Barnett, C. A. and McKibbon, J. B., 1974. Fed. Proc. 33: 1105–1111.
19. Baldwin, D., Jr., Terris, S. and Steiner, D. F., 1980. J. Biol. Chem. 255: 4028–4034.
20. Jacobs, S., Chang, K.-J. and Cuatrecasas, P., 1978. Science 200: 1283–1284.
21. Blundell, T. L., Bedakar, S., Rinderknecht, E. and Humbel, R. E., 1978. Proc. Natl. Acad. Sci. U.S.A. 75: 180–184.
22. Rinderknecht, E. and Humbel, R. E., 1978. J. Biol. Chem. 253: 2769–2776.
23. Blundell, T. L. and Humbel, R. E., 1980. Nature 297: 781–787.
24. Kahn, C. R., Baird, K., Flier, J. S. and Jarrett, D. B., 1977. J. Clin. Invest. 60: 1094–1106.
25. Van Obberghen, E., Spooner, P. M., Kahn, C. R., Chernick, S. S., Garrison, M. M., Karlesson, F. A. and Grunfeld, C., 1979. Nature 280: 500–502.
26. Pitot, H. C., Peraino, C., Morse, P. A. and Potter, J. R., 1964. Natl. Cancer Inst. Monograph 13: 229–242.
27. Steiner, D. F., Chan, C. V., Terris, S. and Hoffman, C., 1978. In: Ciba Foundation Symposium 55, Hepatotrophic Factors, pp. 217–228. Elsevier, Excerpta Medica, North-Holland, Amsterdam.
28. Leffert, H. L. and Koch, K. S., 1978. In: Ciba Foundation Symposium 55, Hepatotrophic Factors, pp. 61–82. Elsevier, Excerpta Medica, North-Holland, Amsterdam.
29. Bucher, N. C. R., Patel, U. and Cohen, S., 1978. In: Ciba Foundation Symposium 55, Hepatotrophic Factors, pp. 95–110. Elsevier, Excerpta Medica, North-Holland, Amsterdam.

30. Steiner, D. F., 1966. Vitamins and Hormones 24: 1-61.
31. Czech, M. P., 1982. Cell 31: 8-10.
32. Massaque, J., Blindermon, L. A. and Czech, M. P., 1982. Fed. Proc. 42: 496.
33. Massague, J. and Czech, M. P., 1982. J. Biol. Chem. 257: 5038-5045.
34. Rosenfeld, R. G., Baldwin, D., Jr., Dollar, L. A., Hintz, R. L., Olefsky, J. M. and Rubenstein, A., 1981. Diabetes 30: 979-982.
35. Megyesi, K., Kahn, C. R., Roth, J., Neville, D. M., Jr., Nissely, S. P., Humbel, R. E. and Froesch, E. R., 1975. J. Biol. Chem. 250: 8990-8496.
36. Rechler, M. M., Zapf, J., Nissely, S. P., Froesch, E. R., Moses, A. C., Podskalny, J. M., Schilling, E. E. and Humbel, R. E., 1980. Endocrinology 107: 1451-1459.
37. Heaton, J. H., Schilling, E. E., Gelehrter, T. D., Rechler, M. M., Spencer, C. J. and Nissely, S. P., 1980. Biochim. Biophys. Acta 632: 192-203.
38. Knight, A. B., Rechler, M. M., Romanus, J. A., Van Obberghen-Schilling, E. E. and Nissely, S. P., 1981. Proc. Natl. Acad. Sci. U.S.A. 78: 2554-2558.

Received 12 December 1982.

Note added in proof

Shortly after submission of this manuscript a report by Massague et al. appeared (J. Biol. Chem. 257: 13958-13963, 1982) demonstrating the interaction of IGF-I and IGF-II with the H-35 cells. Their conclusions regarding the growth response were substantially the same as those described in this article.

Molecular and Cellular Biochemistry 58, 147–156 (1984).

Chromatographic characteristics and subcellular localization of synthase phosphatase, phosphorylase phosphatase and histone phosphatase in human polymorphonuclear leukocytes

Nabeel Nahas, Henning Juhl and Viggo Esmann*
Department of Medicine, Marselisborg Hospital, DK-8000 Aarhus C, Denmark

Summary

Synthase phosphatase, phosphorylase phosphatase and histone phosphatase activity in a leukocyte homogenate were found to have different sedimentation charcteristics: both synthase phosphatase and phosphorylase phosphatase activity are associated with the microsomal fraction, while the majority of histone phosphatase activity (75–85%) was found in the cytosol. Synthase phosphatase, phosphorylase phosphatase and histone phosphatase activities accompanying the microsomal fraction are readily solubilized by 0.3% Triton X-100.

When the solubilized microsomal enzymes were chromatographed on Sephadex G-200, the majority of synthase phosphatase, phosphorylase phosphatase and histone phosphatase activity migrated in single peaks corresponding to apparent molecular weights of 380 000, 250 000 and 68 000, respectively. A minor peak of 30 000, which had phosphatase activity against all three substrates was also obtained.

Ethanol treatment resulted in solubilization and dissociation of the three phosphatase activities. It was found that although ethanol treatment resulted in a 4-fold increase of phosphorylase phosphatase activity, histone phosphatase activity was decreased (by 60%), while synthase phosphatase activity remained stable. Similar results were obtained when ethanol treatment was performed on the 17 000 × **g** supernatant.

Chromatography of the ethanol-treated microsomes (or homogenate) on Sephadex G-200 showed that the phosphatase activity towards synthase D, phosphorylase *a* and phosphohistone coincided a M_r 30 000 species. Heat treatment of the M_r 30 000 peak resulted in dissociation of synthase phosphatase and phosphorylase phosphatase activity.

Synthase phosphatase was inhibited by phosphorylase *a* in a kinetically non-competitive manner while histone phosphatase activity was not inhibited by synthase D (8.5 unit/ml) or by phosphorylase *a* (12 unit/ml).

Introduction

The modification of enzyme function by phosphorylation and dephosphorylation is recognized to play an important regulatory mechanism in the metabolic control of glycogen metabolism. While considerable research efforts have been concentrated on the enzymology and the elucidation of mechanism of protein kinases (1–3), much less is known about the properties of the phosphatases. Recently, extensive studies on the properties of a 'general' phosphoprotein phosphatase prepared from various tissues have been reported (4). The enzyme (M_r 35 000) was highly purified and was shown to dephosphorylate a wide spectrum of phosphorylated proteins, including glycogen synthase and glycogen phosphorylase (5, 6). The low molecular weight protein was purified using harsh procedures which resulted in dissociation of the active subunit from the holoenzyme. The physiological role of the low

* To whom correspondence should be addressed.

molecular weight enzyme was put very much in doubt since such an enzyme cannot be detected in gently prepared homogenates from muscle (7) or liver (8). Multiple specificity of phosphatases has also been reported using purified or semipurified preparations from various tissues (9–11). However, the lack of specificity of phosphatases is not a general finding. Goris et al. (12) reported on a partially purified phosphorylase phosphatase from adrenal cortex and liver that did not activate glycogen synthase. Gratecos et al. (13) reported that their purified phosphorylase phosphatase from muscle dephosphorylates phosphopeptides derived from phosphorylase and the inhibitory component of troponin which were the only substrates for the enzyme. Graves et al. (14) found a phosphatase which acted on phosphorylase and phosphopeptides derived from tryptic digestion of phosphorylase *a*, but was incapable of dephosphorylating other phosphoproteins. Tan and Nuttall (15) reported that different proteins are involved in the dephosphorylation of glycogen synthase, glycogen phosphorylase and phosphohistone in rat liver.

Regarding the variability of the studies and the conflicting results of various reports, it should be noted that few laboratories have examined the subcellular localization of the phosphatases. Hizukuri and Larner (16) provided an important early finding by demonstrating a 'converting' factor of glycogen synthase activity (synthase phosphatase) that sedimented after high speed centrifugation. It was suggested that this fraction was possible of microsomal origin. Margolis et al. (17) have reported that significant quantities of synthase phosphatase and phosphorylase phosphatase are associated with the microsomal fraction. Similar results were also obtained by Tan and Nuttall (15) and Killilea et al. (8).

In this report we describe the subcellular localization, solubilization, chromatographic behaviours and fractionation characteristics of synthase phosphatase, phosphorylase phosphatase and histone phosphatase in human polymorphonuclear leukocytes. These results together with the aging studies of the enzymes as well as the kinetic data illustrate that different proteins are involved in the dephosphorylation of glycogen synthase, glycogen phosphorylase and phosphohistone in human leukocytes. Our results are also indicative of the heterogeneity of the low molecular weight species generated by ethanol treatment.

Materials and methods

Polymorphonuclear leukocytes were obtained from freshly drawn heparinized human peripheral blood as previously described (18). Leukocytes were suspended (1 g/5 ml) in 0.05 M Tris-HCl, pH 7.0, 0.1 mM dithiothreitol (DTT), 0.2 mg/ml soybean trypsin inhibitor, 1 mM benzamidine, 0.1 mM phenylmethyl sulfonyl fluoride (PMSF) and 1 mM EDTA (homogenization buffer). Leukocytes were disintegrated by sonication for 15 sec using a MSE sonifier (20 kHz, amplitude 12) and centrifuged at 17 000 × **g** for 15 min. The supernatant is referred to as the homogenate. The homogenate was centrifuged at 110 000 × **g** for 90 min in a MSE superspeed 65 ultracentrifuge. The precipitate contained the glycogen pellet and the microsomal fraction which was layered above the glycogen pellet. Microsomes were detached from the glycogen pellet by gentle agitation and resuspended in the original volume of buffer using a hand homogenizer or gentle sonication (20 kHz, amplitude 8). The microsmal fraction contained about 90% of the alkaline phosphatase activity present in the leukocyte homogenate. The remaining glycogen pellet was resuspended in buffer by homogenization or sonication. Addition of 0.2% Amberlite MB3-treated glycogen to the homogenate prior to the high speed centrifugation facilitates the removal of the microsomal fraction from the glycogen pellet.

Preparation of substrates for the phosphatases

Leukocyte glycogen synthase I (12 U/mg protein) was purified using concanavalin-A Sepharose and 5′ AMP-Sepharose as described by Juhl (19). Synthase kinase and the catalytic subunit of cAMP-dependent protein kinase were purified from leukocytes according to Juhl (20) and Juhl and Esmann (21), respectively. The phosphorylation of glycogen synthase I was performed at 30 °C in a reaction mixture containing synthase kinase or the catalytic subunit of cAMP-dependent protein kinase. The termination of the reaction and the calculations of ^{32}P incorporated into glycogen synthase were performed as described previously (19).

Rabbit skeletal muscle ^{32}P-labeled phosphorylase was prepared from phosphorylase *b* using [γ-^{32}P]ATP-Mg^{++} and phosphorylase kinase as described by Antoniw et al. (22). The incorporation of ^{32}P into phosphorylase was measured to be 0.97 mole P_i/ mole subunit of the enzyme. ^{32}P-histone IIA was prepared essentially as described by Meisler and Langan (23). The 20% trichloroacetic acid precipitate was washed twice with 50 mM Tris-HCl buffer, pH 7.0, then dissolved in 1.0 ml of buffer and passed through a small column of Sephadex G-25. The ^{32}P-histone obtained after the filtration was dialyzed overnight against 10 mM Tris-HCl, pH 7.0. The dialyzed ^{32}P-histone (50 nmole P_i/ mg) was stored in small portions at -20 °C. Leukocyte glycogen phosphorylase was purified using a 5′ AMP-Sepharose column as described by Sørensen and Wang (24).

Enzyme assays

Muscle and leukocyte phosphorylase *a* activity was determined using [U-^{14}C]glucose-1-phosphate as described by Wang and Esmann (25). For the determination of muscle and leukocyte phosphorylase *a* activity 60 mM and 10 mM glucose-1-phosphate (specific activity 10^6 cpm/ μmole) were used, respectively. Glycogen synthase activity was determined by the incorporation of [U-^{14}C]glucose from UDP-[U-^{14}C]glucose into glycogen (26). The reaction was terminated by spotting 75 μl of the reaction mixture on 2 × 2 cm Whatman 31 ET filter papers, washed three times with 66% ethanol as described by Thomas et al. (27). Total glycogen synthase activity was measured in the presence of 10 mM glucose-6-phosphate. When ^{32}P-labeled synthase or phosphorylase were assayed for activity, the filter papers were in addition washed (3 times) with 5% KOH in 66% ethanol as described by Hutson et al. (28). One unit (U) of glycogen synthase or glycogen phosphorylase is defined as the activity which incorporates 1 μmole of glycosyl units from UDP-glucose or glucose-1-phosphate, respectively into glycogen per minute. Alkaline phosphatase activity was measured according to Kane and Peters (29).

Conditions for phosphatase assays

Glycogen synthase phosphatase incubation mixture contained 50 mM Tris-HCl, pH 7.0, 1 mM dithiothreitol, proteinase inhibitors as in the homogenization buffer, 20–30 mU/ ml (3.0–4.5 nM) of ^{32}P-labeled synthase (subunit MW 85 000) and phosphatase enzyme in a final volume of 200 μl. The sample was added in a volume of 10 μl. The phosphorylase phosphatase incubation mixture was similar to the synthase phosphatase assay system except that 2 U/ ml (63 nM) of ^{32}P-labeled muscle phosphorylase *a* (subunit MW 100 000) was used instead of glycogen synthase. At different times of incubation at 30 °C, 25 μl was removed and mixed with 40 μl 12% trichloro-acetic acid in Eppendorf microtubes. 10 μl of 5 mg/ ml human serum albumin was added as carrier to facilitate precipitation of the labeled protein. Tubes were placed on ice for 20 min and denatured protein was removed by centrifugation for 5 min in an Eppendorf microcentrifuge. 50 μl of the supernatant which contained the released $^{32}P_i$ was counted for radioactivity with Instagel in Packard Liquid Scintillation spectrometer. One unit (U) of synthase phosphatase and phosphorylase phosphatase is defined as the amount of enzyme which hydrolyzes 1 pmol P_i/ min for glycogen synthase and glycogen phosphorylase, respectively. Synthase phosphatase and phosphorylase phosphatase assays were linear until 50–60% of the ^{32}P was released. Paper chromatography of the trichloro-acetic acid supernatant according to Li and Hsiao (30) showed that essentially all the radioactivity released from glycogen synthase and glycogen phosphorylase was in the P_i form. These results eliminate the possibility that proteolytic degradation of the substrates takes place under our assay conditions. Control experiments of Sephadex G-25 gel filtered enzymes excluded influence of low molecular weight metabolites on the enzyme. However, the influence of protein phosphatase inhibitors/ activators cannot be ruled out.

Histone phosphatase activity was determined in a reaction mixture containing 50 mM Tris-HCl, pH 7.0, 1 mM dithiothreitol with proteinase inhibitors as in the homogenization buffer, 0.5 mg/ ml ^{32}P-histone and phosphatase enzyme in a total volume of 100 μl. The sample was added in a volume of 5 μl. After incubation for 10 min at 30 °C, the reaction was terminated by adding 200 μl 50% trichloro-acetic acid followed by 100 μl of 2 mg/ ml human serum albumin. The tubes were placed on ice for 20 min to facilitate precipitation. After centrifugation of the denatured protein, the released ^{32}P in the supernatant was counted for radioactivity. Blank values, i.e. the counts obtained in the absence of phosphatase(s) were substrated in all the assays. One unit of histone phosphatase activity is defined as the amount of enzyme which hydrolyzes 1 nmole P_i/min. Under our assay conditions histone phosphatase activity was linear with time and enzyme concentration until at least 20% of the total radioactivity was released.

Materials

[γ-^{32}P]ATP and other radioactive materials were obtained from the Radiochemical Centre, Amersham, England. Human serum was from OHRA (Beringswerke AG, Germany). Sephadex G-25 fine and G-200 fine were from Pharmacia, Uppsala, Sweden. Instagel was from Packard Instrument Co. (Copenhagen, Denmark). All other chemicals were purchased from Sigma Chemical Co. (St. Louis, U.S.A.).

Results

Subcellular distribution of synthase phosphatase, phosphorylase phosphatase and histone phosphatase activity in human polymorphonuclear leukocytes

Differential centrifugation of the leukocyte homogenate (Table 1), shows that the majority of synthase phosphatase and phosphorylase phosphatase activity are sedimented with the microsomal fraction as judged by the distribution of alkaline phosphatase (31). The glycogen pellet and the 110 000 × **g** supernatant contained small amounts of the enzymes. The majority of histone phosphatase activity was found in the cytosol. Before assaying, samples were preincubated at 30 °C for 20 min to convert endogenous substrates to products and to minimize interference of endogenous metabolites in the assay as suggested by Tan and Nuttall (15). After treatment of the homogenate with salivary α-amylase until glycogen was completely degraded, as indicated by the absence of iodine staining by the sensitive method of Krisman (32), 85 to 90% of the synthase phosphatase and phosphorylase activity remained associated with the microsomal fraction. The association of synthase phosphatase with the microsomal fraction was found irrespective of whether the ^{32}P-synthase was prepared using synthase kinase or the catalytic subunit of cAMP-dependent protein kinase. However, glycogen synthase phosphorylated by synthase kinase is a significantly (about 30%) better substrate for synthase phosphatase than synthase prepared by using the catalytic subunit of protein kinase. In order to examine whether the particulate associated phosphatases could be related to co-sedimented lysosomal contamination, lysosomes were isolated using the method of Tsung et al. (33) and found to contain negligible synthase phosphatase and phosphorylase phosphatase activity.

Table 1. Distribution of synthase phosphatase, phosphorylase phosphatase, histone phosphatase and alkaline phosphatase activities in leukocyte subfractions.

Fraction	Synthase phosphatase U/ml		Phosphorylase phosphatase U/ml	Histone phosphatase U/ml	Alkaline phosphatase mμ/ml
	CS	SK			
Homogenate	2.06	2.61	11.39	1.35	10.58
110 000 × **g** supernatant	0.65	0.87	2.91	1.05	1.12
Microsomes	1.68	2.23	9.2	0.36	8.76
Glycogen	0.16	0.43	1.37	0.08	0.64

Leukocyte homogenate (3.5 mg protein/ml) obtained as described in Methods was centrifuged at 110 000 × **g** for 90 min. The microsomal fraction which was sedimented above the glycogen pellet was removed and dispersed in the original volume of buffer (0.9 mg protein/ml) using a hand homogenizer. The remaining glycogen was suspended in the same volume of buffer by gentle sonication (0.3 mg protein/ml). The fractions obtained were checked for synthase phosphatase, phosphorylase phosphatase and histone phosphatase activities. ^{32}P-labeled glycogen synthase (1.4 P_i/subunit) was prepared either by the catalytic subunit of protein kinase (CS) or with synthase kinase (SK). The incubation mixture contained 28 mU ^{32}P-synthase/ml. Phosphorylase phosphatase and histone phosphatase activities were measured as described in 'Methods' after incubation for 5 and 10 min at 30 °C, respectively. Alkaline phosphatase was measured according to Kane and Peters (29).

The glycogen pellet which contains more than 95% of glycogen synthase and glycogen phosphorylase activity was found to contain less than 10% of synthase phosphatase and phosphorylase phosphatase activity. This activity might be attributed to slight contamination by microsomes. A similar distribution pattern of synthase phosphatase has been obtained using rabbit skeletal muscle glycogen synthase D, purified according to Nimmo et al. (34), as substrate.

We have obtained similar results with rat liver and skeletal muscle homogenates (not shown). Mature human erythrocytes, which lack endoplasmic reticulum, were found to contain a considerable amount of histone phosphatase activity (7.5 U/g hemoglobin) while exhibiting negligible activity towards glycogen synthase and glycogen phosphorylase. Our results confirm the lack of synthase phosphatase in human erythrocytes as reported by Moses et al. (35). It is worthwhile mentioning that vigorous sonication of the leukocyte suspension resulted in a higher proportion of synthase phosphatase and phosphorylase phosphatase activity in the soluble fraction. However, even after extensive sonication of the leukocytes the specific activity of the microsomal synthase phosphatase and phosphorylase, based on protein content, was at least 6 times higher than in the 110 000 × **g** supernatant.

Effect of ethanol and trypsin treatment on the phosphatase activity

Liver, heart and skeletal muscle phosphorylase phosphatase activity was increased several-fold by ammonium sulphate precipitation followed by ethanol treatment which resulted in dissociation of the active subunit from the native enzyme (4). The effect of ethanol treatment on the activity of phosphatases in a leukocyte homogenate and in the microsomal fraction was examined (Table 2) using the conditions of Brandt et al. (4). It was found that although a four-fold increase in the phosphorylase phosphatase activity was obtained as a result of ethanol treatment of the microsomes and the homogenate, histone phosphatase activity was decreased (by 60%) and synthase phosphatase activity was unchanged. The ethanol treatment of the microsomes also resulted in solubilization of the microsomal associated phosphatases. The same results were obtained if the ammonium sulphate step was omitted. The dissociation of phosphorylase phosphatase and histone phosphatase activities by ethanol treatment has also been described in rat liver (15). Trypsin treatment, which was reported to activate phosphorylase phosphatase from various tissues (36), was less effective (1.6-fold increase) than ethanol in activating leukocyte phosphorylase phosphatase. However, much of the synthase phosphatase activity (60–70%) of the leukocyte homogenate (prepared without proteinase inhibitors) was lost by trypsinization.

Solubilization of the microsomal phosphatases

In order to clarify whether the phosphatases associated with the microsomal fraction are bound to

Table 2. Effect of ethanol and trypsin treatment on synthase phosphatase, phosphorylase and histone phosphatase activities.

Fraction	Treatment	Synthase phosphatase U/ml	Phosphorylase phosphatase U/ml	Histone phosphatase U/ml
Homogenate	None	1.92	12.17	1.43
	Ethanol	1.85	45.63	0.66
	Trypsin	0.62	18.93	–
Microsomes	None	1.34	8.63	0.36
	Ethanol	1.32	30.86	0.21

Leukocyte homogenate and microsomes were subjected to ammonium sulphate precipitation followed by ethanol treatment as described by Brandt et al. (4). After centrifugation of the denatured proteins, samples were desalted using small Sephadex G-25 columns, dialyzed for 5 hr and checked for activity.

For trypsin treatment, homogenates were prepared without proteinase inhibitors. 100 μg/ml of trypsin (diphenyl carbamyl chloride treated) was added. After incubation at 30 °C/20 min, 0.5 mg/ml soybean trypsin inhibitor was added and the homogenate checked for synthase phosphatase and phosphorylase phosphatase activity.

^{32}P-glycogen synthase (1.4 P_i/subunit) used in this experiment was phosphorylated by the catalytic subunit of cAMP-dependent protein kinase.

Table 3. The effect of storage and heat treatment on synthase phosphatase and phosphorylase phosphatase activities.

Treatment	Synthase phosphatase U/ml	% inhibition	Phosphorylase phosphatase U/ml	% inhibition
Before storage	1.31	–	9.60	–
24 hr/4 °C	0.59	55	9.22	4
48 hr/4 °C	0.21	74	8.16	15
54 °C/5 min	0.63	42	7.81	19

Synthase phosphatase and phosphorylase phosphatase activities were measured in the microsomal Triton extract following storage at 4 °C and heat treatment. Synthase phosphatase activity was measured using ^{32}P-glycogen synthase (1.1 P_i/subunit) phosphorylated by the catalytic subunit of cAMP-dependent protein kinase.

the membranes, an experiment was performed in which microsomes were incubated 5 min at 30 °C with different detergents such as Triton X-100 and sodium deoxycholate. The particulate material obtained after high speed centrifugation contained only 5–10% of the original microsomal phosphatase activity while 83–95% of the phosphatases were recovered in the supernatant (not shown). These results suggest that the phosphatases associated with the microsomal fraction are bound to the microsomal membranes. In the present work Triton X-100 was routinely used for solubilization of the phosphatases.

Effect of storage and heat treatment on synthase phosphatase and phosphorylase phosphatase activity

Although ethanol treatment of the homogenate and the microsomal fraction was found to have different effects on the activity of the phosphatases, we have tried to obtain more information about the susceptibility of synthase phosphatase and phosphorylase phosphatase to different inactivation treatments. Table 3 shows that when the solubilized microsomal phosphatases were stored at 4 °C for 24 hr, a considerable loss of synthase phosphatase activity had occurred while phosphorylase phosphatase activity remained unaffected. Further storage of the enzyme (48 hr) resulted in continuing decline in synthase phosphatase activity (70% loss of activity) with only modest inactivation of phosphorylase phosphatase. Heat treatment of the Triton X-100 extracted enzymes at 54 °C for 5 min shows different heat susceptibilities of synthase phosphatase and phosphorylase phosphatase activities.

The effect of substrate competition on the phosphatases

Synthase phosphatase activity measured in the Triton-solubilized microsomal extract was inhibited by purified leukocyte phosphorylase *a* in a kinetically non-competitive manner (Fig. 1). Similar results were obtained when intact microsomes were used as the source of phosphatases. On the other hand phosphorylase *b*, up to 2.5 U/ml had no effect on the dephosphorylation of ^{32}P-labeled glycogen synthase. Histone phosphatase activity, measured

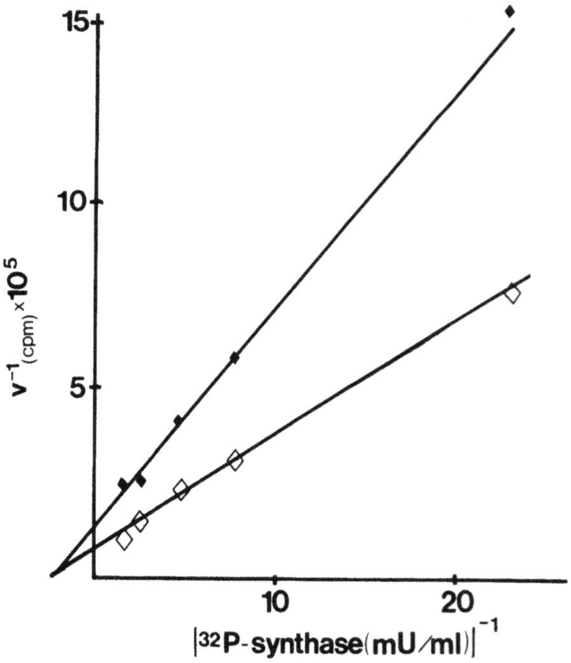

Fig. 1. The effect of phosphorylase *a* on synthase phosphatase activity. Lineweaver-Burk plot of synthase phosphatase activity with (♦) and without (◇) 0.1 U/ml leukocyte phosphorylase *a*.

in the 110 000 × **g** supernatant, was not affected by the addition of either leukocyte synthase D (8.5 U/ml) or phosphorylase *a* (12 U/ml). However, glycogen synthase D also inhibited the dephosphorylation of ^{32}P-labeled phosphorylase in a non-competitive manner (not shown).

Column chromatography of the solubilized microsomal phosphatases

The microsomal Triton extract which contained most of the leukocyte synthase phosphatase and phosphorylase phosphatase activity was subjected to Sephadex G-200 gel chromatography (Fig. 2A). Most of the synthase phosphatase, phosphorylase phosphatase and histone phosphatase activity migrated in large molecular weight forms corresponding to M_r 380 000, 250 000 and 68 000, respectively. Estimation of the molecular weight of protein has been performed using adequate marker enzymes according to Andrews (37). Although a considerable overlap between synthase phosphatase and phosphorylase phosphatase elution profiles was found, the activity of the peaks did not coincide. Chromatography of the solubilized enzymes in the presence of 0.1–0.3% Triton X-100 in the running buffer had no effect on the elution profile of the enzymes. Concentration of the high molecular weight forms of phosphorylase phosphatase and synthase phosphatase by vacuum dialysis (fractions 3–19), resulted in 60% loss of synthase phosphatase but 20–30% of phosphorylase phosphatase activity. Rechromatography of the concentrated high molecular weight synthase phosphatase and phosphorylase phosphatase resulted in the same elution pattern as shown in Fig. 2A. These results rule out

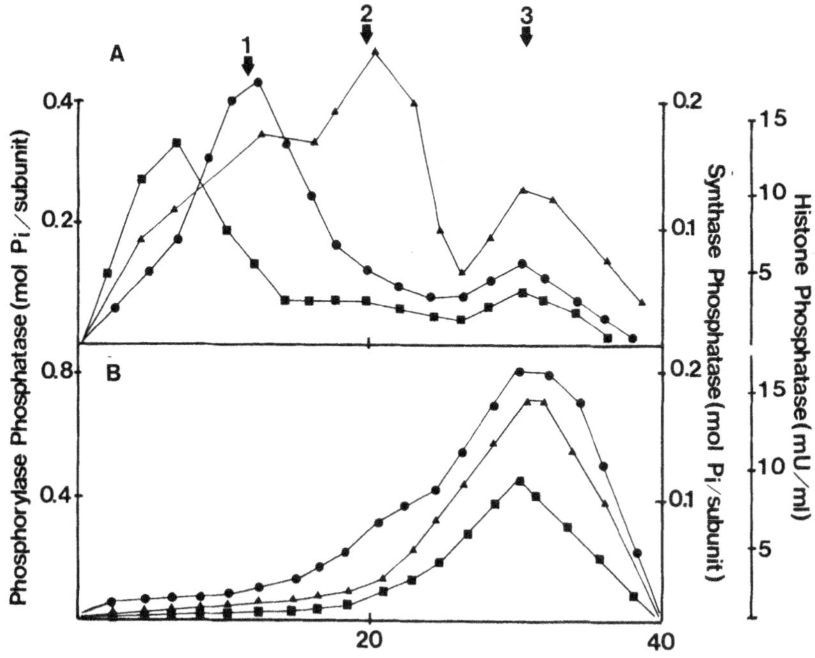

Fig. 2. Gel chromatography of the microsomal phosphatases and the ethanol treated microsomes. The microsomal fraction was incubated 5 min at 30 °C with 0.3% Triton X-100 as described in Table 3. The solubilized fraction obtained after centrifugation of the particulate protein (110 000× **g** for 90 min) was treated with Biobeads SM-X to adsorb the detergent then subjected to Sephadex G-200 column (0.9 × 110 cm) chromatography (A). Fractions of 15 drops were collected and checked for synthase phosphatase (■), phosphorylase phosphatase (●) and histone phosphatase (▲) activity (incubation for 30 min at 30 °C). 2 ml of the microsomal fraction (1.4 mg protein/ml) were treated with ethanol following ammonium sulphate precipitation (Brandt et al. (4)). After removal of the denatured protein (15 000 × **g** for 15 min) 1.0 ml of the supernatant was subjected to gel chromatography (B). The fractions collected were examined for the phosphatase activity. Chromatography was carried out at 4 °C at flow rate of 4.5 ml/hr (V_0 = 33.5 ml). The arrows 1, 2 and 3 denote the elution position of liver catalase (M_r 240 000), bovine serum albumin (M_r 68 000) and carbonic anhydrase (M_r 30 000), respectively.

the possibility that the loss of activity by concentration of the phosphatases is due to aggregation of the enzymes. Addition of 15% glycerol before concentrating the fractions protected only phosphorylase phosphatase activity and not synthase phosphatase activity.

A minor peak of M_r 30 000, which constitutes activity of the three phosphatases, was also eluted from the column. When the microsomal fraction was treated with ethanol and subjected to gel chromatography, the M_r 30 000 peak having phosphatase activity against glycogen synthase, glycogen phosphorylase and phosphohistone was the only form eluted from the column (Fig. 2B).

The dissociation of a low molecular weight, catalytically active, protein from the native enzyme upon ethanol treatment has been observed in rat liver phosphorylase phosphatase (4, 6, 7). However, the main point in our results is that synthase phosphatase and histone phosphatase, which are distinct proteins as indicated by gel chromatography, are also dissociated to M_r 30 000 species. The fractions containing the low molecular weight enzyme obtained after gel chromatography were pooled and subjected to heat treatment (Table 4). The interesting results of different heat susceptibilities of the low molecular weight proteins of synthase phosphatase and phosphorylase phosphatase activity was obtained. These results indicate that the low molecular weight protein, obtained after ethanol treatment, maintains tertiary structure with intramolecular hydrogen bonds which are susceptible to heat. This concept might be supported by the fact

Table 4. The effect of heat treatment on the low molecular weight phosphatases.

	Synthase phosphatase		Phosphorylase phosphatase	
	U/ml	% inhibition	U/ml	% inhibition
Control	0.29	-	1.74	-
54 °C/5 min	0.24	17	0.78	55

Fractions with the highest activity obtained after gel chromatography of the ethanol-treated microsomes were pooled and subjected to heat treatment at 54 °C for 5 min. Activity was checked before and after treatment. Synthase phosphatase activity was determined with ^{32}P-synthase phosphorylated by the catalytic subunit of cAMP-dependent protein kinase (1.1 P_i/subunit). Synthase phosphatase and phosphorylase phosphatase activities were measured after incubation at 30 °C for 30 and 20 min, respectively.

that the low molecular weight enzyme is responsive to different allosteric modifiers such as Mn^{++} and Mg^{++}. 3 mM $MnCl_2$ and 3 mM $MgCl_2$ have stimulated synthase phosphatase activity by 36% and 23% respectively, but decreased phosphorylase phosphatase activity by 58% and 40% respectively. These cations are believed to modify the phosphatase activity rather than affecting the substrates (38, 39). It is noticed that while the high molecular weight synthase phosphatase is more susceptible to heat than phosphorylase phosphatase, the opposite is true for the low molecular weight enzymes. It could be assumed that the 'regulatory subunits' of the enzymes have a role in heat stability of the protein.

Discussion

Studies of phosphatase activity in liver and muscle extracts, using a variety of phosphoprotein substrates, have led to two general findings: the existence of multiple enzyme forms (40, 41) and evidence for multiple specificity of the phosphatase (5, 6, 9, 36). Our study of the phosphatases in human polymorphonuclear leukocytes, provides several lines of evidence against the concept of multiple specificity of proteins having synthase phosphatase, phosphorylase phosphatase and histone phosphatase activity. In our work we have routinely used potent proteinase inhibitors to protect the enzymes against proteolytic modifications, which are known to give rise to multiple enzyme forms (8). This is especially important in working with leukocyte homogenates, since leukocytes by their very nature, as phagocytic cells, are expected to contain high levels of proteinases.

The subcellular localization of synthase phosphatase and phosphorylase phosphatase was found to be mainly in the microsomal fraction, while the majority of histone phosphatase activity was found in the soluble fraction. Almost all investigators, working on the fine structure of cells, have shown the close morphological association of the smooth endoplasmic reticulum and the glycogen particles in hepatocytes (17, 42–44). Cardell (43) and Babcock and Cardell (42) provided an important observation of an increase in the amount of free microsomes in hepatocytes during glycogen synthesis. This finding may provide a link between the morphological observations of a close association between microsomes and glycogen particles and the increase of free microsomes during glycogen synthesis. Our study, which indicates the association of synthase phosphatase and phosphorylase phosphatase with the microsomal membranes, may simplify the implication of these findings. It is possible that the microsomes and the glycogen particles form a microenvironment in the cell for synthesis and breakdown of glycogen. The subcellular location of microsomes places the phosphatases in close proximity to their substrates, i.e. glycogen synthase and glycogen phosphorylase. This microenvironment is thus rich in both key enzymes of glycogen metabolism (glycogen synthase and glycogen phosphorylase) and their converting enzymes, i.e. synthase phosphatase and phosphorylase phosphatase.

Treating the microsomes with detergents resulted in solubilization of the microsomal attached phosphatases. Storage of the solubilized phosphatases at 4 °C showed that great loss of synthase phosphatase activity had occurred while phosphorylase phosphatase activity was less susceptible to storage. Column chromatography of the microsomal enzymes on Sephadex G-200 resulted in the separation of the phosphatases, although a considerable overlap between synthase phosphatase and phosphorylase phosphatase activity was obtained. However, the peaks did not coincide. We regard this finding of particular importance since it provides direct evidence that different proteins are involved in the dephosphorylation of glycogen synthase, glycogen phosphorylase and phosphohistone. A minor peak of M_r 30 000, which had phosphatase activity towards the three substrates was also present after gel chromatography.

Our results are supported by those published by the group of Tsuiki. The high molecular weight synthase phosphatase, phosphorylase phosphatase and histone phosphatase activity in rat liver were resolved in three separate fractions (47) and the phosphorylase phosphatase and histone phosphatase enzymes were purified extensively (48, 49).

Ethanol treatment of the microsomes resulted in solubilization of the membrane-bound phosphatases. Subsequent chromatography showed that the activities of all the phosphatases were present in the M_r 30 000 species. However, the low molecular weight enzyme which has activity against the three substrates examined, is likely to have originated from different proteins as judged by gel filtration of the microsomal Triton extract. Heat treatment of the M_r 30 000 species eluted from the column after ethanol treatment indicates the heterogeneity of the low molecular weight enzymes. It seems that the 'active subunit' of synthase phosphatase and phosphorylase phosphatase, although having similar molecular weights, have different identities. A conformational change of a single M_r 30 000 species followed by a change in substrate specificity is a possibility that cannot be ruled out. No further studies have been done to differentiate between the 'active subunits' of the enzymes.

It might be that the difference between the active subunits of synthase phosphatase and phosphorylase phosphatase resides in only a few amino acids composing the polypeptide, which makes it difficult to separate them by conventional methods. Our results thus might explain the data obtained in liver studies by Killilea et al. (6), which showed co-purification of the enzymes only after the ethanol precipitation step. Khandelwal et al. (5) have also emphasized the dissociation of larger molecular weight phosphatases to low molecular weight forms without change in the activity characteristics.

The kinetic data on the effect of phosphorylase *a* on the activation of glycogen synthase shows a non-competitive inhibition of synthase phosphatase by phosphorylase *a*. Similar results have been obtained by Tan and Nuttall (15) with the enzyme from the rat liver homogenate. Other authors have reported competitive or mixed type inhibition kinetics of phosphatase(s) with different substrates (5, 9). Their data were used as evidence for the identity of the phosphatases. It is difficult to explain the non-competitive inhibition obtained in our studies, but it is probably due to protein–protein interaction.

Khatra and Soderling (45) showed that dephosphorylation in the N-terminal end of glycogen synthase was predominantly responsible for the activation of glycogen synthase. The phosphatase dephosphorylated the N-terminal end three times more rapidly than the C-terminal end. Similarly, Juhl et al. (46) have demonstrated a close correlation between activation of glycogen synthase and dephosphorylation of a chymotryptic liberated peptide (c-A) from the N-terminal region of leukocyte glycogen synthase when using leukocyte synthase phosphatase associated with the microsomal fraction.

Acknowledgements

The authors are greatly indebted to Mrs Mary Nahas for her expert technical assistance and to Mrs Lone Manner-Jakobsen for typing the manuscript. This work was supported by the Danish Research Council, Grant No. 512-15484 and by grants from NOVO's Fond, Fonden til laegevidenskabens Fremme and The Institute for Experimental Clinical Research, Aarhus University.

References

1. Soderling, T. R., Jett, F. M., Hutson, N. J. and Khatra, S. B., 1977. J. Biol. Chem. 252: 7517–7524.
2. Nimmo, H. G., Proud, C. G. and Cohen, P., 1976. Eur. J. Biochem. 68: 31–44.
3. Juhl, H. and Esmann, V., 1980. Mol. Cell. Biochem. 30: 131–139.
4. Brandt, H., Killilea, S. D. and Lee, E. Y. C., 1974. Biochem. Biophys. Res. Commun. 66: 598–604.
5. Khandelwal, R. L., Vandenheede, J. R. and Krebs, E. G., 1976. J. Biol. Chem. 251: 4850–4858.
6. Killilea, S. D., Brandt, H., Lee, E. Y. C. and Whelan, W. J., 1976. J. Biol. Chem. 251: 2363–2368.
7. Mellgren, R. L., Aylward, J. H., Killilea, S. D. and Lee, E. Y. C., 1979. J. Biol. Chem. 254: 648–652.
8. Killilea, S. D., Mellgren, R. L., Aylward, J. H., Methieh, M. E. and Lee, E. Y. C., 1979. Arch. Biochem. Biophys. 193: 130–139.
9. Nakai, C. and Thomas, J. A., 1974. J. Biol. Chem. 249: 6459–6467.
10. Kato, K. and Bishop, J. S., 1972. J. Biol. Chem. 247: 7420–7429.
11. England, P. J., Stull, J. T. and Krebs, E. G., 1972. J. Biol. Chem. 247: 5275–5277.
12. Goris, J., Kalala, L. R. and Merlevede, W., 1974. Arch. Int. Physiol. Biochem. 82: 988–990.
13. Gratecos, D., Detwiler, T. and Fisher, E. H., 1973. In: Metabolic Interconversions of Enzymes (Fisher, E. H., Krebs, E. G., Neurath, H. and Stadtman, E. R., eds.), pp. 43–52. Springer-Verlag, New York.
14. Graves, D. H., Fisher, E. H. and Krebs, E. G., 1960. J. Biol. Chem. 235: 805–809.
15. Tan, A. W. H. and Nuttall, F. Q., 1978. Biochim. Biophys. Acta 522: 139–150.
16. Hizukuri, S. and Larner, J., 1964. Biochemistry 3: 1983–1788.
17. Margolis, R. N., Cardell, R. R. and Curnow, R. T., 1979. J. Cell. Biol. 83: 348–356.
18. Esmann, V., 1964. Metabolism 13: 354–360.
19. Juhl, H., 1981. Mol. Cell. Biochem. 35: 93–101.
20. Juhl, H., 1979. Mol. Cell. Biochem. 26: 19–27.
21. Juhl, H. and Esmann, V., 1979. Mol. Cell. Biochem. 26: 3–18.
22. Antoniw, J. F., Nimmo, H. G., Yeaman, S. J. and Cohen, P., 1977. Biochem. J. 162: 423–433.
23. Meisler, M. H. and Langan, T. A., 1969. J. Biol. Chem. 244: 4961–4968.
24. Sørensen, N. B. and Wang, P., 1975. Biochem. Biophys. Res. Commun. 67: 883–887.
25. Wang, P. and Esmann, V., 1972. Anal. Biochem. 47: 495–500.
26. Sølling, H., 1979. Eur. J. Biochem. 94: 231–242.
27. Thomas, J. A., Schlender, K. K. and Larner, J. 1968. Anal. Biochem. 25: 486–499.
28. Hutson, N. J., Khatra, B. S. and Soderling, T. R., 1977. J. Biol. Chem. 253: 2540–2545.
29. Kane, S. P. and Peters, T. J., 1974. Clin. Sci. Mol. Med. 49: 179–182.
30. Li, H. C. and Hsiao, K. J., 1977. Arch. Biochem. Biophys. 179: 147–156.
31. Grustin, G. J. S., Wilson, P. D. and Peters, T. J., 1979. J. Cell Sci. 36: 401–412.
32. Krisman, C. R., 1962. Anal. Biochem. 4: 17–23.
33. Tsung, P. K., Sakamoto, T. and Weissmann, G., 1975. Biochem. J. 145: 437–448.
34. Nimmo, H. G., Proud, C. G. and Cohen, P., 1976. Eur. J. Biochem. 68: 21–30.
35. Moses, S. W., Bashan, N. and Gutman, A., 1972. Eur. J. Biochem. 30: 205–210.
36. Lee, E. Y. C., Brandt, H., Capulong, Z. L. and Killilea, S. D., 1976. Advan. Enzyme Regul. 14: 467–490.
37. Andrews, P., 1965. Biochem. J. 96: 595–606.
38. Martensen, T. M., Brotherton, J. E. and Graves, D. J., 1973. J. Biol. Chem. 248: 8323–8328.
39. Khatra, B. S. and Soderling, T. R., 1978. Biochem. Biophys. Res. Commun. 85: 647–654.
40. Goris, J. and Merlevede, W., 1972. Arch. Int. Physiol. Biochim. 80: 967–969.
41. Kobayashi, M., Kato, K. and Sato, S., 1975. Biochim. Biophys. Acta 377: 343–355.
42. Babcock, M. B. and Cardell, R. R., Jr., 1975. Am. J. Anal. 143: 399–438.
43. Cardell, R. R., Jr., 1977. Int. Rev. Cyt. 48: 221–279.
44. Margolis, R. N., Garfield, S. A. and Cardell, R. R., Jr., 1979. Endocrinology 104: 1722–1723.
45. Khatra, B. S. and Soderling, T. R., 1978. J. Biol. Chem. 253: 5247–5250.
46. Juhl, H., Nahas, N. and Esmann, V., 1982. Biochim. Biophys. Acta 704: 509–514.
47. Kikuchi, K., Tamura, S., Hirafa, A. and Tsuiki, S., 1977. Biochem. Biophys. Res. Commun. 75: 29–37.
48. Tamura, S., Kikuchi, H., Kikuchi, K., Hiraga, A. and Tsuiki, S., 1980. Eur. J. Biochem. 104: 347–355.
49. Tamura, S. and Tsuiki, S., 1980. Eur. J. Biochem. 111: 217–224.

Received 12 December 1982.

Molecular and Cellular Biochemistry 58, 157–163 (1984).

Glycosyl transferases in mouse and human milk fat globule membranes

Armando J. Parodi,[1,2] Edward W. Blank,[1] Jerry Peterson[1] and Roberto Ceriani[1]
[1] *Bruce Lyon Memorial Research Laboratory, Children's Hospital Medical Center, 51st and Grove Streets, Oakland, CA 94609, U.S.A.;* [2] *Department of Microbiology, The Wellcome Research Laboratories, Research Triangle Park, 27709, U.S.A.*

Summary

Membranes isolated from mouse and human milk fat globules were found to contain the enzymes responsible for the synthesis of dolichol monophosphate mannose and dolichol monophosphate glucose as well as those involved in the transference of the glycosyl residues from the two dolichol derivatives to dolichol diphosphate oligosaccharides. The levels of most of the enzymes were comparable to those found in mouse mammary gland microsomes. The presence of enzymes involved in protein glycosylation via dolichol derivatives in the milk fat globule membrane provides evidence in favor of an outward flow of membrane components from the rough endoplasmic reticulum, where these enzymes are active *in vivo*, towards the cell surface.

Introduction

When dolichol, a lipid containing 17–21 isoprene units, is linked to various oligosaccharides, the derivatives obtained are involved as intermediates in the synthesis of glycoproteins of the asparagine linkage type (for a review see Ref. 1). Dolichol diphosphate N-acetylglucosamine (Dol-P-P-GlcNAc) is formed upon reaction of dolichol monophosphate (Dol-P) and uridine diphosphate N-acetylglucosamine (UDP-GlcNAc). Another N-acetylglucosamine and several mannose and glucose residues are then succesively added. The final product is a dolichol diphosphate (Dol-P-P) derivative containing 2 N-acetylglucosamine, 9 mannose and 3 glucose units. The donors of these monosaccharides are in some cases the respective sugar nucleotides and in others dolichol monophosphate mannose (Dol-P-Man) or dolichol monophosphate glucose (Dol-P-Glc) which in turn are formed upon reaction of guanosine diphosphate mannose (GDP-Man) or uridine diphosphate glucose (UDP-Glc) and Dol-P. The complete oligosaccharide is then transferred *en bloc* to asparagine residues in proteins (2). In the case of the synthesis of the 'complex' type glycoproteins, the glucose and 6 mannose residues are trimmed and N-acetylglucosamine, galactose, sialic acid and fucose unis are transferred to the glycoproteins directly from the respective sugar nucleotides (3). When glycoproteins of the 'high mannose' type, that is, those containing only 2 N-acetylglucosamine and between 5 and 9 mannose residues are to be synthesized, only the glucose and fewer mannose residues than in the complex type case are excised (4).

All the enzymes involved in this pathway have been found to be membrane bound. The transference of the oligosaccharide from the Dol-P-P derivative to proteins occurs in the rough endoplasmic reticulum before the synthesis of the polypeptide chain has been completed whereas the addition of the more distal monosaccharides directly from the sugar nucleotides takes place in the Golgi apparatus (1).

We report here that membranes from human and mouse milk fat globules contain several of the glycosyl transferases involved in the synthesis of dolichol derivatives and that the levels of these enzymes are in many cases comparable to those found in mouse mammary gland microsomes. The interest of this study lies in the fact that the membranes used can be considered as almost pure plasma membranes from mammary epithelial cells.

Materials and methods

Materials

UDP-[^{14}C]Glc (223 Ci/mol); GDP-[^{14}C]Man (210 Ci/mol) and UDP-[^{14}C]GlcNAc (235 Ci/mol) were from New England Nuclear. Dol-P was prepared by chemical phosphorylation of liver dolichol as reported by Behrens and Leloir (5). Unlabeled Dol-P-P-oligosaccharides were extracted and partially purified from pig liver as described before for the similar compounds from rat liver (6). Dol-P-[^{14}C]Glc was synthesized with UDP-[^{14}C]Glc and Dol-P by rabbit liver microsomes as reported previously (5).

Membrane preparations

Crude membranes from milk fat globules were prepared as follows: freshly drawn milk was diluted about three-fold with 0.9% NaCl and centrifuged for 20 min at 27 500 **g** in a swinging bucket rotor. The upper creamy layer was withdrawn with a spatula and diluted with a 0.25 M sucrose, 5 mM EDTA (pH 7.4) solution (8 ml of cream to 60 ml of total volume). The suspension was freezed, thawed and sonicated four times for 20 sec each at half maximum intensity in a Biosonik III sonicator. The suspension was then brought to 37 °C, homogenized with a glass-Teflon homogenizer and centrifuged at 95 000 **g** for 60 min. The pellets were resuspended in 10 ml of the 0.25 M sucrose, 5 mM EDTA (pH 7.4) solution and recentrifuged as before. For the enzyme assays, the final pellet was homogenized in about 1 ml of the above mentioned sucrose-EDTA solution at protein concentrations of 10–20 mg/ml. All procedures were performed at 0–4 °C except where otherwise stated.

Washing of the fat globules was omitted because no Dol-P-dependent glycosyltransferases could be detected in milk serum. Moreover, in all organisms studied so far these glycosyltransferases have been found to be membrane bound (1).

The milk fat globule membranes were further purified as described by Powell et al. (7). The same nomenclature employed in that paper (M3′ and N3′) is used here. It should be noted that fractions M3′ and N3′ have been shown to have the morphology and the marker enzymes of pure milk fat globule membranes (7).

Mouse mammary gland microsomes were prepared following as close as possible the procedure employed for the preparation of crude membranes from milk fat globules. Mammary glands were withdrawn from 10 mice 5 days after delivery, freezed and thawed, suspended in 60 ml of 0.25 M sucrose, 5 mM EDTA (pH 7.4), disintegrated in a Waring blender, sonicated as above, brought to 37 °C, homogenized with a glass-Teflon homogenizer, and centrifuged for 15 min at 17 000 **g** at 4 °C. The supernatant was centrifuged at high speed as described above for the crude milk fat globule membranes. The pellet was finally resuspended in the same sucrose-EDTA solution at protein concentrations of 40–60 mg/ml.

Incubation mixtures

The incubation mixtures contained in a total volume of 100 μl:50 mM Tris-maleate buffer (pH 7.5), 10 mM $MgCl_2$, 0.05 μCi of radioactive sugar nucleotides and about 0.6 mg of membrane proteins. Dol-P and pig liver Dol-P-P oligosaccharides (3.7 nmol and 0.75 μmol of total phosphorous, respectively) were added where indicated. Triton X-100 concentrations were 0.6% in mixtures devoid of the pig liver lipids and 1.2% in the ones containing them. The detergent was omitted where indicated. The amounts of all the components were doubled when the synthesis of oligosaccharide derivatives of dolichol was being investigated. Where indicated in the text, Dol-P-[^{14}C]Glc (30 000 cpm) was substituted for UDP-[^{14}C]Glc plus Dol-P. In this case the incubation mixtures contained 5 mM EDTA (pH 7.4) and no $MgCl_2$.

When mouse mammary gland microsomes were employed, 12 mM adenosine-5′-monophosphate was added to the incubation mixture to inhibit possible nucleoside pyrophosphate sugars pyrophosphatases present in the microsomal preparation. It was found later, however, that omission of this substance did not modify the results obtained.

After 30 min at 37 °C the reactions were stopped by the successive addition of 0.6 ml of methanol, 0.9 ml of chloroform, 0.25 ml of water and 1.5 mg of rabbit liver microsomal proteins. The latter were used as carrier. The lower phase obtained upon low speed centrifugation was washed twice 0.35 ml of chloroform–methanol–4 mM $MgCl_2$ (1:16:16). The proteinaceous interphase was washed twice with 1.5 ml of water and extracted three times with 1 ml of chloroform–methanol–water (1:1:0.3). The extracts and the lower phase were pooled where indicated.

Methods

The mild acid hydrolysis was performed as follows: the lipid-soluble reaction products were dried and heated in 1 ml of 0.02 M HCl for 15 min at 100 °C. Two ml of methanol and 3 ml of chloroform were added. The upper phase obtained upon low speed centrifugation contained the water soluble substances produced by the acid treatment. These substances were spotted on Whatman No. 1 paper and developed with solvent A: 1-butanol–ethanol–water (10:1:2) or B: 1-butanol–pyridine–water (4:3:4). The papers were cut into 1 cm pieces and counted in a toluene-based scintillation solution. Protein concentrations were measured as described by Lowry et al. (8) with bovine serum albumin as standard. RNA was measured as described by Campbell and Sargent (9). Total phosphorous in Dol-P and pig liver Dol-P-P-oligosaccharides was measured as described by Chalvardjian and Rudnicki (10) and by Ames (11), respectively. Unlabeled standards in paper chromatographies were revealed by the alkaline silver nitrate procedure (12).

Experimental

Synthesis of monosaccharide derivatives of dolichol by crude membranes

Crude membranes from mouse or human milk fat globules as well as mouse mammary gland microsomes were incubated with UDP-[^{14}C]Glc, GDP-[^{14}C]Man or UDP-[^{14}C]GlcNAc in the absence or presence of synthetic Dol-P and the labeled substances soluble in the lower phase of a chloroform–methanol–water (3:2:1) partition isolated. Dolichol derivatives containing short saccharide moieties are soluble in the lower, organic phase of the above-mentioned mixture of solvents. Unreacted sugar nucleotides or their possible decomposition products are soluble in the upper, aqueous phase (5). As seen in Table 1 there was a small incorporation into material soluble in the organic phase when exogenous lipid derivative was omitted whereas upon its addition a several-fold enhancement of sugar transfer occurred. The lipid-soluble substances synthesized in the presence of Dol-P and either one of the three sugar nucleotides were submitted to mild acid hydrolysis followed by chloroform–methanol–water (3:2:1) partitions. In all cases the label appeared in the upper, aqueous phases. The radioactive substances thus made water soluble were chromatographed on paper with solvent A. The compounds migrated as the sugar moieties of the sugar nucleotides originally added to the reaction mixtures (Fig. 1A-C). As the reactions were stimulated by synthetic Dol-P and the products were labile under mild acid conditions it can be concluded that the enzymes responsible for the synthesis of Dol-P-Glc, Dol-P-Man and Dol-P-P-GlcNAc are present in the crude membranes from mouse and human milk fat globules. All these enzymatic levels were roughly similar to those present in mouse mammary gland microsomes except for that of the enzyme involved in the synthesis of Dol-P-P-GlcNAc in mouse membranes which was much lower.

Purification of milk fat globule membranes

As there have been several reports about trapping of endoplasmic reticula debris by milk fat globules, the membranes were purified according to Powell et al. (9). The material thus obtained was completely free from RNA. The ratio RNA/protein was 0.135 for mouse mammary gland microsomes, indicative of a rough-rich preparation, and less than 0.001 for purified milk fat globule membranes. Nevertheless, as seen in Table 1, the levels of the enzymes involved in the synthesis of Dol-P-Man and Dol-P-Glc in fractions M3′ and N3′ from both human and mouse milk fat globule membranes were comparable to those of the crude preparations. An exception was the enzyme that catalyzes the formation of Dol-P-P-GlcNAc. No activity could be detected in the purified fractions from mice which is not surprising because the crude preparation was almost completely devoid of it. In

Table 1. Synthesis of monosaccharide derivatives of dolichol[a].

Sugar nucleotide	Enzyme source[b]	Not incubated	Incubated minus Dol-P	Incubated plus Dol-P
UDP-[^{14}C]Glc	H	16	263	11113
	H-M3′	56	61	3420
	H-N3′	410	364	5830
	M	15	155	4124
	M-M3′	212	270	10113
	M-N3′	73	195	4828
	MM	25	1641	6467
GDP-[^{14}C]Man	H	30	956	20883
	H-M3′	402	333	33202
	H-N3′	1662	1238	45884
	M	20	226	12852
	M-M3′	470	506	7084
	M-N3′	303	328	1811
	MM	24	1119	8953
UDP-[^{14}C]GlcNAc	H	11	95	4117
	H-M3′	14	5	10
	H-N3′	1363	801	1602
	M	8	12	117
	M-M3′	293	119	141
	M-N3′	141	108	165
	MM	22	280	751

[a] The monosaccharide derivatives of dolichol were synthesized and purified as described under Materials and methods. The numerical values shown correspond to the cpm in the lower phases of the chloroform–methanol–water (3:2:1) partitions. They were counted in a dioxane-based scintillation mixture. As not exactly the same amount of membrane proteins were used in all incubations, the results have been adjusted to 0.6 mg of protein per tube.

[b] H, H-M′, and H-N3′ stand respectively for crude human milk fat globule membranes and for fractions M3′ and N3′ of the same membranes purifed according to Ref. 7, M, M-M3′ and M-N3′ have a similar meaning as above but referred to material obtained from mice. MM stands for mouse mammary gland microsomes.

addition, although the crude human membranes had high enzymatic levels, appreciable activity could be detected in purified fraction N3′ but not in M3’.

Synthesis of mannose-labeled oligosaccharide derivatives of dolichol

Membranes from mouse and human milk fat globules also contain the enzymes involved in the transference of mannosyl residues from Dol-P-Man to Dol-P-P-oligosaccharides. It must be mentioned that the latter compounds that contain rather large saccharide moieties are insoluble in the lower phase of a chloroform–methanol–water (3:2:1) partition of the reaction mixture and remain together with the proteinaceous interphase. They are soluble, however, in chloroform–methanol–water (1:1:0.3). The above-mentioned membranes were incubated with GDP-[^{14}C]Man and Dol-P and the substances soluble in the lower phase of a chloroform–methanol–water (3:2:1) partition and those extracted from the interphase with chloroform–methanol–water (1:1:0.3) were pooled, dried, and submitted to a mild acid hydrolysis. This procedure releases the oligosaccharides from Dol-P-P. The water-soluble compounds thus generated were chromatographed on paper with solvent B. In all cases there was a labeled substance present that migrated with a R_{Glc} of 1.09, that is, as internal mannose standards did (Fig. 2A-C). This labeled monosaccharide came from the hydrolysis of Dol-P-[^{14}C] Man synthesized in the reaction mixtures. No compounds that migrated with R_{Glc} values smaller than 1.09, that is, as oligosaccharides, were formed when the mouse milk fat globule mem-

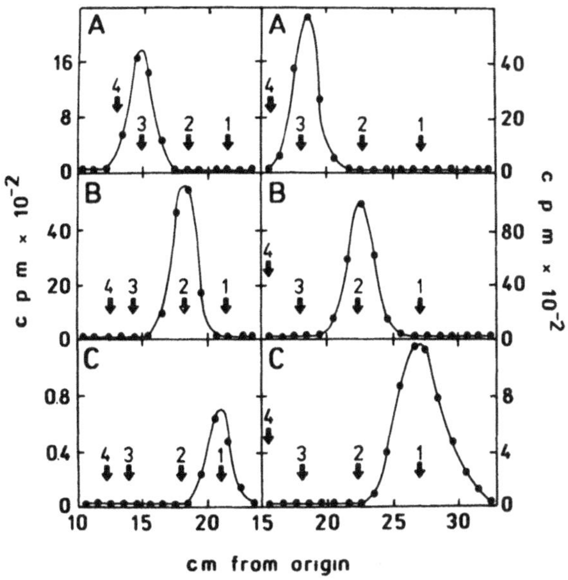

Fig. 1. Paper chromatography of the sugar moieties of the monosaccharide derivatives of dolichol. The monosaccharide derivatives of dolichol were synthesized and purified as described under Materials and methods. The lower phases of the chloroform–methanol–water (3:2:1) partitions were dried and submitted to mild acid hydrolysis followed by paper chromatography with solvent A. The products obtained with mouse and human milk fat globule membranes are represented in the left and right side columns respectively. Letters A, B and C correspond to the compounds synthesized with UDP-[^{14}C]Glc, GDP-[^{14}C]Man and UDP-[^{14}C]GlcNac respectively. Fractions not shown had no radioactivity. Numbers 1, 2, 3 and 4 correspond to the position where N-acetyl-glucosamine, mannose, glucose and galactose standards migrated, respectively.

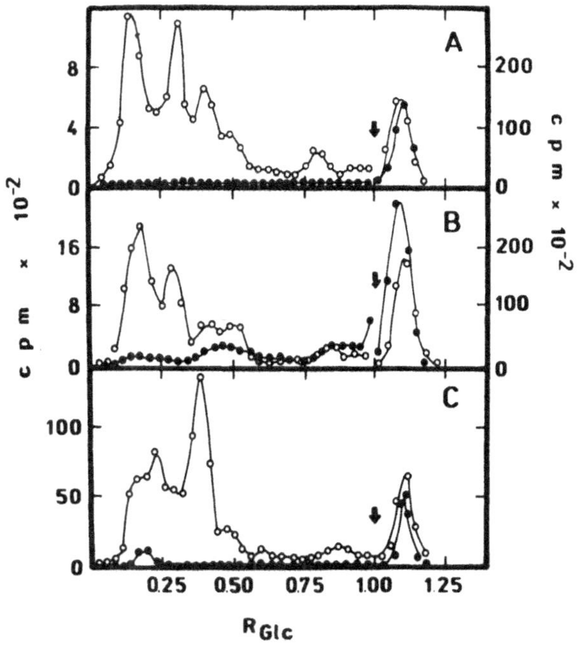

Fig. 2. Paper chromatography of the sugar moieties of the mannose-labeled oligosaccharide derivatives of dolichol. The mannose-labeled oligosaccharide derivative of dolichol were synthesized in the absence (full circles) or presence (empty circles) of pig liver Dol-P-P-oligosaccharides by mouse (A) or human (B) milk fat globule membranes or by mouse mammary gland microsomes (C). The reaction products soluble in the lower phases of chloroform–methanol–water (3:2:1) partitions or extracted with chloroform–methanol–water (1:1:0.3) from the interphases were mixed, dried and subjected to mild acid hydrolysis. The water-soluble products thus generated were run on paper chromatography with solvent B. The arrows indicate the position where unlabeled glucose standards migrated. In figures A and B fractions with R_{Glc} values smaller than 1.00 are represented with the ordinate scales at the left of the figures. The ones at the right are for the rest of the fractions.

branes were used (Fig. 2A).

A very small amount of such compounds was found when the human or mouse mammary gland membranes were analyzed (Fig. 2B-C). These results could be due to a low level of the corresponding mannosyl transferases or of endogenous acceptors. In order to test these possibilities, unlabeled Dol-P-P-oligosaccharides from pig liver were added to the incubation mixtures containing mouse and human milk fat globule membranes and mouse mammary gland microsomes and the reaction products processed as described above. In all cases oligosaccharides were formed and in roughly the same amounts (Fig. 2A-C). The formation of mannose-labeled Dol-P-P-oligosaccharides involves the transfer of mannosyl residues from GDP-Man and from Dol-P-Man to the Dol-P-P derivative. The sugar nucleotide is the substrate for the addition of the first 5 mannose units whereas Dol-P-Man is involved in the transfer of the last 4 residues (13). The oligosaccharides containing from 6 to 9 mannose units in addition to the 2 N-acetylglucosamine residues migrate with R_{Glc} values smaller than 0.35.

As seen in Fig. 2A-C mainly the synthesis involving Dol-P-Man as intermediate were stimulated. As mentioned before the enzyme catalyzing the formation of Dol-P-Man was present in purified milk fat globule membranes.

Synthesis of glucose-labeled oligosaccharide derivatives of dolichol

The enzymes involved in the transference of glucosyl residues from Dol-P-Glc to Dol-P-P-oligosaccharides are also present in the membranes from mouse and human milk fat globules. However, in this case the relative levels of the transferases could not be ascertained because the addition of detergents to the incubation mixtures produced a marked inhibition of the enzymes. Detergents are required when substances as Dol-P-Glc or Dol-P-P-oligosaccharides, which are normally insoluble in pure aqueous systems, are added to the assays (1). The presence of exogenous Dol-P-P derivatives is required in order to evaluate the enzymatic levels regardless of the amounts of endogenous acceptors. Human milk fat globule membranes were incubated with UDP-[14C]Glc plus Dol-P or with Dol-P-[14C]Glc and the labeled substances that remained with the interphase in a chloroform–methanol–water (3:2:1) partition of the reaction mixture but that were soluble in chloroform–methanol–water (1:1:0.3) were subjected to mild acid hydrolysis and paper chromatography. Oligosaccharides that migrated with R_{Glc} values of 0.08–0.12 were formed (Fig. 3A-B). It is known that the acceptors for the glucosyl residues are endogenous Dol-P-P derivatives (1).

Addition of exogenous unlabeled Dol-P-P-oligosaccharides from pig liver to incubation mixtures similar to those described in the legends to Fig. 3A-B did not enhance the synthesis of the glucose-labeled Dol-P-P derivatives. No such compounds were formed when membranes from mouse milk fat globules were incubated with UDP[14C]Glc, Dol-P and detergent. This result was apparently due to the absence of endogenous acceptors because [14C]-glucosyl residues were transferred to Dol-P-P derivatives when the pig liver lipids were added to the incubation mixtures (Fig. 3C).

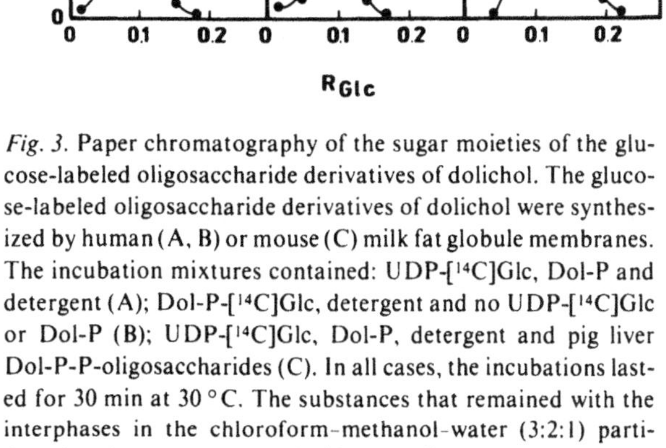

Fig. 3. Paper chromatography of the sugar moieties of the glucose-labeled oligosaccharide derivatives of dolichol. The glucose-labeled oligosaccharide derivatives of dolichol were synthesized by human (A, B) or mouse (C) milk fat globule membranes. The incubation mixtures contained: UDP-[14C]Glc, Dol-P and detergent (A); Dol-P-[14C]Glc, detergent and no UDP-[14C]Glc or Dol-P (B); UDP-[14C]Glc, Dol-P, detergent and pig liver Dol-P-P-oligosaccharides (C). In all cases, the incubations lasted for 30 min at 30 °C. The substances that remained with the interphases in the chloroform–methanol–water (3:2:1) partitions but that were extracted with chloroform–methanol–water (1:1:0.3) were subjected to mild acid hydrolysis followed by paper chromatography with solvent B.

No transfer to endogenous proteins could be detected when [14C]-glucose-labeled Dol-P-P-oligosaccharides were incubated with mouse or human milk fat globule membranes.

Discussion

The milk fat globule membrane has been intensively studied as it constitutes a source of almost pure apical plasma membrane from mammary epithelial cells (14). It should be noted, however, that recent studies using the freeze-etching technique have shown that some of the intrinsic proteins of the cell plasma membrane are lost during the budding of the fat globule (15). Therefore, although all the proteins present in the milk fat globule membranes were initially in the cell plasmalemma, the opposite could not be true.

Our findings seem to conform to the proposal that membrane components flow from the innermost organelles towards the surface membrane (16). The enzymes whose existence in the membranes from milk fat globules is described here participate *in vivo* in the glycosylation of proteins in the rough endoplasmic reticulum. In this organelle the oligosaccharide is transferred from the Dol-P-P derivative to uncompleted polypeptide chains (17). It is not surprising that the levels of the enzymes under study were analogous in many cases to those found in liver microsomes. The lactating mammary epithelial cells are relentlessly exporting fat globules surrounded by portions of the apical plasma membranes. It has been calculated that the bovine cells must replace an amount of plasma membrane equivalent to their apical surface in 8–10 hours (18). This replacement is brought about by the continuous flow of material from the inner organelles to the plasma membranes.

Glycosyl transferases involved in the synthesis of dolichol derivatives have been found also in human erythrocyte and rabbit reticulocyte plasma membranes, thus supporting our present results (19, 20). These enzymes were shown to be intrinsic and not peripheral proteins of the erythrocyte plasma membrane (19). Further, highly purified rat liver plasma membrane preparations seem to be able to synthesize a mannolipid, presumably Dol-P-Man when incubated with GDP-[^{14}C]Man (21). Other enzymes present in the endoplasmic reticulum of mammary epithelial cells like acetylcholinesterase, glucose-6-phosphatase, NADH dehydrogenase and xantine oxidase have been found also as constituents of the milk fat globule membrane (22-24). The latter from bovine and human milk contain a galactosyl transferase responsible for the attachment of one of the distal monosaccharides of the 'complex' type glycoproteins (7, 23). Intracellularly the enzyme is concentrated in the Golgi apparatus and no lipid intermediates are involved in this glycosyltransfer reaction.

No evidence for the *in vivo* glycosylation of proteins in the plasma membrane has been found. For instance, it has been reported that enzymatically desialylated glycoproteins of the plasma membrane are not repaired *in situ* but replaced by new, intact species at the same rate as other similar components of the membrane (25).

Nevertheless, it has been postulated that glycosyl transferases could play a role in cell to cell adhesion as they interact with their respective substrates present in neighboring cells (26, 27). The results presented in this study show that this hypothesis is untenable at least for the glycosyl transferases involved in the synthesis of dolichol derivatives. The mouse milk fat globule membranes appeared to be completely devoid of Dol-P-P-oligosaccharides and thus of one of the components of the hypothetical anchoring system. It is therefore suggested that all glycosyl transferases present in plasma membranes are only derivatives of the glycosylation apparatus of the cell which have been transferred to the surface by the continuous vectorial flow from the inner organelles to the plasma membranes.

Acknowledgements

This work was supported in part by Grant No. CA 19455 from the National Cancer Institute, Biomedical Research Support Grant No. RR 05467 from the National Institutes of Health, DHEW, and Grant No. DPT-99 from the American Cancer Society.

References

1. Parodi, A. J. and Leloir, L. F., 1979. Biochim. Biophys. Acta 559: 1–37.
2. Parodi, A. J., Behrens, N. H., Leloir, L. F. and Carminatti, H., 1972. Proc. Natl. Acad. Sci. U.S.A. 69: 3268–3272.
3. Tabas, I., Schlesinger, S. and Kornfeld, S., 1978. J. Biol. Chem. 253: 716–722.
4. Robbins, P. W., Hubbard, S. C., Turco, S. J. and Wirth, D. F., 1977. Cell 12: 893–900.
5. Behrens, N. H. and Leloir, L. F., 1970. Proc. Natl. Acad. Sci. U.S.A. 66: 153–159.
6. Parodi, A. J., 1979. J. Biol. Chem. 254: 8343–8352.
7. Powell, J. T., Jarlfors, V. and Brew, K., 1977. J. Cell. Biol. 72: 617–627.
8. Lowry, O. H., Rosenbrough, N. J., Farr, A. L. and Randall, R. J., 1951. J. Biol. Chem. 193: 265–275.
9. Campbell, P. N. and Sargent, J. R., 1967. In: 'Techniques in Protein Biosynthesis' (Campbell, P. N. and Sargent, J. R., eds.), Vol. I, pp. 301–303, Academic Press, New York.
10. Chalvardjian, A. and Rudnicki, E., 1970. Anal. Biochem. 36: 225–226.
11. Ames, B. N., 1966. Methods Enzymol. 8: 115–118.
12. Trevelyan, W. E., Procter, D. P. and Hamilton, J. S., 1950. Nature 166: 444–445.
13. Chapman, A., Fujimoto, K. and Kornfeld, S., 1980. J. Biol. Chem. 255: 4441–4446.
14. Patton, S. and Keenan, T. W., 1975. Biochim. Biophys. Acta 415: 273–309.
15. Peixoto de Menezes, A. and Pinto da Silva, P., 1978. J. Cell. Biol. 76: 767–778.
16. Morré, D. J., Kartenbeck, J. and Franke, W. W., 1979. Biochim. Biophys. Acta 559: 71–152.
17. Bergman, L. W. and Kuehl, W. M., 1978. Biochemistry 17: 5174–5180.
18. Franke, W. W., Lüder, M. R., Kartenbeck, J., Zerban, H. and Keenan, T. W., 1976. J. Cell. Biol. 69: 173–195.
19. Martín-Barrientos, J. and Parodi, A. J., 1977. Mol. Cell. Biochem. 16: 111–117.
20. Parodi, A. J. and Martín-Barrientos, J., 1977. Biochim. Biophys. Acta 500: 80–88.
21. Merrit, W. D., Morré, D. J., Franke, W. W. and Keenan, T. W., 1977. Biochim. Biophys. Acta 497: 820–824.
22. Dowben, R. M., Brunner, J. R. and Philpot, D. E., 1967. Biochim. Biophys. Acta 135: 1–10.
23. Martel, M. B. and Got, R., 1972. FEBS Lett. 21: 220–222.
24. Kitchen, B. J., 1974. Biochim. Biophys. Acta 356: 257–269.
25. Buck, C. A. and Warren, L., 1976. J. Cell. Physiol. 89: 187–199.
26. Shur, B. D. and Roth, S., 1975. Biochim. Biophys. Acta 415: 473–512.
27. Shur, B. D., 1982. J. Biol. Chem. 257: 6871–6878.

Received 12 December 1982.

Molecular and Cellular Biochemistry 58, 165–171 (1984).

Cyclic AMP-dependent regulation of activities of synthetase and phosphodiesterase of 2′,5′-oligoadenylate in NIH 3T3 cells

Alexander V. Itkes, Kyril T. Turpaev, Olga N. Kartasheva, Constantin A. Kafiani and Evgenii S. Severin
Institute of Molecular Biology, U.S.S.R. Academy of Sciences, Moscow 117984, U.S.S.R.

Summary

Treatment of NIH 3T3 cells with adenylate cyclase activator adrenaline (10^{-6} M) or cAMP phosphodiesterase inhibitor theophylline (10^{-3} M) was shown to lead to intracellular cAMP elevation followed by a 2.0- to 2.5-fold increase in the 2′,5′-oligoadenylate synthetase activity. This process was blocked by actinomycin D. The rise in the intracellular cAMP level was also followed by a 3–4-fold decrease in the activity of 2′-phosphodiesterase. Propranolol prevented this inhibition but actinomycin D produced only a negligible effect on the process. Incubation of the cell homogenate with purified catalytic subunit of cAMP-dependent protein kinase and ATP also resulted in a decrease of 2′-phosphodiesterase activity. These results indicate that cAMP is involved in the regulation of enzymes of the 2′,5′-oligoadenylate system. The possibility that certain biological functions of cAMP are implemented via 2′,5′-oligoadenylate-dependent processes is discussed.

Introduction

2′,5′-oligoadenylate (2′,5′-oligo(A)), i.e. oligomer consisting of several AMP residues (usually 3 or 4) connected through 2′,5′-phosphodiester bonds, is known to be an important regulator of metabolic activity. Recent extensive studies have shown that the main mechanism of 2′,5′-oligo(A) action involves activation of a specific ribonuclease which in the presence of the above activator hydrolyzes single-stranded RNA (1–3). This leads to a substantial decrease of mRNA pool in the cell and inhibition of protein synthesis that is observed both in the cell-free system (1, 4–9) and *in vivo* (10–12). Two enzymes regulate the 2′,5′-oligo(A) concentration in the cell, synthetase that synthesizes 2′,5′-oligo(A) from ATP (its activity, at least *in vitro*, is exhibited only in the presence of double-stranded RNA), and 2′-phosphodiesterase that hydrolyzes 2′,5′-oligo(A) to give AMP and ATP (reviewed in Refs. 13, 14). Changes in the activities of these enzymes have been detected under a number of effects on the cell causing either increase or decrease in the intensity of 2′,5′-oligo(A)-dependent processes.

Among the processes involving 2′,5′-oligo(A), the development of the interferon-induced antiviral resistance of cell is most thoroughly studied (15–17). Interferon causes a significant increase in the 2′,5′-oligo(A) synthetase activity (4–9, 13, 18). It is essential that the role of the synthetase activator, at least in some cases, is likely to be played by double-stranded regions of viral RNA that cause a preferential hydrolysis of single-stranded regions of RNA which are located in the immediate proximity to double-stranded regions, i.e. to the sites of 2′,5′-oligo(A) synthesis (19). However, 2′,5′-oligo(A)-insensitive viral propagation has also been reported; in such cases, the effect of interferon is probably mediated by other biochemical processes (20).

In a large number of studies devoted to 2′,5′-oligo(A), its antiproliferative effect was investigated. It was found that the direct introduction of 2′,5′-oligo(A) into the cell (10, 12), as well as the addition of the so-called core oligonucleotide, i.e.

2′,5′-oligo(A) lacking the triphosphate residue at 5′-terminal of molecule, into the culture medium (21, 22) produces a substantial antiproliferative action. It appears therefore that it is 2′,5′-oligo(A) that produces a cytostatic effect observed in the interferon-treated cells. Moreover, the 2′,5′-oligo(A) level in the cells was shown to correlate with inhibition of cell proliferation also in the absence of interferon. Thus, in the confluent cells the activity of 2′,5′-oligo(A) synthetase is two times higher, and that of 2′-phosphodiesterase four times lower, than in rapidly growing cells (23). Thus, the elevation of intracellular concentration of 2′,5′-oligo(A) at the reduced rate of cell division is presumably a widely occurring process.

In our previous studies on the regulation of 2′,5′-oligo(A) synthetase activity (24, 25), we demonstrated that the addition into the culture medium of cAMP phosphodiesterase inhibitors was accompanied by a rise in the intracellular cAMP level followed by an elevation of 2′,5′-oligo(A) synthetase activity. This effect was observed when NIH 3T3 cells (mouse fibroblasts) and CaOV cells (human ovary carcinoma) were treated with two standard cAMP phosphodiesterase inhibitors, 1-methyl-3-isobutylxanthine and theophylline.

This study presents evidence that a rise in the intracellular cAMP level due to adenylate cyclase activation by adrenaline as well as by cAMP phosphodiesterase inhibition with theophylline leads to an increase in the activity of 2′,5′-oligo(A) synthetase. Also data are presented on inhibition of phosphodiesterase of 2′,5′-oligo(A) (2′-phosphodiesterase) caused by cAMP elevation.

Materials and methods

Cell culture

NIH 3T3 cells were cultivated in Eagle's medium containing 10% calf serum (Serva, B.R.D.). To rapidly growing cells (48 hr before confluence), 10^{-3} M theophylline (Merck, B. R. D.) or 10^{-7}–10^{-5} M adrenaline (Sigma, U.S.A.) was added into the culture medium and the cells were incubated for 1 to 16 hr. When incubation was carried out in the presence of actinomycin D (Reanal, Hungary), the latter was added into the medium to 1 μg/ml, simultaneously with adrenaline. When propranolol (Sigma, U.S.A.) was used for incubation, it was introduced into the medium at a concentration of 10^{-6} M, 40 min before adrenaline.

2′,5′-oligo(A) synthetase assay

Preparation of the cell homogenate, obtaining of the homogenate fraction devoid of 2′-phosphodiesterase activity and assay of 2′,5′-oligo(A) synthetase were performed as described earlier (24). The core trimer of 2′,5′-oligo(A) used as standard in the experiments was a kind gift of Dr S. N. Mikhailov and Prof. M. Ya. Karpeisky (Institute of Molecular Biology, U.S.S.R. Acad. Sci., Moscow). The enzyme activity was calculated as the number of AMP residues incorporated into the 2′,5′-oligo(A) trimer.

Preparation of 2′,5′-oligo(A)

The homogenate fraction devoid of 2′-phosphodiesterase activity was incubated with [^{3}H]ATP as in the synthetase assay (24), but [^{3}H]ATP concentration in the incubation mixture was 10^{-4} M ($5 \cdot 10^{13}$ cpm/mol). After a 20 hr incubation, an equal volume of water was added, and the mixture was applied to a DEAE-cellulose DE-52 column (1 × 1 cm) (Whatman, U.K.). The column was washed with 90 mM KCl, 20 mM Tris-HCl (pH 7.5). The 2′,5′-oligo(A) fraction was eluted with 500 mM KCl, 20 mM Tris-HCl (pH 7.5), applied to a Sephadex G-15 column (0.8 × 50 cm) (Pharmacia, Sweden) and eluted with water. The fraction of 2′,5′-oligo(A) trimer was collected and lyophylized. Thin-layer chromatography on Silufol UV-254 plates (Č.S.S.R.) developed with the technique described (24) showed that the 2′,5′-oligo(A) trimer fraction contained 80–90%, and the ATP fraction, 10–20% of the total radioactivity in the preparation.

2′-phosphodiesterase assay

Samples (30 μl) containing $3 \cdot 10^{-7}$ M [^{3}H]-oligo(A) (1 500 to 2 000 cpm), 20–40 μl of cell homogenate protein, 125 mM K-acetate, 10 mM KCl, 5 mM $MgCl_2$, 1 mM dithiothreitol, 20 mM Tris-HCl (pH 7.5) were incubated for 2 hr at 37 °C. The reaction was terminated by heating to 90 °C for 10 min. Then 5 μl of 100 mM Tris and 5 μl of *E. coli* alkaline phosphatase solution (Boehringer, B.R.D.)

were added (0.01 unit per sample) and the mixture was incubated for 6 to 8 hr at 37 °C. Thin-layer chromatography was run as described (24); the enzyme activity was calculated as the decrease in the number of AMP residues in the substrate, i.e. 2′,5′-oligo(A) trimer fraction.

Treatment of cell homogenate with protein kinase

The catalytic subunit of cAMP-dependent protein kinase from pig brain was isolated as described (26, 27). 10 μl of the solution of the catalytic subunit (1 mg/ml in 10 mM $MgCl_2$, 20 mM Tris-HCl, pH 7.2) was added per 90 ml of cell homogenate. ATP was added simultaneously to 1 mM. [^{32}P]ATP (Amersham, U.K.) was added, where indicated ($7.2 \cdot 10^5$ cpm per 100 μl of the mixture). Samples were incubated for 2–8 hr at 37 °C; 2 hr before the end of incubation, 2′,5′-oligo(A) was added, and the protein kinase and 2′-phosphodiesterase reactions proceeded simultaneously. 2′-Phosphodiesterase activity was measured as described above. The control homogenate was incubated in the absence of the catalytic subunit of protein kinase, with ATP added to 1 mM. [^{32}P]-incorporation into the homogenate proteins was measured as described (26).

Determination of cAMP concentration

cAMP concentration was determined as described (24). cAMP content was calculated per total amount of cell protein.

Experimental

Effects of theophylline and adrenaline on the 2′,5′-oligo(A) synthetase activity

With rapidly growing NIH 3T3 cells, 2′,5′-oligo(A) synthetase activity assayed in the cell homogenate fraction free from 2′-phosphodiesterase activity (24), was 95 ± 15 pmol/mg · hr, i.e. 32 pmol/mg · hr if calculated per total protein amount in the homogenate.

After a 16 hr incubation in the presence of 1 mM theophylline, 2′,5′-oligo(A) synthetase activity was increased 2- to 3-fold (Table 1). The addition into the medium of 1 μg/ml of actinomycin D completely prevented the effect of adrenaline on 2′,5′-oligo(A) synthetase activity (Table 1).

Table 1. 2′,5′-oligo(A) synthetase activity in theophylline- and adrenaline-treated NIH 3T3 cells.

Treatment	Reagent concentration (M)	2′,5′-oligo(A) synthetase activity (% of initial level)
Control	–	100 ± 20
Theophylline	10^{-3}	230 ± 20
Adrenaline	10^{-7}	140 ± 20
Adrenaline	10^{-6}	220 ± 50
Adrenaline	10^{-5}	240 ± 40
Adrenaline + 1 μg/ml of actinomycin D	10^{-6}	90 ± 30

The enzyme activity was assayed 16 hr after addition of theophylline or adrenaline into a culture medium. The given values are the mean (± SE) of 4 parallel experiments.

Figure 1A demonstrates that after theophylline and adrenaline treatment, a transient 2.5- to 2.8-fold rise in cAMP level occurred as compared to the initial level (7.7 ± 0.8 pmol/mg cell protein), followed by a fall to the normal level.

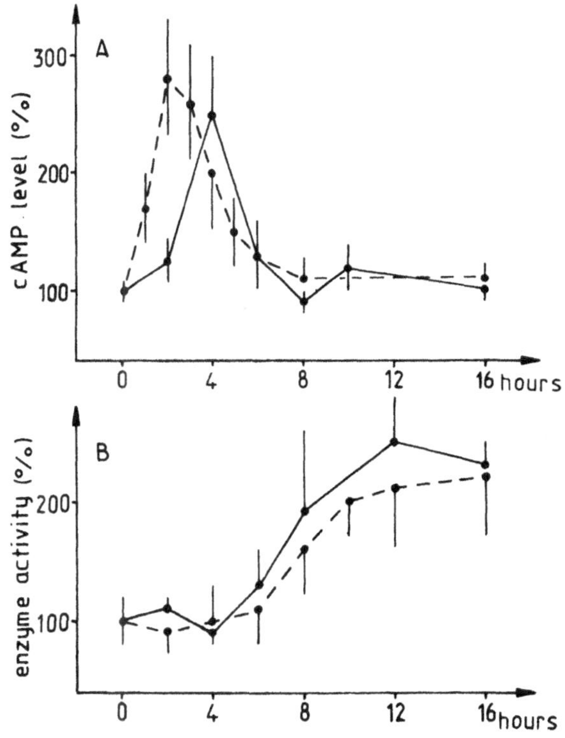

Fig. 1. cAMP level (A) and 2′,5′-oligo(A) synthetase activity (B) in NIH 3T3 cells vs time after addition into a culture medium of 10^{-3} M theophylline (—) and 10^{-6} M adrenaline (---). The given values are the mean (±SE) of 4 parallel experiments. cAMP level and 2′,5′-oligo(A) synthetase activity in control cells were at the initial level for at least 16 hr.

The increase in 2′,5′-oligo(A) synthetase activity in both cases began a few hours after the cAMP elevation. The maximum activity of the enzyme exceeded the initial level 2.2 to 2.5 times (Fig. 1B). Thus, irrespective of its mechanism, the intracellular cAMP elevation led to a rise in 2′,5′-oligo(A) synthetase activity. It is essential that the curves of cAMP concentration in the two cases were similar and the extent of the synthetase activity increase, following cAMP elevation, was the same. This finding supports the suggestion that 2′,5′-oligo(A) synthetase activity rises as a result of the cAMP elevation. Inhibition of the described process by actinomycin D favours the proposal that in our experiments an induction of the enzyme occurred rather than its cAMP-dependent modification.

Effect of theophylline and adrenaline on the 2′-phosphodiesterase activity

2′-Phosphodiesterase activity in NIH 3T3 cell homogenate was 200 ± 40 pmol/mg · hr at 2′,5′-oligo(A) concentration $3 \cdot 10^{-7}$ M (see Materials and methods). In the presence of 1 mM theophylline, the enzyme activity decreased by 40%. This fall began 8 hr after the inhibitor addition into the culture medium (Fig. 2). A fall in 2′-phosphodiesterase activity was also observed in the presence of 10^{-6} M adrenaline. However, in this case the decrease started almost simultaneously with the increase of cAMP level (Fig. 1 and 2).

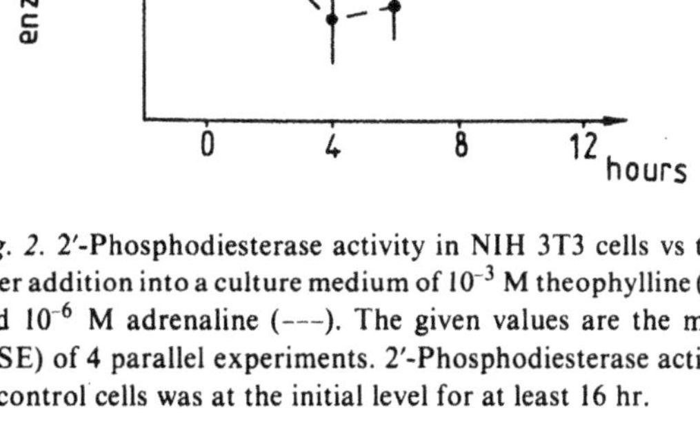

Fig. 2. 2′-Phosphodiesterase activity in NIH 3T3 cells vs time after addition into a culture medium of 10^{-3} M theophylline (—) and 10^{-6} M adrenaline (---). The given values are the mean (±SE) of 4 parallel experiments. 2′-Phosphodiesterase activity in control cells was at the initial level for at least 16 hr.

Thus, cAMP elevation in the cells after treatment with both theophylline and adrenaline led to a fall in the 2′-phosphodiesterase activity. However, as the time of activity decrease in the above cases was different (Fig. 2), it was not clear whether it was the enhancement of cAMP level that caused the observed fall in the enzyme activity. In order to ascertain that the effect of the decrease in the 2′-phosphodiesterase activity was due to the rise in cAMP concentration, incubation of the cells with 10^{-6} M adrenaline was conducted in the presence of 10^{-6} M propranolol. Under these experimental conditions, no decrease in 2′-phosphodiesterase activity was observed (Table 2). Propranolol is known to be highly specific reagent blocking β-adrenergic receptors and preventing adenylate cyclase activation by adrenaline. Thus, these results argue that the adrenaline-dependent decrease in 2′,5′-oligo(A) phosphodiesterase activity was mediated by adenylate cyclase activation and cAMP elevation.

Table 2. 2′-Phosphodiesterase activity in NIH 3T3 cells treated with adrenaline in the presence of actinomycin D and propranolol.

Treatment	2′-Phosphodiesterase activity (% of initial level)
Adrenaline	30 ± 10
Adrenaline + 1 μg/ml actinomycin D	50 ± 20
Adrenaline + 10^{-6} M propranolol	100 ± 20

The enzyme activity was assayed 4 hr after addition of reagents into a culture medium. Adrenaline concentration in a culture medium was 10^{-6} M. The given values are the mean (±SE) of 4 parallel experiments.

In the cells incubated in the presence of 10^{-6} M adrenaline and 1 μg/ml actinomycin D, a decrease in the 2′-phosphodiesterase activity was also detected; the enzyme activity after 4 hr incubation amounted to 50% of the initial level (Table 2). It appears, therefore, that the observed decrease in the 2′-phosphodiesterase activity in the cells treated with adrenaline does not require mRNA synthesis and is not due to the induction of a protein inhibitor of this enzyme. Another possible cause of the cAMP-induced decrease in the 2′-phosphodiesterase activity can be the modification of the enzyme by a cAMP-dependent protein kinase that would be activated by the rise of the intracellular cAMP following the theophylline or adrenaline treatment of the cells.

Effect of protein kinase treatment of cell homogenate on 2′-phosphodiesterase activity

To test the above proposal we carried out a model experiment in which we studied the effect of phosphorylation of the cell homogenate on its 2′-phosphodiesterase activity. In this experiment we used the purified catalytic subunit of the pig brain cAMP-dependent protein kinase.

The catalytic subunit and ATP were added to the cell homogenate and the 2′-phosphodiesterase activity was assayed as described in Materials and methods, as a function of time of incubation of the mixture. Figure 3 shows that in the control homogenate (i.e. without the addition of the kinase catalytic subunit), the 2′-phosphodiesterase activity remained unchanged for at least 4 hr, and then decreased to 70% of the initial level. In contrast, the 2′-phosphodiesterase activity decreased rapidly in the homogenate incubated with the added catalytic subunit of the protein kinase. [^{32}P]-incorporation into the proteins of the kinase-treated cell homogenate after 6 hr incubation over that in the control

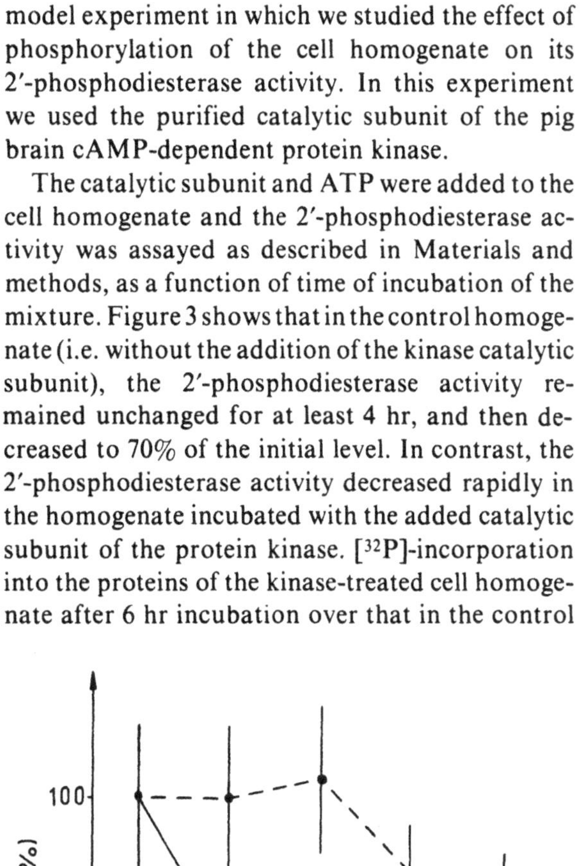

Fig. 3. 2′-Phosphodiesterase activity in NIH 3T3 cells homogenate vs time of incubation of the homogenate with catalytic subunit of cAMP-dependent protein kinase (—) and in the control homogenate (---). The given values are the mean (±SE) of 4 parallel experiments.

was 4.3 mol phosphate per 1 mol protein (calculated per average MW of protein of 50 000).

Thus, phosphorylation by cAMP-dependent protein kinase of the homogenate proteins was accompanied with an inhibition of 2′-phosphodiesterase. We believe that the catalytic subunit of cAMP-dependent protein kinase from NIH 3T3 cells produces an effect similar to that of the catalytic subunit of the pig brain kinase. This is supported by the fact that catalytic subunits of protein kinases from different sources are very similar in their structures and properties (28). Our results on phosphorylation of proteins in the cell homogenate *in vitro* and the data on 2′-phosphodiesterase activity and cAMP level in theophylline- and adrenaline-treated cells allow us to suggest that the cAMP elevation and subsequent activation of cAMP-dependent phosphorylation in the cells are the cause of the observed decrease in 2′-phosphodiesterase activity.

Discussion

We described the cAMP-dependent regulation of activities of 2′,5′-oligo(A) synthetase and phosphodiesterase in NIH 3T3 cell culture. The data allow us to conclude that intracellular cAMP elevation results in an increase of 2′,5′-oligo(A) synthetase activity. The data reported in the literature suggest that this mechanism is functional, at least partially, when various effects on a cell cause the enzyme activity to rise. Thus, interferon treatment of the cell leads to a cAMP elevation (29–32); the degree of this increase (2- to 3-fold as compared to the initial cAMP content) corresponds to the elevation reported in this paper. It is likely that the increase in cAMP concentration contributes to the interferon-induced elevation of 2′,5′-oligo(A) synthetase activity. This proposal fits with the finding that cAMP elevation results in an enhancement of 2′,5′-oligo(A) synthetase and the development of antiviral resistance of the cells (25).

It should be noted, however, that in the interferon-treated cells, the rise in 2′,5′-oligo(A) synthetase activity is substantially stronger than that observed in this study (13, 14). This suggests the existence of additional factor(s) that cause an induction of the enzyme in the case of interferon action. Furthermore, it was shown that cAMP level rises in the cells during their transition from the active division state to the quiescent state (33–35). The increased 2′,5′-oligo(A) synthetase activity was also detected in the quiescent cells (23). In this case, a 2- to 3-fold increase in the enzyme activity was observed (23) in agreement with the increase observed in our experiments. It seems possible that cAMP is also involved in the process of enzyme elevation in quiescent cells. The mechanism of increasing the activity of this enzyme presumably involves transcription of the corresponding gene. This idea agrees with the fact that the interferon-induced elevation of 2′,5′-oligo(A) synthetase activity is transcription-dependent (36). However, the final elucidation of the mechanism of cAMP-dependent increase of synthetase activity requires further investigation.

The experimental data also suggest that a rise in the intracellular cAMP level and an accompanying activation of the cAMP-dependent phosphorylation are the cause of the observed decrease in 2′-phosphodiesterase activity. It is essential that in the quiescent cells that are characterized by the enhanced cAMP level (33–35), 2′-phosphodiesterase activity decreases 4-fold (23) that fits well with our results.

On the whole, the results of this study support our supposition (24, 25) about cAMP involvement in the induction of 2′,5′-oligo(A) synthetase and give evidence of the existence of one more regulatory function of cAMP, viz. cAMP-dependent inactivation of 2′-phosphodiesterase. Moreover, a suggestion is possible that one of the biochemical mechanisms of antiproliferative cAMP action reported for a whole number of cell cultures (reviewed in Refs. 30–32) is connected with intracellular 2′,5′-oligo(A) elevation resulting from cAMP-dependent activation of its synthetase and inhibition of its phosphodiesterase. This leads to 2′,5′-oligo(A)-dependent inhibition of protein synthesis and decrease in cell division. These observations render it reasonable to consider cAMP and 2′,5′-oligo(A) to be components of a system involved in the regulation of the main metabolic cell processes.

References

1. Schmidt, A., Zilberstein, A., Shulman, L., Federman, P., Berissi, H. and Revel, M., 1978. FEBS Lett. 25: 257–264.
2. Chernajovsky, Y., Kimchi, A., Schmidt, A., Zilberstein, A. and Revel, M., 1979. Eur. J. Biochem. 96: 35–41.
3. Slattery, E., Ghosh, N., Samanta, H. and Lengyel, P., 1979. Proc. Natl. Acad. Sci. U.S.A. 76: 4778–4782.
4. Clemens, M. J. and Williams, B. R. G., 1978. Cell 13: 565–572.
5. Baglioni, C., Minks, M. A. and Maroney, P. A., 1978. Nature 273: 684–687.
6. Ratner, W., Wiegand, R.C., Farell, P. J., Seng, G. C., Cabrer, B. and Lengyel, P., 1978. Biochem. Biophys. Res. Commun. 81: 947–951.
7. Williams, B. R. G., Kerr, I. M., Gilbert, C. S., White, C. N. and Ball, L. A., 1978. Eur. J. Biochem. 92: 455–462.
8. Nilsen, T. W. and Baglioni, C., 1979. Proc. Natl. Acad. Sci. U.S.A. 76: 2600–2604.
9. Kerr, I. M. and Broun, R. E., 1978. Proc. Natl. Acad. Sci. U.S.A. 75: 256–260.
10. Williams, B. R. G., Golgher, R. R. and Kerr, I. M., 1979. FEBS Lett. 105: 47–52.
11. Hovanessian, A. G., Wood, J. N., Meurs, E. and Montagnier, L., 1979. Proc. Natl. Acad. Sci. U.S.A. 76: 3261–3265.
12. Hovanessian, A. G. and Wood, J. N., 1980. Virology 101: 81–90.
13. Baglioni, C., 1979. Cell 17: 255–264.
14. Revel, M., 1979. In: Regulation of Macromolecular Synthesis by Low-Molecular-Weight Mediators (Koch, G. and Riehter, D., eds.), pp. 341–359. Academic Press, New York.
15. Baglioni, C., Maroney, P. A. and West, D. K., 1979. Biochemistry 18: 1765–1770.
16. Kimchi, A., Shulman, L., Schmidt, A., Chernajovsky, Y., Fradin, A. and Revel, M. 1979. Proc. Natl. Acad. Sci. U.S.A. 76: 3208–3212.
17. Williams, B. R. G., Golgher, R. R., Broun, R. E., Gilbert, C. S. and Kerr, I. M., 1979. Nature 282: 582–586.
18. Ball, L. A. and White, C. N., 1978. Proc. Natl. Acad. Sci. U.S.A. 75: 1167–1171.
19. Nilsen, T. W., Weissman, S. G. and Baglioni, C., 1980. Biochemistry 19: 5574–5579.
20. Hovanessian, A. G., Meurs, E. and Montagnier, L., 1981. J. Interferon Res. 1: 179–190.
21. Williams, B. R. G. and Kerr, I. M., 1978. Nature 276: 88–89.
22. Kimchi, A., Shure, H. and Revel, M., 1979. Nature 282: 849–851.
23. Kimchi, A., Shure, H. and Revel, M., 1981. Eur. J. Biochem. 114: 5–10.
24. Itkes, A. V., Turpaev, K. T., Kartasheva, O. N., Vagonis, A. I., Kafiani, C. A. and Severin, E. S., 1982. Biochem. Int. 5: 15–21.
25. Itkes, A. V., Krispin, T. I., Shloma, D. V., Balandin, I. G., Tunitskaya, V. L. and Severin, E. S., 1982. Biochem. Int. 5: 389–398.
26. Nesterova, M. V., Sashchenko, L. P., Vasiliev, V. Yu. and Severin, E. S., 1975. Biochim. Biophys. Acta 377: 271–281.
27. Severin, E. S., Sashchenko, L. P., Kochetkov, S. N. and Kurochkin, S. N., 1978. Adv. Cycl. Nucl. Res. 9: 171–184.
28. Flockhart, D. A. and Corbin, J. D., 1982. Critical Rev. Biochem. 12: 133–186.
29. Weber, J. M. and Stewart, R. B., 1975. J. Gen. Virol. 28: 363–372.

30. Meldolesi, M. F., Friedman, R. M. and Kohn, L. D., 1977. Biochem. Biophys. Res. Commun. 79: 239–246.
31. Goldberg, S. J., O'Shaugnessy, M. V. and Stewart, R. B., 1980. J. Gen. Virol. 48: 377–381.
32. Schneck, J., Rager-Zisman, B., Rosen, O. M. and Bloom, B. R., 1982. Proc. Natl. Acad. Sci. U.S.A. 79: 1879–1883.
33. Abell, C. W. and Monahan, T. M., 1973. J. Cell Biol. 59: 549–558.
34. Pastan, I. H., Johnson, G. S. and Anderson, W. B., 1975. Ann. Rev. Biochem. 44: 491–522.
35. Friedman, D. L., Johnson, R. A. and Zeilig, C. E., 1976. Adv. Cycl. Nucl. Res. 7: 69–114.
36. Shulman, L. and Revel, M., 1980. Nature 288: 98–100.

Received 12 December 1983.

Molecular and Cellular Biochemistry 58, 173–181 (1984).

Biological maturation and β-adrenergic effectors: development of β-adrenergic receptors in rabbit heart

William Schumacher*, Bernard L. Mirkin and J. R. Sheppard**
Division of Clinical Pharmacology, Departments of Pharmacology and Pediatrics, Medical School, Department of Genetics and Cell Biology, Dight Institute for Human Genetics, University of Minnesota, Minneapolis, MN 55455, U.S.A.

Summary

The β-adrenergic receptor, transduction processes and catalytic activity of the adenylate cyclase enzyme complex have been investigated in rabbit heart at different stages of biological maturation. The binding of $[^3H]$-dihydroalprenolol to a washed membrane preparation isolated from rabbit ventricular muscle was used to characterize β-adrenergic receptors. Significant age-related differences were noted in β-receptor affinity (K_d) and density (RD) of neonatal and adult animals; the adult K_d was 3.7-fold greater and the RD 2-fold higher than the neonates. No significant differences in these parameters were detected among the 27-day old fetus and the 1- and 7-day old neonates. Age-dependent differences in agonist isoproterenol affinity for the receptor were not observed in contrast to the significant changes in antagonist (DHA) affinity.

Age-related changes in receptor affinity were also quantitated by determining the inhibitory potency of alprenolol on isoproterenol stimulated adenylate cyclase enzyme activity. A decreased affinity of the β-adrenergic receptor for alprenolol in the adult heart was indicated by a 3.7-fold greater K_i for the adult than the 1-day old neonate. Ontogenic variations in the coupling efficiency between the receptor and catalytic components of the adenylate cyclase complex were also evaluated. The K_d of the β-adrenergic receptor for isoproterenol and the EC50 for adenylate cyclase stimulation were determined under similar conditions. The corresponding coupling index (K_d/EC50) was found to be 2.4-fold greater in the 1-day old neonate than adult, suggesting that for a given percentage increase in adenylate cyclase activity, a lower percentage of β-adrenergic receptor sites need be occupied in the neonate. These data extend previous studies (29) and indicate all components of the rabbit heart adenylate cyclase enzyme complex (i.e., the β-adrenergic receptor, the GTP-dependent transduction event, and the catalytic subunit) exhibit significant developmental changes.

Introduction

The major functional components of the adenylate cyclase (AC) enzyme complex consist of a receptor that selectively recognizes and binds specific hormones, a catalytic subunit that promotes formation of adenosine 3′:5′-monophosphate (cAMP) from ATP and a transducer or coupling system that determines the efficiency with which an occupied receptor stimulates or inhibits the catalytic activity (26). Numerous studies have described the effects of biological maturation on AC alone (15); however, few have examined the functional integration of the components during development (29).

In the present investigation, several β-adrenergic receptor-related parameters (i.e., receptor-ligand

* *Current address*: William A. Schumacher, Ph.D., Department of Pharmacology, University of Michigan, M 6322 Medical Science Building, Ann Arbor, MI 48109, U.S.A.

** *Reprint address*: J. R. Sheppard, Dight Institute for Human Genetics, 400 Church Street S. E., Minneapolis, MN 55455, U.S.A.

binding, coupling efficiency and AC activity) have been concurrently evaluated in the 27-day old fetal, 1- and 7-day old neonatal and adult rabbit hearts. These data suggest that changes in β-adrenergic receptor affinity and density as well as coupling efficiency occur in the rabbit heart during the post-natal period.

Methods

Washed particulate membrane (WPM) preparations

New Zealand White Rabbits were obtained from the Oak Crest Rabbitry in Hanover, Minnesota. Average weights (mean ± S.D.) were: 35 ± 12 g (27-day old fetus), 59 ± 13 g (1-day old), 113 ± 29 g (7-day old), and 2.8 ± 0.4 kg (adult). Animals were sacrificed by a stunning blow to the head, the hearts excised within 2–3 minutes and placed in ice-cold buffer *A* (10 mM Tris-HCl, pH 7.4 at 4 °C with 2 mM dithiothreitol). The number of hearts prepared simultaneously varied according to the age of the animal. In general, the ensuing schedule was used: 27-day old fetus (9-10 hearts), 1- and 7-day old neonate (4–5 hearts) and adult (1 heart). The pooled hearts were always obtained from littermates. Atria and vessels were removed and the resulting ventricular muscle minced with a scissors. All tissue preparation was carried out at 4 °C.

The ventricular tissue yields were approximately 1.0 gram (27-day old fetus), 1.0–1.5 grams (1-day old), 1.5–2.0 grams (7-day old), and 2–4 grams (adult). The tissue mince was suspended in buffer *A* (10 volumes/wt) and homogenized at 4 °C in a Waring blender for 6–16 seconds duration, depending upon the amount of cardiac tissue used. Further homogenization was carried out in a Tekmar SDT tissuemizer using a 10-sec duration at setting 70 and a 2-sec duration at setting 100. The tissue suspension was then filtered through a 250 μ nylon mesh and the filtrate collected.

Washed particulate membranes (WPM) were prepared by subjecting this filtrate to the following series of low speed centrifugations using a Sorvall SS-34: 600 **g** for 15 min,, 400 **g** for 10 min, 200 **g** for 10 min, and 200 **g** for 10 min. The pellet was retained after each centrifugation and resuspended in 5 volumes of buffer *A* with 6 strokes of a motor-driven Teflon pestle homogenizer. The pellet from the 600 **g** centrifugation was gently washed with buffer *A* to remove the fluffy lipid layer which was discarded. Following the final 200 **g** centrifugation the pellet was resuspended in 10 volumes of buffer *B* (buffer *A* + 0.25 M sucrose) and centrifuged at 30 000 **g** for 15 min. The resulting pellet was washed three times with 5 ml of buffer *C* (5 mM $MgCl_2$ and 50 mM Tris-HCl pH 7.4, at 37 °C) and resuspended with a motor-driven teflon pestle homogenizer in an equal volume of buffer *C* to yield approximately 2 mg protein/ml. This suspension was aliquoted into 0.5–2.0 ml volumes, frozen immediately in dry ice-acetone and stored at -70 °C. Membranes could be stored in this manner for up to 9 months without loss of receptor binding or adenylate cyclase activity; however, they were typically used within 3 months. This preparation was similar to that described by Snyder and Drummond (31), the major alteration being the elimination of the KCl extraction of contractile proteins. Proteins were determined by the Lowry method (19) using a standard comprised of bovine serum albumin.

Adenylate cyclase assay

The enzymatic reaction was carried out in an incubation volume of 0.5 ml that contained 50 mM Tris-HCl (pH 7.4), 1.8 mM ATP, 2.5 μg R020 (phosphodiesterase inhibitor), 0.25 mM EGTA, 2 mM mercaptoethanol, 1.9 units pyruvate kinase (PK), 5 mM phosphoenolpyruvate (PEP), 5.5 mM KCl, 1.0 μM guanyl-5′-yl imidodiphosphate (Gpp(NH)p, a GTP analog), 2.5 mM $MgCl_2$ and approximately 0.1 mg membrane protein. Reactions were initiated by addition of membrane suspensions to prewarmed tubes that were maintained in an oscillating water bath set at 37 °C. Incubations were carried out for 8-min periods during which time enzyme velocities were shown to be linear. All reactions were terminated by inserting the test tube in a boiling water bath for 3 min. Samples were stored at 4 °C until assayed for cAMP by a competition binding procedure (4, 29).

β-Adrenergic receptor binding assay

The ligand binding reaction was performed at 37 °C in an oscillating water bath. All reactions were carried out in triplicate. This process was initiated by adding 0.08 ml of membrane preparation (0.1–0.3 mg protein) to a reaction mixture (0.2 ml final volume) contained in polystyrene tubes. The incubation media were similar to those used in the adenylate cyclase enzyme assay and included: 50 mM Tris-HCl (pH 7.4), 1.8 mM ATP, 2.5 μg R020, 2.5 mM EGTA, 2 mM mercaptoethanol, 5 mM PEP, 1.9 units PK, 5.5 mM KCl, 1 mM $MgCl_2$ and [^{3}H]-DHA. Non-specific binding was determined by addition of 10 μM propranolol to duplicate tubes. The reaction was terminated after 13 min by rapid addition of 3 ml ice-cold buffer *C*. The entire mixture was then filtered over a Gelman type AE glass fiber filter paper held in a millipore filtration manifold and washed with 12 ml of cold buffer *C*. A continuous vacuum sufficient to produce a flow rate of 1 ml/sec was applied during the entire filtration which was completed within 20 sec. The filters were air dried and placed in a vial containing 10 ml of scintillation cocktail (toluene with 0.04% 2a70 omniflour). Samples were counted in a scintillation spectrometer at an efficiency of 45% for 10 min.

Statistical analysis

Tests of equality between the means of groups were performed using the Student's *t*-test. When more than two groups were compared, interactions were determined by the one factor analysis of variance (ANOVA) or two factor ANOVA with repeated measures; if interactions were found, Duncan's least square test was used to determine which group means differed significantly. Least squares linear regression analysis was employed in Scatchard and Schild plot analyses.

Materials

[^{3}H]-cyclic AMP (40–60 Ci/mmole) and [^{3}H]-DHA (32–49 Ci/nmole) were purchased from the New England Nuclear Co. and (d-, l-) Gpp(NH)p from ICN Pharmaceuticals. 2-Mercaptoethanol was obtained from Eastman Organic, 2a70 from Research Products International and Norit SG Extra charcoal from the Baker Company. ATP, phosphoenolpyruvate, pyruvate kinase, DTT, BSA, d,l- and l-isoproterenol, l-phenylephrine and d- and l-norepinephrine were purchased from Sigma. Propranolol and l-alprenolol were generously provided by the Ayerst Pharmaceutical Co. and RO-20-1724 was a gift from H. Sheppard (Hoffman-LaRoche).

Results

[^{3}H]-DHA binding to washed particulate membranes (WPM): characteristics of the ligand–receptor interaction

The binding properties of [^{3}H]-DHA to WPM was initially studied in preparations derived from adult rabbit hearts. The specific binding of [^{3}H]-DHA to receptor sites was saturated with increasing ligand concentration and Scatchard analyses of these data (Fig. 1) demonstrated a K_d of 11.3 nM and a RD of 176 fmol/mg protein. Specific binding at 10 nM [^{3}H]-DHA was 70% of the total counts bound. About half of the non-specific binding was present on the filter discs.

Stereospecificity of the ligand–receptor interaction was determined by comparing the displacement of bound [^{3}H]-DHA in the presence and absence of l- and d-isomers of several β-adrenergic agonists (Fig. 2). The compounds employed in these experiments exhibited a high order of selec-

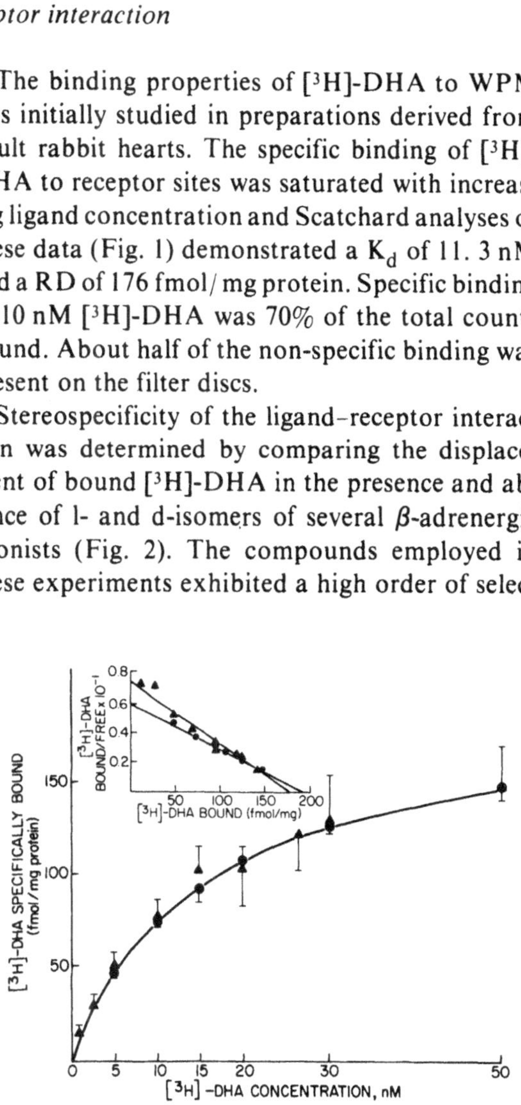

Fig. 1. Specific binding of [^{3}H]-DHA to myocardial membranes (WPM) from adult rabbits. WPM (▲, ●) fractions were incubated as described in Methods. Specific binding, that displaceable by 10^{-5} M dl-propranolol has been plotted as a function of [^{3}H]-DHA concentration. The insert is a Scatchard plot (27) of the data from which the K_d and RD were estimated from the slopes and abscissa intercepts, respectively. Results were averaged from 2 experiments on separate membrane preparations. Data expressed as mean ± SEM.

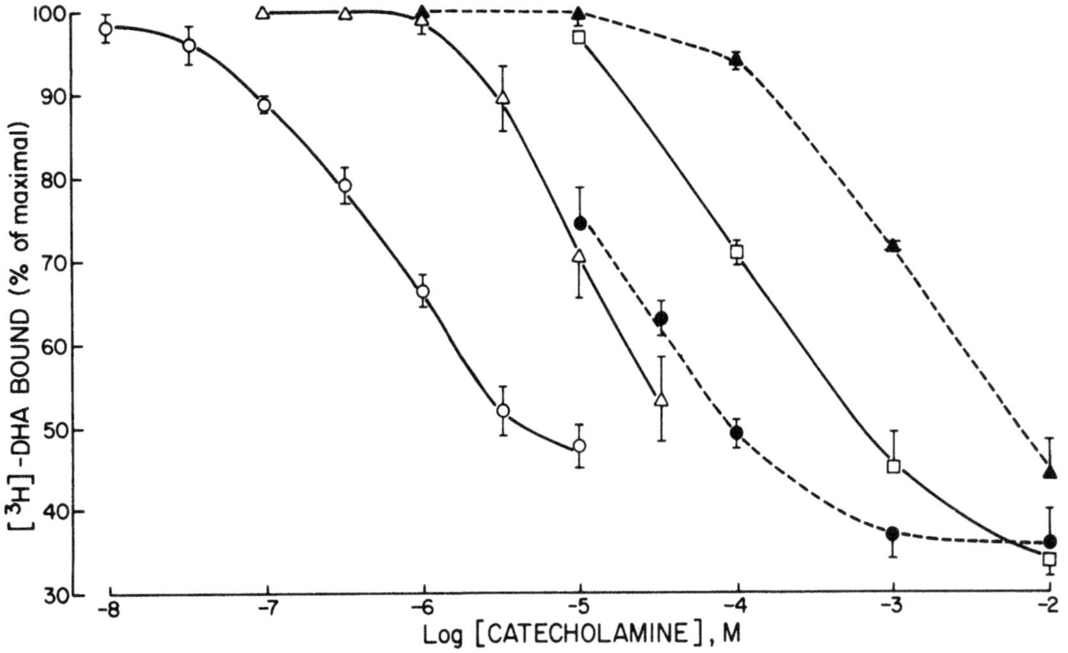

Fig. 2. Effect of β-adrenergic agonists on [^{3}H]-DHA binding in adult rabbit myocardial membranes. WPM from three adult rabbits were incubated as described in Methods with 10 nM [^{3}H]-DHA and various concentrations of the following β-adrenergic agonists: l-isoproterenol (○), d-isoproterenol (●), l-norepinephrine (Δ), d-norepinephrine (▲), l-phenylephrine (□). Binding was expressed as the percentage of total [^{3}H]-DHA bound and was plotted against agonist concentration. Specific binding averaged 75% of the total counts bound. Data expressed as mean ± SEM ($n = 3$). K_d values were calculated from these data using the equation $K_d = IC_{50}/1 + (S)/K_m$ where IC_{50} represents the concentration of drug inhibiting specific binding by 50%, S equals the concentration of [^{3}H]-DHA and K_m the K_d of [^{3}H]-DHA determined from Scatchard analysis. The resulting K_d (μM) values were: 0.3 (l-isoproterenol), 8.5 (d-isoproterenol), 4.9 (l-norepinephrine), 550 (d-norepinephrine), 55 (l-phenylephrine).

tively and the following order of potency: l-isoproterenol > l-norepinephrine > l-phenylephrine. The relative potency of the l- and d-stereoisomers l/d in this regard was 11 for norepinephrine, and 28 for isoproterenol.

Kinetic analyses of specific [^{3}H]-DHA binding to WPM demonstrated that equilibrium was achieved within 2 min and remained stable for 15 min. When 10 μM d-, l-propranolol was added to the reaction mixture, after equilibrium had been established, about 90% of the specifically bound [^{3}H]-DHA was dissociated within 3 min (data not shown).

Effect of biological maturation on binding of [^{3}H]-DHA to washed particulate membranes

The relationship between [^{3}H]-DHA concentration and specific binding of WPM derived from 27-day old fetal, 1- and 7-day old neonatal and adult rabbits was analyzed (Fig. 3). Specific binding in these experiments was approximately 70% of the total binding at 10 nM [^{3}H]-DHA. Saturation was more clearly observable in WPM isolated from the 27-day old fetus and adult than in the 1- or 7-day old neonate over a broad concentration is ligand used (1–50 nM, [^{3}H]-DHA). Scatchard analyses (27) of these data were linear for the adult and 27-day old fetus, suggesting that a single class of binding sites dominated in these preparations. In contrast, the 1- and 7-day old neonates generated curvilinear plots, reflecting a lack of saturation between 30–50 nM [^{3}H]-DHA (Fig. 4).

The K_d and RD values obtained for the high affinity binding sites are summarized in Table 1. There were no significant differences between the 27-day old fetus and 1-or 7-day old neonate with respect to either parameter. However, the adult K_d was 3- to 4-fold greater and the RD approximately 2-fold higher than any of the younger age groups.

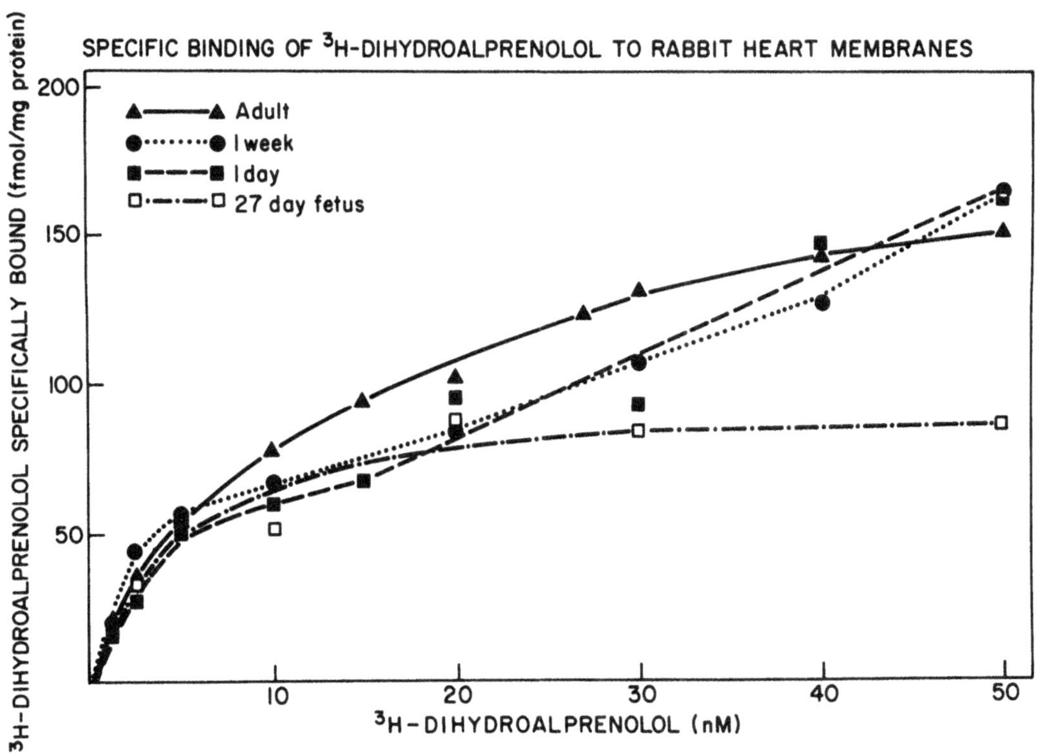

Fig. 3. Specific binding of [^{3}H]-DHA to rabbit myocardial membranes as a function of biological maturation. WPM from 27-day old fetus (□), 1-day old neonatal (■), 1-week old neonatal (○), and adult (▲) were incubated with various concentrations of [^{3}H]-DHA and specific binding determined as described in the Methods section. Each point constitutes an average of experiments performed on separate membrane preparations: 27-day old fetus ($n = 4$–6), 1-day old neonatal ($n = 5$), 1-week old neonatal ($n = 4$–10) and adult ($n = 4$–8).

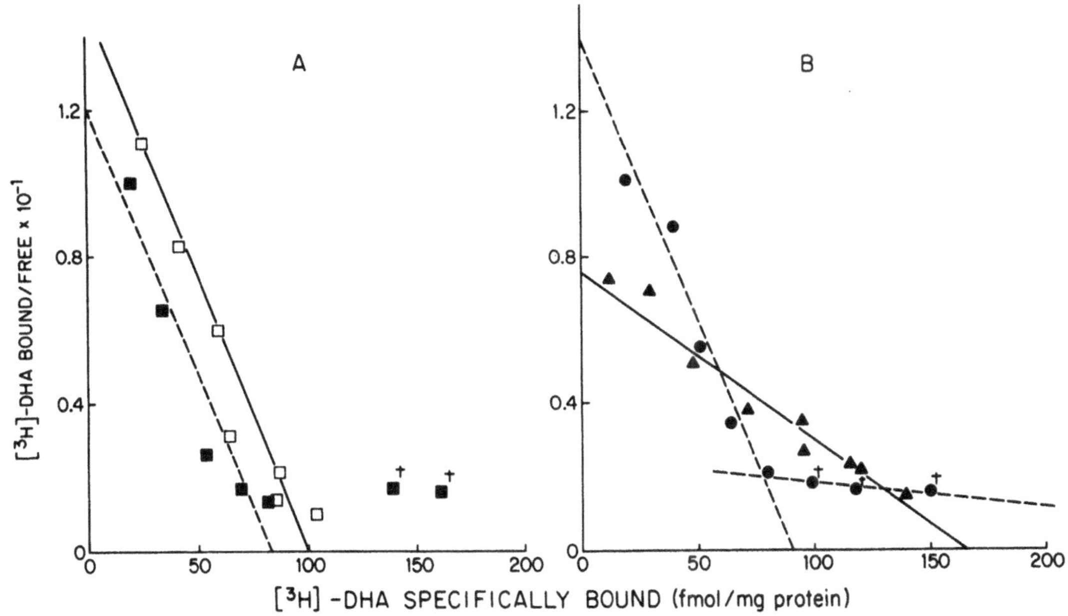

Fig. 4. Scatchard analysis of [^{3}H]-DHA specific binding to myocardial membranes as a function of biological maturation. Data analyzed were obtained from the experiments described in Fig. 3 and contrast the 27-day old fetus (□) and 1-day old neonatal (■) in panel A and the 7-day old neonatal (○) with adult (▲) in panel B. The correlation value (r) for the least square regression analysis were as follows: (□) = 0.986, (■) = 0.940, (●) = 0.980, (▲) = 0.920. The (+) data for the 1- and 7-day old neonatals were omitted from the analysis since they appeared to comprise another binding component of low affinity.

Table 1. Scatchard analysis of individual membrane preparations.

	r	K_d (nM)	B_{max} (fmol/mg)
Adult	-0.958	8.69	188
	-0.849	9,57	143
	-0.964	9.75	137
	-0.916	23.20	227
	-0.972	22.69	276
	-0.966	15.10	218
	-0.989	8.95	160
	-0.979	7.75	137
	Ave ± SEM	13.2 ± 2.3	186 ± 18.0
7-day old neonatal	-0.931	5.70	154
	-0.923	4.69	152
	-0.933	4.02	96
	-0.855	2.61	115
	-0.824	2.60	68
	-0.940	2.63	77
	Ave ± SEM	[a]3.8 ± 0.5	[a]111 ± 12.7
1-day old neonatal	0.850	2.85	88
	0.826	5.34	68
	0.827	2.68	50
	0.866	3.41	99
	0.945	4.30	124
	Ave ± SEM	[a]3.7 ± 0.5	[a]86± 13
27-day old fetus	0.976	2.24	74
	0.841	4.70	131
	0.921	2.54	70
	0.940	2.17	80
	0.915	3.84	104
	0.924	3.44	122
	Ave ± SEM		[a]3.2 ± 0.4 [a][illegible]

[a] Less than adult, $p < 0.001$.

Data were obtained by the method indicated in Fig. 4 except that individual rather than pooled data have been presented.

Inhibition of WPM adenylate cyclase activity by l-alprenolol

The K_i of l-alprenolol was determined by determining its potency as an inhibitor of isoproterenol-stimulated adenylate cyclase activity in WPM prepared from adult and 1-day old rabbits. Increasing concentrations of l-alprenolol produced parallel shifts in the l-isoproterenol dose-response curve (data not shown). The competitive nature of this inhibition was confirmed by slope values of unity (range 0.991 to 1.074) and the ability of high isoproterenol concentrations to overcome alprenolol inhibition. Schild analyses (28) of these data indicated that the average K_i for the adult (3.50 ± 0.17 nM) was 3.7-fold greater ($p < 0.05$) than of the 1-day old neonate (1.01 ± 0.22 nM) (Fig. 5). These observations suggest that the affinity of the receptors in the neonatal WPM have a greater affinity for β-adrenergic antagonists than the adult.

Relationship between receptor occupation and adenylate cyclase activation by l-isoproterenol: coupling efficiency of reaction

The IC_{50} of isoproterenol for [^{3}H]-DHA occupied receptor sites was used to determine the K_d of the agonist in WPM prepared from 1-day old and adult hearts. The average K_d of the neonate (0.32 μM) did not differ significantly from that of the adult (0.30 μM). The average EC_{50} for isoproterenol-induced stimulation of adenylate cyclase was 0.047 μM in the 1-day old and 0.108 μM in the adult ($p < 0.001$). In these experiments the GTP analog, Gpp(NH)p, was always present in these experiments at a concentration of 1.0 μM. It failed to elicit any shift in the EC_{50} of isoproterenol, although the enzymatic activity of the preparation was increased two-fold in the presence of this compound. The coupling efficiency index (K_d/EC_{50}) determined from these data was found to be greater in the 1-day old neonate (0.34 μM/0.047 μM = 6.8) than in adult (0.30 μM/0.108 μM = 2.8).

Discussion

The binding of [^{3}H]-DHA to membrane fractions prepared from rabbit heart membranes has been shown to fulfill accepted criteria for the identification and characterization of β-adrenergic receptors. The lack of a well-defined saturation pattern for [^{3}H]-DHA specific binding was noted in WPM prepared from 1- and 7-day old neonatal myocardium. The resulting concave-upward Scatchard plot may indicate the existence of two classes of binding sites, a two step reaction process, or negative cooperativity among sites. Quantitation and interpretation of the K_d was, therefore, made less precise by the non-linearity of these plots. Precedents for these observations have been previously published. Krawietz et al. (17) described a similar

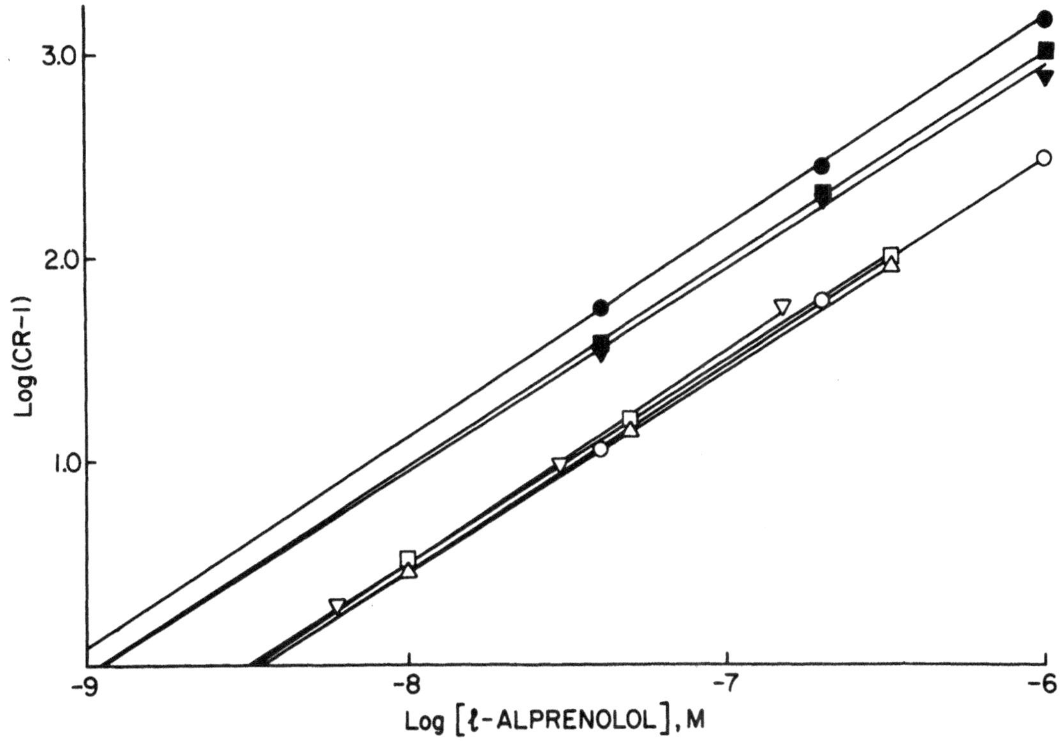

Fig. 5. Schild analysis of l-alprenolol inhibition of l-isoproterenol stimulated myocardial adenylate cyclase. Data were obtained from the experiments using 1-day old neonatal (▼, ■, ●) or adult (Δ, ○, □, ∇) preparations. Concentration ratios (CR) were calculated by dividing the ratio of the EC50 for l-isoproterenol determined in the presence of alprenolol by EC50 determined in its absence. Log (CR-1) was plotted against alprenolol concentration. The abscissa intercepts were used to estimate the individual K_i values. The individual K_i (nm) values were: (▼) 1.19, (■) 1.07, (●) 0.76, (∇) 1.074, (○) 3.65, (Δ) 3.61, (□) 3.28.

curvilinear Scatchard analysis for [^{3}H]-DHA binding to guinea pig heart membranes and suggested this was due to negative cooperativity among receptors. Winek and Bhalla (34) reported that [^{3}H]-DHA concentrations in excess of 5 nM, bound to a second class of low affinity sites in rat myocardium.

Due to the uncertainty of the K_d values that could be estimated by this approach, we sought verification by determining the K_i of adenylate cyclase with l-alprenolol. The null hypothesis for these studies was based on the assumption that the two forms of the antagonist (DHA and [^{3}H]-DHA) lack intrinsic activity and possess equal affinity for the receptor as claimed by Mukherjee et al. (23); consequently, both parameters, K_d and K_i, should be equal. These data revealed the K_d's to be about 3.5-fold greater than the K_i's for WPM derived from either the 1-day old neonate or adult. Several lines of evidence favor the validity of the K_i values. The K_i obtained for the adult (3.5 nM) was close to that reported by Kaumann and Birnbaumer (14) for the inhibition of myocardial adenylate cyclase activity (4.2 nM) and papillary muscle contraction (1.7 nM) by alprenolol. In addition, the Schild analysis conformed better to theoretical expectations than the Scatchard analysis.

Discrepancies between the K_d and K_i of an antagonist have been noted elsewhere in the literature. Coleman et al. (8) measured K_d and K_i values for various derivatives of phenoxypropranolamines in cardiac membranes. The relative K_d values as estimated from linear regression analysis of these data were about 3-fold greater than the K_i's for cardiac membranes. Lucas et al. (20) found the K_d for alprenolol and [^{3}H]-DHA to be two times greater than the K_i in C_6-glioma cells. Other investigators employing adult cardiac membrane preparations comparable to those used in this study have estimated K_d's for [^{3}H]-DHA to range between 0.3 and 10.8 nM (median = 6 nM) and RD's between 35–350 fmol/mg protein (median = 89 fmol/mg protein (1, 3, 5, 6, 11, 16, 17, 18, 33, 34). The only previous report on rabbit heart membranes (9) estimated a [^{3}H]-DHA K_d of 6.1 nM as compared to 13.2 nM in the present study.

We found that Schild and Scatchard analyses predicted a 3.7-fold increase in the K_d of the adult receptor when compared to the neonate, indicating that Scatchard analysis accurately reflected changes in K_d and RD occurring during development. In contrast to the differences in K_d observed with alprenolol, the K_d of isoproterenol determined in 1-day old and adult tissues were similar. These observations suggested the existence of developmentally regulated distinctions between agonist and antagonist binding to the β-adrenergic receptor. Other investigators using cardiac membrane preparations have indicated an absence of maturationally related changes in the K_d of [^{3}H]-DHA. These studies employed rats (3,32), and mice (5).

The demonstration of an increase in β-adrenergic receptor density in adult as compared to neonate cardiac membranes is also contrary to previous reports. Baker and Potter (3) and Whitsett and Darovec-Beckerman (32) observed a decrease in the RD (per g of tissue) of adult rats when compared to 2-week old animals. Chen et al. (5) described a progressive 4-fold increase in RD (relative to tissue weight) between the 17-day old fetus and 3-day old neonate, which was followed by a 50% reduction in RD at 3–5 months of age. While species-specific differences may account for the lack of correlation between these reports and the current study, variations in membrane preparation were also evident.

Changes in the reactivity of β-adrenergic effectors during development have been associated with numerous biological events including growth of sympathetic nerve fibers (24, 30), hormonal influences (12), as well as cellular proliferation and differentiation (7). Adrenergic innervation to the rabbit heart develops primarily during extrauterine life. Friedman et al. (10) reported that norepinephrine tissue levels reach adult values three weeks after parturition, with the major increase occurring between days 2 and 14. Roffi and Motelica-Heino (25) confirmed these data and noted a rapid increase in myocardial norepinephrine within 4 hours of birth, followed by little or no change between 4 and 72 hours. This was presumed to result from an enhancement of sympathetic activity following delivery. We were unable to detect any changes in receptor function or number during this phase of extrauterine development, which was marked by rapid increases in tissue catecholamine concentration. Our data suggest that variations in endogenous tissue norepinephrine levels may not be a significant modulator of RD during perinatal development. This contrasts with accompanying changes in RD observed following denervation supersensitivity (35) and agonist-induced desensitization (15).

It is well established that sympathetic induced myocardial responses are at least in part mediated by cAMP (22). Our results on the coupling efficiency of the AC system as estimated from the K_d/EC_{50} ratio for isoproterenol indicate that while fewer β-adrenergic receptors per mg protein are present in the neonate as compared to the adult, a smaller fraction need be occupied for maximal adenylate cyclase enzyme activation in the neonate.

The analysis of developmentally related changes in the adenylate cyclase enzyme complex is difficult and requires a broad perspective. This study has

Table 2. Characteristics of the β-adrenergic receptor, transduction and catalytic subunits of the 1-day old neonatal and adult rabbit heart adenylate cyclase complex

Subunit	Parameter	Age of rabbit	
		1-day old	adult
Receptor	Receptor density (fmol/mg prot)	86 ± 13	186 ± 18
	K_d of H^3-dihydroalprenolol (nM)	3.7 ± 0.5	13.2 ± 2.3
	K_d of l-isoproterenol (nM)	340	300
Transduction	Coupling efficiency ($K_{d/EC_{50}}$		
	for l-isoproterenol	6.8	2.8
	Gpp(NH)p preincubation (pmol/min/mg)	25 ± 1.8	493 ± 64
Catalytic	Basal activity (pmol/min/nm)	23 ± 0.8	18 ± 0.3
	NaF stimulated activity (pmol/min/mg)		
	Activity after 5 hr (4 °C)	221 ± 28	451 ± 31

demonstrated that β-adrenergic receptor density (RD) in the heart increases approximately two-fold from birth to adult life. Receptor affinity for the β-adrenergic agonist, isoproterenol, does not appear to change; the coupling efficiency index of isoproterenol is greater in the neonate, potentially compensating for the decreased number of receptors present at this time (Table 2).

Sodium fluoride activation and Gpp(NH)p preincubation studies have also revealed a major increase in the catalytic capacity of adenylate cyclase with advancing age (29). In light of cyclic AMP's biological role as a second messenger, these alterations in the adenylate cyclase enzyme may be related to the other ontogenic changes which occur in cardiac muscle physiology.

Acknowledgement

This investigation was partially supported by funds from USPHS Grants GM-07466 and GM-07397, the Hoffmann La Roche Foundation and the Leukemia Task Force.

References

1. Alexander, W. R., Williams, L. T. and Lefkowitz, R. J., 1975. Proc. Natl. Acad. Sci. U.S.A. 72: 1564–1568.
2. Aurback, G. P., Fedak, S. A., Woodward, C. J., Palmer, J. S., Hauser, D. and Troxler, F., 1974. Science 186: 1223–1224.
3. Baker, S. D. and Potter, L. T., 1980. Br. J. Pharmacol. 68: 65–70.
4. Brown, B. L., Albano, J. D. M., Ekins, R. P. and Sgherzi, A. M., 1971. Biochem. J. 121: 561–562.
5. Chen, F. C. M., Yamamura, H. I. and Roeske, W. R., 1979. Eur. J. Pharmacol. 58: 255–264.
6. Ciaraldi, T. and Marinetti, G. V., 1977. Biochem. Biophys. Res. Commun. 74: 984–991.
7. Claycomb, W. C., 1976. J. Biol. Chem. 251: 6082–6089.
8. Coleman, A. J., Paterson, D. S. and Somerville, A. R., 1979. Biochem. Pharmacol. 128: 1011–1013.
9. Coleman, A. J. and Somerville, A. R., 1977. Br. J. Pharmacol. 59: 83–93.
10. Friedman, W. F., Pool, P. E., Jacobowitz, D., Seagren, S. C. and Braunwald, E., 1968. Circ. Res. 23: 25–32.
11. Glaubiger, G., Tsai, B. S. and Lefkowitz, R. J., 1978. Nature 273: 240–242.
12. Green, T. D., O'Dea, R. F. and Mirkin, B. L., 1979. Ann. Rev. Pharmacol. Toxicol. 19: 285–322.
13. Howlett, A. C., VanArsdale, P. M. and Gilman, A. G., 1978. Mol. Pharmacol. 14: 531–539.
14. Kaumann, A. J. and Birnbaumer, L., 1973. Acta Phys. Lat. Amer. 23: 619–620.
15. Krall, J. F., Connelly, M. and Tuck, M. L., 1980. J. Pharmacol. Exp. Ther. 214: 554–560.
16. Krawietz, W., Poppert, D., Erdmann, E., Glossmann, H., Struck, C. J. and Konrad, C., 1976. Naunyn Schmiedeberg Arch. Pharmacol. 295: 215–224.
17. Krawietz, W., Weinsteiger, M., Pruchniewski, M. and Erdmann, E., 1979. Biochem. Pharmacol. 28: 2999–3007.
18. Limas, C. and Limas, C. J., 1978. Biochem. Biophys. Res. Commun. 83: 710–714.
19. Lowry, O. H., Rosebrough, W. J. and Randall, R. J., 1951. J. Biol. Chem. 193: 265–289.
20. Lucas and Bockaert, J., 1977. Mol. Pharmacol. 13: 314–329. 314–329.
21. Maguire, M. E., Ross, E. M. and Gilman, A. G., 1977. Adv. Cyclic Nucleotide Res. 8: 1–74.
22. Mayer, S. E., 1972. J. Pharmacol. Exp. Ther. 181: 116–125.
23. Mukherjee, C., Caron, M. G. and Coverstone, M., 1975. J. Biol. Chem. 250: 4869–4876.
24. Pappano, A. J., 1977. Pharmacol. Rev. 29: 3–33.
25. Roffi, J. and Motelica-Heino, I., 1975. Experientia (Basel) 31: 194–195.
26. Ross, E. M. and Gilman, A. G., 1977. Proc. Natl. Acad. Sci. U.S.A. 74: 3715–3719.
27. Scatchard, G., 1949. Ann. N.Y. Acad. Sci. 51: 660–672.
28. Schild, H. O., 1947. J. Pharmacol. 2: 189–206.
29. Schumacher, W., Sheppard, J. R. and Mirkin, B. L., 1982. J. Pharmacol. Exp. Ther. 223: 587–593.
30. Slotkin, T. A., Smith, P. G., Lau, C. and Bareis, D. L., 1980. In: Biogenic Amines in Development (Parves, S. and Parves, H. eds.), Elsevier, North Holland, Amsterdam.
31. Snyder, F. F. and Drummond, G. I., 1978. Arch. Biochem. Biophys. 185: 116–125.
32. Whitsett, J. A. and Darovec-Beckerman, L., 1981. Ped. Res. 15: 1363–1369.
33. Williams, L. T. and Lefkowitz, R. J., 1977. J. Biol. Chem. 252: 2787–2789.
34. Winek, R. and Bhalla, R., 1979. Biochem. Biophys. Res. Commun 91: 200–206.
35. Yamada, S., Yamamura, H. I. and Roeske, W. R., 1980. Mol. Pharmacol. 18: 185–192.

Received 1 February 1983.

Molecular and Cellular Biochemistry 58, 183–186 (1984).

Protein patterns of the nuclear matrix in differently proliferating and malignant cells

S. N. Kuzmina, T. V. Buldyaeva, S. B. Akopov and I. B. Zbarsky
Biochemistry Laboratory, N. K. Koltzov Institute of Developmental Biology, Academy of Sciences of the U.S.S.R, 26 Vavilov Street, 117334 Moscow, U.S.S.R.

Summary

SDS-polyacrylamide gel electrophoresis of the nuclear matrix proteins showed a predominance of high molecular weight and low molecular weight polypeptides in the nuclear matrices of hepatomas and cultured Chinese hamster fibroblasts as compared to quiescent and regenerating rat liver. These features were more prominent in solid hepatoma 27 than in Zajdela ascites hepatoma or in cultured cells. In proliferating cells (tumors, regenerating liver, log phase cell culture) a polypeptide band of 150 kD and lamin B were conspicuous at the expence of lamins A and C.

Introduction

Manifold functions in regulation of metabolic processes in the nucleus, such as DNA replication, transcription, nucleo-cytoplasmic RNA transport etc. are ascribed to the nuclear skeleton structure or the nuclear matrix (NM) along with its role in maintenance of the nuclear shape and higher orders of chromatin and chromosome organization (1–6).

Neither in different phases of HeLa cell mitotic cycle (7), nor in regenerating liver (8, 9) any significant alteration in protein pattern of the NM was found. However, in the NM of solid rat hepatoma 27 (8, 10, 11) and Zajdela ascites hepatoma (12) high molecular weight polypeptides absent or very faint in normal or regenerating liver were reported; in early stages of tumor growth low molecular weight polypeptides were also prominent (6, 8). A decrease in a lamin (protein with molecular weight of 69 kilodaltons (kD)) and an increase in a protein with molecular weight of 100 kD in Zajdela hepatoma NM were also described (13).

In this study the NM protein patterns of cultured cells in log and lag phases, of normal and regenerating rat liver and solid ascites rat hepatomas were compared.

Materials and methods

The nuclei and the NMs were isolated from livers, regenerating livers and hepatomas of male white rats weighing 150–200 g taken from Central Animal House of the Academy of Medical Sciences of the U.S.S.R. as described earlier (12, 14). Regenerating livers were taken at 20 hr after partial hepatectomy according to Higgins and Anderson (15). Hepatoma 27 was used at 6 weeks after subcutaneous transplantation. Zajdela hepatoma – in 5 days after intraperitoneal inoculation. Chinese hamster *Cricetulus griseus* fibroblasts, strain B77 FAF 28, clone 431 were grown in Eagle medium (40%), medium 199 (40%) and calf serum (20%) with penicillin and streptomycin, 50 U/ml each, at 37 °C in air containing 5% CO_2. The cells were collected in 2.5 days (log phase) and 5.5 days (lag phase) of growth. The fibroblasts were homogenized for 90 sec and shaken for 30 min at 0 °C in 0.25 M sucrose containing 1.5 mM $CaCl_2$, 5 mM $MgCl_2$, 0.5% Nonidet P 40 and 0.5% Triton X-100, then the nuclei were sedimented at 250 g for 10 min and washed with the same medium without detergents.

Electrophoresis of the NM proteins was performed in slab 12% SDS-polyacrylamide gel at

4 °C for 5 hr or in tubes (for radioactivity studies) during 3–3.5 hr (12).

The incorporation of labeled amino acids was studied by incubation of Zajdela hepatoma cells in ascitic fluid with 1 μCi/ml [^{14}C] Chlorella protein hydrolysate (Prague, Chechoslovakia, specific activity 1 000 μCi/g). Then, the disc electrophoretogram of the NM was cut to 48 equal slices. The proteins were eluted from each slice of the gel with 0.01 N Tris-HCl buffer, pH 7.2, containing 0.1% SDS at 37–40 °C for 18 hr. Then, the proteins were precipitated by aceton: 0.1 N HCl (6:1, v/v) and the radioactivity was counted in liquid scintillator 'Intertechnique SL 30'. Protein was determined by the method of Lowry et al. (16).

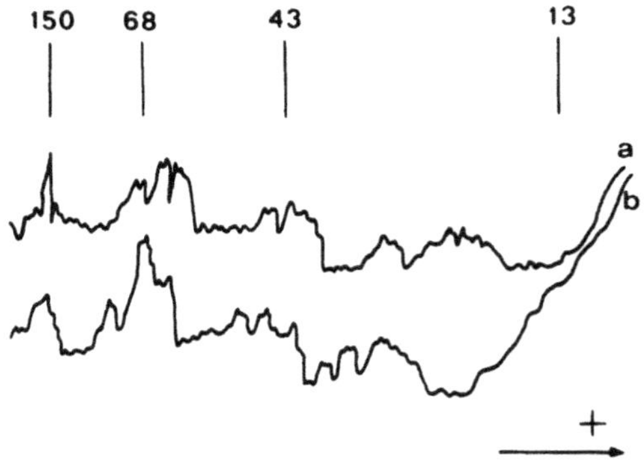

Fig. 2. Densitometric tracing of electrophoretograms of nuclear matrix of Chinese hamster fibroblasts in log phase (a) and lag phase (b) of growth. Details of electrophoresis in Fig. 1.

Experimental and discussion

Electrophoretic study revealed about 30 similar peaks in the NMs of quiescent and regenerating liver (Fig. 1). However, electrophoretic profile of regenerating liver differed from that of its quiescent counterpart in decreased highest and lowest peaks of polypeptide 'triplet' at 65–70 kD molecular weight (lamins A and C), while the intermediate peak (lamin B) was enhanced. Three peaks in the region of about 30 kD were also reduced.

In the NM of cultured fibroblasts (Fig. 2) in log phase compared to the lag phase a peak at 150 kD molecular weight was prominent, groups of polypeptides at 55–60 kD and 20–30 kD were also enhanced, while the 'triplet' region and a region of 30–50 kD as well as low molecular weight polypeptides were reduced.

In the NMs of both hepatomas the lamin B was also prominent at the expence of lamins A and C (Fig. 3); the regions of high molecular weight (150–200 kD) and low molecular weight (13 kD and lower) polypeptides were increased, while the regions of 40–50 kD and about 30 kD were diminished. These features were more prominent in the case of solid hepatoma 27, while in ascites Zajdela hepatoma mainly a peak of about 150 kD was conspicuous.

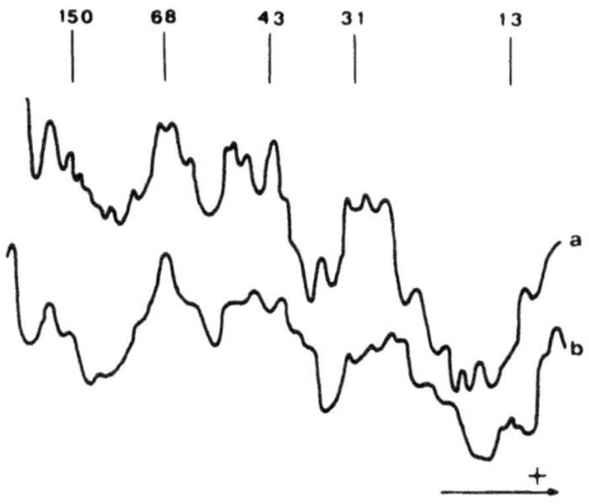

Fig. 1. Densitometric tracing of nuclear matrix electrophoretograms of quiescent liver (a) and regenerating liver (b). Electrophoresis was run in SDS-12% polyacrylamide gel. Molecular weight standards: β unit of bacterial RNA-polymerase (150 kD), bovine serum albumin (68 kD), ovalbumin (43 kD), DNase I (31 kD), cytochrome *c* (13 kD).

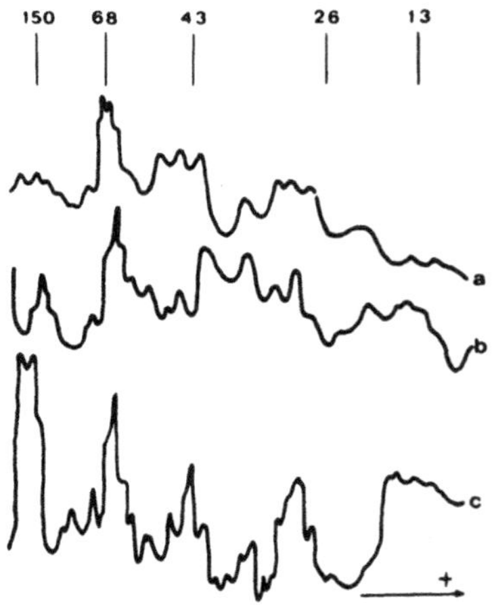

Fig. 3. Densitometric tracing of electrophoretograms of the nuclear matrix of rat liver (a), Zajdela hepatoma (b) and hepatoma 27 (c). Details of electrophoresis in Fig. 1.

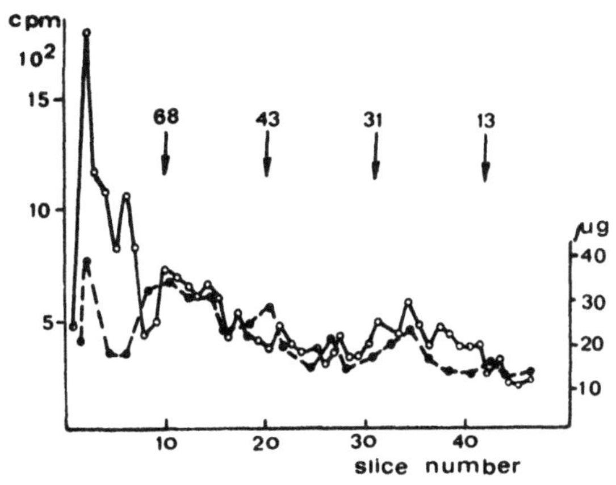

Fig. 4. Contents (µg) and radioactivity (cpm) of Zajdela hepatoma nuclear matrix proteins separated by SDS-polyacrylamide electrophoresis after 30 min incubation with [^{14}C] Chlorella protein hydrolysate. Details of electrophoresis in Fig. 1. ●---●---●--- protein (µg); ○—○—○— radioactivity (cpm).

Thus, a conspicuous middle component of the 'triplet' - the lamin B and a peak with molecular weight of about 150 kD in the NMs may be regarded as characteristic of rapidly proliferating cells and are common to regenerating liver, cultured cells and hepatomas. It is noteworthy that both, a polypeptide of 150 kD and the lamin B might be regarded to be characteristic of the nuclear pore complexes (17–19). Really, the reported increase in nuclear pore density in regenerating rat liver (20) might be important for proliferation.

In the protein profile of the NMs of both hepatomas, along with the features mentioned above, an accumulation of high molecular weight polypeptides (150–200 kD) and low molecular weight polypeptides (13 kD and lower) as well as some decrease in the region of 45–50 kD polypeptides is also characteristic. Those peculiarities of the tumor NM were reported by us earlier (6, 8, 10, 11). Similar features in the high molecular weight region were revealed in nuclear alkali-insoluble residual protein as well (21). These characteristics are more prominent in solid hepatoma 27 than in ascites Zajdela hepatoma. They are also detected in cultured fibroblasts, but are absent from regenerating liver (Figs. 1–3). Thus, these features may be regarded as peculiar of dedifferentiated tumor growth. A weak manifestation of these features in cultured fibroblasts might demonstrate their spontaneously transformed state.

The high molecular weight group of polypeptides turned out to be most active in the incorporation of labeled amino acids in Zajdela hepatoma cells (Fig. 4). The lowest specific radioactivity was revealed in the region of polypeptides with molecular weight between 30 and 70 kD, while the specific activity of low molecular weight group exceeded 1.5-fold, and that of high molecular weight polypeptides 3–3.5-fold the radioactivity of the intermediate group. Thus, these high molecular weight polypeptides prevailing in the tumor NM appeared also to be most active metabolically.

References

1. Berezney, R., 1979. In: The Cell Nucleus (Busch, H., ed.), Vol. 7, pp. 63–100. Academic Press, New York.
2. Faiferman, I. and Pogo, A. O., 1975. Biochemistry 14: 3808–3816.
3. Georgiev, G. P., Nedospasov, S. A. and Bakaev, V. V., 1978. In: The Cell Nucleus (Busch, H., ed.), Vol. 6, pp. 3–34. Academic Press, New York.
4. Miller, T. E., Huang Chen-Ya and Pogo, A. O., 1978. J. Cell. Biol. 76: 675–691.
5. Shaper, J. H., Pardoll, D. M., Kaufmann, S. H., Barrak, E. R., Vogelstein, B. and Coffey, D. S., 1979. In: Adv. Enzyme Regul. (Weber, G., ed.), Vol. 17, pp. 213–248. Pergamon Press, Oxford.
6. Zbarsky, I. B., 1981. Molec. Biol. Rep. 7: 139–148.
7. Hodge, L. D., Mancini, P., Davis, F. M. and Heywood, P., 1977. J. Cell. Biol. 72: 194–208.
8. Buldyaeva, T. V., Kuzmina, S. N. and Zbarsky, I. B., 1978. Dokl. Akad. Nauk SSSR 241: 1461–1464.
9. Berezney, R., Basler, J., Kaplan, S. C. and Hughes, B. B., 1979. Eur. J. Cell Biol. 20: 139–142.
10. Zbarsky, I. B., Kuzmina, S. N. and Buldyaeva, T. V., 1979. In: Macromolecules in the Functioning Cell (Salvatore, F., Marino, G. and Volpe, P., eds.), 1st Soviet-Italian Symposium, Capri, May 24–27, 1978, pp. 15–29. Plenum Press, New York.
11. Vokurkovà, N. and Zbarsky, I. B., 1982. Voprosy Med. Khimii 28: 113–116.
12. Kuzmina, S. N., Buldyaeva, T. V. and Zbarsky, I. B., 1980. Biokhimiya 45: 1417–1424.
13. Berezney, R., Basler, J., Hughes, B. B. and Kaplan, S. C., 1979. Cancer Res. 39: 3031–3039.
14. Kuzmina, S. N., Buldyaeva, T. V., Troitskaya, L. P. and Zbarsky, I. B., 1981. Eur. J. Cell Biol. 25: 225–232.
15. Higgins, G. M. and Anderson, R. M., 1931. Arch. Pathol. 12: 186–202.
16. Lowry, O. H., Rosebrough, N. J., Farr, A. L. and Randall, R. J., 1951. J. Biol. Chem. 193: 265–275.
17. Krohne, G., Franke, W. W. and Scheer, U., 1978. Exp. Cell Res. 116: 85–102.
18. Zbarsky, I. B., Virtanen, I., Lehto, V.-P., Buldyaeva, T. V. and Kuzmina, S. N., 1982. Tsitologiya 24: 1424–1429.

19. Gerace, L., Ottaviano, Y. and Kondor-Koch, C., 1982. J. Cell Biol. 95: 826–837.
20. Kuzmina, S. N., Troitskaya, L. P., Mirkhamidova, F. A., Buldyaeva, T. V., Zbarsky, I. B. and Grigoriev, V. B., 1979. Tsitologiya 21: 883–889.
21. Zbarsky, I. B., Perevoshchikova, K. A., Filatova, L. S. and Popova, V. S., 1980. Voprosy Med. Khimii 26: 406–410.

Received 10 June 1983.

Molecular and Cellular Biochemistry 58, 187–191 (1984).

Current thinking

Is there a general paradigm of cyclic AMP action in eukaryotes?

Martin L. Pall*
Program in Genetics and Cell Biology, and Program in Biochemistry/Biophysics, Washington State University, Pullman, WA 99164-4350, U.S.A.

Summary

The cyclic AMP control system in eukaryotes has been highly conserved evolutionarily in four of its central properties. Such conservation suggests conservation of the regulatory function of cyclic AMP. Conservation is seen in the properties of adenylate cyclase, cyclic AMP-dependent protein kinase and, among diverse lower eukaryotes, the control of endogenous cyclic AMP levels. A conserved regulatory response to cyclic AMP is the stimulation of glycolysis and inhibition of gluconeogenesis. The control of glycolysis and gluconeogenesis is proposed to be evidence of general pattern of cyclic AMP action in many lower and higher eukaryotic cells.

Introduction

Cyclic AMP has been reported to influence an amazing variety of cellular properties in animal cells and in microorganisms. Any general pattern of action that can be discerned should be of great value in understanding the literature on this central regulatory molecule. The only proposed general paradigm of cyclic AMP action which has received wide attention is based on the role of this molecule as a mediator of carbon catabolite repression in bacteria, such as *Escherichia coli*. In *E. coli*, the availability of carbon components regulates cyclic AMP levels which, in turn, regulate a variety of functions which are said to be catabolite repressible. This has led to the suggestion that cyclic AMP may have a general role as a carbon starvation signal (1, 2).

The thrust of this paper is that: 1) the *E. coli* pattern of function appears not to be applicable to eukaryotes; 2) certain important properties of this regulatory system are conserved among diverse lower and higher eukaryotes; 3) a general pattern of cyclic AMP action can be seen in its action in several lower eukaryotes involving stimulation of glycolysis and inhibition of gluconeogenesis; 4) this lower eukaryotic pattern may also be relevant to cyclic AMP action in higher animal cells.

The possibility that cyclic AMP mediates carbon catabolite repression in eukaryotes (as in *E. coli*) has been tested recently using mutants of the fungus, *Neurospora crassa*, and the yeast, *Saccharomyces cerevisiae*. A glucose transport system which shows carbon catabolite repression in *Neurospora* shows normal control in mutants which have little or no adenylate cyclase or cyclic AMP (3). In mutants of yeast which have increased permeability to cyclic AMP, exogenous cyclic AMP fails to overcome carbon catabolite repression of galactokinase (4). It may be inferred that cyclic AMP does not act as a general mediator of carbon catabolite repression in these lower eukaryotes. This difference between *E. coli* and eukaryotes should not be surprising. The intracellular receptor for cyclic AMP in bacteria is a protein which regulates transcription, whereas in lower eukaryotes and in animals, it is a cyclic AMP-dependent protein kinase. It may well be that the evolutionary gap between prokaryotes and eukaryotes in this area, as in so many others, will remain enigmatic.

* To whom requests for offprints should be addressed.

We have looked for evidence of evolutionarily conserved properties for the cyclic AMP control system in lower eukaryotes where perhaps more basic properties and functions may be more easily discerned than in higher animal cells.

Four areas of eukaryotic conservation

There are several areas of the cyclic AMP control system which show substantial conservation in eukaryotic evolution. Such conservation suggests that the cyclic AMP control function may have been substantially conserved, as well. Four of these areas are listed in Table 1.

Table 1. Areas of conservation of the cyclic AMP control system in eukaryotes.

1. Properties of protein kinase including regulatory subunit and catalytic specificity.
2. Separate catalytic and regulatory components of a guanine nucleotide-stimulated adenylate cyclase.
3. Similar control of endogenous cyclic AMP levels in diverse lower eukaryotes.
4. Stimulation of glycolysis and inhibition of gluconeogenesis by cyclic AMP.

One area of striking conservation is in the receptor for cyclic AMP, the cyclic AMP-dependent protein kinase. It was demonstrated some time ago that the catalytic specificity for different protein substrates of the yeast cyclic AMP-dependent protein kinase is strikingly similar to that of the animal enzyme (5). The regulatory subunits of these enzymes in lower eukaryotes are also quite similar to those of mammals (5–10). One strikingly similar set of comparisons of regulatory subunits are shown in Table 2.

Table 2. Properties of regulatory component of cyclic AMP-dependent protein kinase.

	Mammalian type I	*Neurospora*
Estimated molecular weight (SDS gels)	47 000	47 000
Aggregation state	Dimer	Dimer
Isoelectric point	5.35–5.57	5.4–5.5
Moles of cyclic AMP bound/subunit	2	2
Cyclic AMP binding sites with high affinity for 8 bromocyclic AMP	1	1
Stokes radius	44Å	46Å

Neurospora data from Ref. 6.

It has also been recently shown that a second critical enzyme of the control system shows signs of conservation. The adenylate cyclase in *Neurospora*, yeast, and possibly *Dictyostelium* have separate catalytic and regulatory components (11–14). As in the animal enzyme (15), the regulatory components are required for high ATP-Mg^{2+}-dependent, guanine-nucleotide-stimulated activity. It has recently been reported that the regulatory component of the animal enzyme can interact with and regulate the catalytic component of the *Neurospora* enzyme (16).

A third area of conservation is in the control of endogenous cyclic AMP levels. Several diverse lower eukaryotes show similar cyclic AMP increases in treatment with uncouplers of oxidative phosphorylation or with certain pore-forming antibiotics (3, 17–19). These organisms are thought to be only very distantly related, *Neurospora* and yeast, for example are estimated to have diverged from a common origin about 800 million years ago (20) with *Mucor* and *Coprinus* being still more distantly related. The results suggest that as yet undetermined mechanisms which control endogenous cyclic AMP levels in these organisms have been conserved through much of the eukaryotic evolution.

A fourth area of conservation has direct implications on cyclic AMP function.

Cyclic AMP stimulation of glycolysis and inhibition of gluconeogenesis

It was first proposed about two and a half years ago that cyclic AMP may act to stimulate glycolysis in many lower eukaryotes (21). Since then, substantial new evidence has been obtained, not only for stimulation of glycolysis, but also for the inhibition of gluconeogenesis. The pattern of evidence available is diagrammed in Fig. 1. Both *in vitro* and *in vivo* effects of cyclic AMP are shown.

The two main storage carbohydrates in fungi are glycogen and trehalose. *In vitro* studies of the glycogen control system in *Neurospora, Coprinus* and in yeast (22–28) show that glycogenolysis is stimulated by cyclic AMP through its influence on protein phosphorylation. Studies in yeast also demonstrate that trehalose degradation is stimulated by cyclic AMP (28–31). Thus, cyclic AMP should provide an increase in hexose phosphates

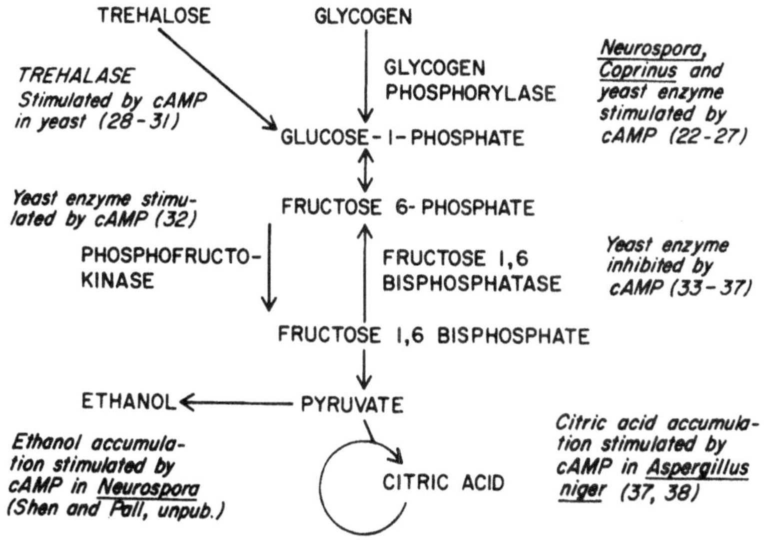

Fig. 1.

for the organisms and obviously the most likely fate of these compounds would be catabolism via the glycolytic pathway.

The major control point in glycolysis is thought to be phosphofructokinase. This enzyme from yeast is activated by phosphorylation via cyclic AMP-dependent protein kinase (32). The protein kinase used in these studies was mammalian as the yeast cyclic AMP-dependent protein kinase was not available at that time. However, because it is known that the catalytic specificity of the yeast cyclic AMP-dependent protein kinase is very similar to the mammalian enzyme (5), it seems likely that the yeast kinase should have a similar effect.

Gluconeogenesis is generally regulated in reciprocal fashion with glycolysis, thus helping avoid futile cycling (and consequent ATP hydrolysis) which might otherwise occur if both processes were active simultaneously. An important control point in gluconeogenesis is at fructose 1,6-bisphosphatase. Fructose 1,6-bisphosphatase from two yeasts *Saccharomyces* and *Kluvermyces* is inhibited by cyclic AMP-dependent phosphorylation (28, 33–38). Fructose 1,6-bisphosphatase also appears to be rendered more sensitive to proteolytic degradation by cyclic AMP-dependent phosphorylation (36, 37). It seems likely from those studies that cyclic AMP inhibits gluconeogenesis in these organisms.

A number of *in vivo* studies provides confirming evidence that cyclic AMP activates glycolysis and inhibits gluconeogenesis. Citric acid accumulation in *Aspergillus niger* is thought to be rate limited by the glycolytic rate of the culture and is found to be stimulated by exogenous cyclic AMP (38, 39). In *Neurospora,* ethanol accumulation is thought to reflect glycolytic rates (40) and ethanol accumulation is very low in adenylate cyclase-deficient mutants and is greatly stimulated by two cyclic AMP analogs (P.-H. Shen and M. L. Pall, unpublished data). The adenylate cyclase-deficient mutants of *Neurospora* grow relatively poorly on a number of carbon sources but grow better than wild-type on acetate (41). Growth on acetate requires high levels of gluconeogenesis. I suggest that in these mutants without cyclic AMP, gluconeogenesis activities may be higher than in wild-type, providing an explanation for the higher growth rate on acetate.

The evidence for cyclic AMP stimulation of glycolysis and inhibition of gluconeogenesis suggests that these represent highly conserved responses to cyclic AMP among these various lower eukaryotes. It falls short, however, of telling us the function of cyclic AMP in these organisms. In order to understand its function, we must know what the major parameters are that control cyclic AMP levels and also what other cellular responses may be directly controlled by cyclic AMP. Thus, we can see evidence for a highly conserved function, but cannot fully describe that function.

One question which should be raised is: Are the

responses to cyclic AMP in lower eukaryotes really very different from those described in *E. coli?* After all, in both systems, cyclic AMP acts to control carbon metabolism. There are, however, very substantial differences between the two systems. In *E. coli* and related bacteria, cyclic AMP acts to activate functions required to utilize various exogenous carbon sources whereas in lower eukaryotes it is the utilization of endogenous reserves and subsequent intermediary metabolism which is regulated. The one specific function which has been tested in both systems shows opposite cyclic AMP effects. Cyclic AMP inhibits glycogenolysis in *E. coli* (42, 43) whereas, as noted above, it activates glycogenolysis in lower eukaryotes and in higher animal cells.

Applicability to higher animal cells

Is the cyclic AMP action proposed here from studies of lower eukaryotes also applicable to higher animal cells? Certainly cyclic AMP stimulation of glycogenolysis in animal cells was the first response found for cyclic AMP and has since been reported to occur in many types of animal cells. However, most studies of cyclic AMP on metabolism of animal cells have used the biochemist's favorite tissue, liver. In liver, cyclic AMP stimulates both glycogenolysis and gluconeogenesis but inhibits glycolysis (44, 45). This makes good functional sense because liver is a storage organ which functions to regulate blood glucose levels by releasing or taking up glucose. Cyclic AMP stimulates glucose release from liver by stimulating both glycogenolysis and gluconeogenesis, but such a pattern of control would not make functional sense in a non-storage tissue. There have been reports that cyclic AMP may stimulate glycolysis in such diverse tissue as *Ascaris* and liver fluke muscle (46, 47), thyroid tissue (48), maturing rat oocytes (49) and guinea pig heart muscle (50). It may well be that cyclic AMP stimulation of glycolysis and inhibition of gluconeogenesis may be a common pattern throughout most of the eukaryotic kingdom.

Acknowledgement

The author's research has been supported by NIH grant GM 24368.

References

1. Jost JP, Rickenberg HV: Cyclic AMP. Annu Rev Biochem 40:741–773, 1971.
2. Tomkins GM: The metabolic code. Science 189:760–763, 1975.
3. Pall ML, Trevillyan JM, Hinman N: Deficient cyclic adenosine 3′,5′-monophosphate control in mutants of two genes of Neurospora crassa. Mol Cell Biol 1:1–8, 1981.
4. Matsumoto K, Uno I, Toh-e A, Ishikawa T, Oshima Y: Cyclic AMP may not be involved in catabolite suppression in Saccharomyces cerevisiae:evidence from mutants capable of utilizing it as an adenine source. J. Bacteriol. 150:277–285, 1982.
5. Takai Y, Sakai K, Morishita Y, Yamamura H, Nishizuka Y: Functional similarities of yeast and mammalian adenosine 3′,5′-monophosphate dependent protein kinase. Biochem Biophys Res Commun 59:646–652, 1974.
6. Trevillyan JM, Pall ML: Isolation and properties of a cyclic AMP-binding protein from Neurospora. Evidence for its role as the regulatory subunit of cyclic AMP-dependent protein kinase. J Biol Chem 257:3978–3986, 1982.
7. Hixson CS, Krebs EG: Characterization of a cyclic AMP binding protein from baker's yeast:identification as a regulatory subunit of cyclic AMP-dependent protein kinase. J Biol Chem 255:2137–2145, 1980.
8. Takai Y, Yamamura H, Nishizuka Y: Adenosine 3′:5′-monophosphate-dependent protein kinase from yeast. J Biol Chem 249:530–535, 1974.
9. Sy J, Roselle M: Phosphorylation of the regulatory subunit of yeast cAMP-dependent protein kinase. Proc Natl Acad Sci USA 79:2874–2877, 1982.
10. Leichtling BH, Spitz E, Rickenberg HV: A cAMP-binding protein from Dictyostelium discoideum regulates mammalian protein kinase. Biochem Biophys Res Commun 100: 515–522.
11. Rosenberg GB, Pall ML: Characterization of an ATP-Mg^{2+}-dependent guanine nucleotide-stimulated adenylate cyclase from Neurospora crassa. Arch Biochem Biophys 221:243–253, 1983.
12. Rosenberg GB, Pall ML: Reconstitution of adenylate cyclase in Neurospora from two components of the enzyme. Arch Biochem Biophys 221:254–260, 1983.
13. Casperson GF, Walker N, Brasier AR, Bourne HR: A guanine nucleotide-sensitive adenylate cyclase in the yeast Saccharomyces cerevisiae. J Biol Chem 258:7911–7914, 1983.
14. Leichtling BH, Coffman DS, Yaeger ES, Rickenberg HV, al-Jumaliy W, Haley BE: Occurrence of the adenylate cyclase 'G protein' in membranes of Dictyostelium discoideum. Biochem Biophys Res Commun 102:1187–1195, 1981.
15. Ross EM, Gilman AG: Biochemical properties of hormone-sensitive adenylate cyclase. Ann Rev Biochem 49:533–564, 1980.
16. Flawia MM, Kornblihtt AR, Reis JR, Torruella M, Torres HN: Reconstitution of a hormone-sensitive adenylate cyclase with membrane extracts from Neurospora and avian erythrocytes. J Biol Chem 258:8255–8259, 1983.
17. Trevillyan JM, Pall ML: Control of cyclic adenosine 3′,5′-monophosphate levels by depolarizing agents in fungi. J Bacteriol 138:397–403, 1979.

18. Uno I, Ishikawa T: Control of adenosine 3′,5′-monophosphate level and protein phosphorylation by depolarizing agents in Coprinus macrorhizus. Biochim Biophys Acta 672:108–113, 1981.
19. Jaynes PK, McDonough JP, Mahler HR: Properties and possible functions of the adenylate cyclase in plasma membranes of Saccharomyces cerevisiae. Molec Cell Biol 2: 1481–1491, 1982.
20. Fitch WM, Margoliash E: Construction of phylogenetic trees. Science 155:279–284, 1967.
21. Pall ML: Adenosine 3′,5′-phosphate in fungi. Microbiol Rev 45:462–480, 1981.
22. Tellez-Inon MT, Torres HN: Interconvertible forms of glycogen phosphorylase in Neurospora crassa. Proc Natl Acad Sci USA 66:459–463, 1970.
23. Tellez-Inon MT, Torres HN: Regulation of glycogen phosphorylase *a* phosphatase in Neurospora crassa. Biochim Biophys Acta 297:399–412, 1973.
24. Uno I, Ishikawa T: Effect of cyclic AMP on glycogen phosphorylation in Coprinus macrorhizus. Biochim Biophys Acta 452:112–120, 1976.
25. Uno I, Ishikawa T: Effect of cyclic AMP on glycogen synthetase (EC-2.4.1.11) in Coprinus macrorhizus. J Gen Appl Microbiol 24:193–197, 1978.
26. Mishra C: Evidence for the occurrence of glycogen synthase phosphatases and kinases in yeast. FEMS Microbiol Lett 18:25–29, 1983.
27. Pohlig G: Wingerden-Drissen R, Becker JU: Characterization of phosphorylase kinase activities in yeast. Biochem Biophys Res Commun 714:331–338, 1983.
28. Ortiz CH, Maia JCC, Tenan MN, Braz-Padrao GR, Mattoon JR, Panek AD: Regulation of yeast trehalase by a monocyclic, cyclic AMP-dependent phosphorylation-dephosphorylation cascade system. J Bacteriol 153:644–651, 1983.
29. Van der Platt JV, Van Solingen P: Cyclic 3′, 5′ adenosine monophosphate stimulates trehalose degradation in baker's yeast. Biochem Biophys Res Commun 56: 580–587, 1974.
30. Van Solingen P, Van der Platt JV: Partial purification of the protein system controlling the breakdown of trehalose in baker's yeast. Biochem Biophys Res Commun 62: 553–560, 1975.
31. Wiemkin A, Schellenberg M: Does a cyclic AMP-dependent phosphorylation initiate the transfer of trehalase from the cytosol into the vacuoles in Saccharomyces cerevisiae? FEBS Lett 150:329–331, 1982.
32. Vinuela E, Salas ML, Salas M, Sols A: Two interconvertable forms of yeast phosphofructokinase with different sensitivity to end product inhibition. Biochem. Biophys Res Commun 15:243–249, 1964.
33. Cancedo JM, Mazon MJ, Cancedo C: Fructose 2,6-bisphosphate activates the cAMP-dependent phosphorylation of yeast fructose-1,6-bisphosphatase activity in vitro. J Biol Chem 258:5998–5999, 1983.
34. Gangedo JM, Mazon MJ, Gangedo C: Kinetic differences between two intraconvertible forms of fructose 1,6-bisphosphatase from Saccharomyces cerevisiae. Arch Biochem Biophys 218:478–482, 1982.
35. Londesborough J: Cyclic nucleotide-dependent inactivation of yeast fructose 1,6-bisphosphatase by ATP. FEBS Lett 144:269–272, 1982.
36. Tortora P, Burlini N, Leoni F: Guerritore, A. Dependence on cyclic AMP of glucose-induced inactivation of yeast gluconeogenic enzymes. FEBS Lett 155:39–42, 1983.
37. Toyoda Y, Sy J: Catabolite inactivation of yeast fructose 1,6-bisphosphatase (abstract). Fed Proc 42:2050, 1983.
38. Wold WSM, Suzuki I: Demonstration in Aspergillus niger of adenyl cyclase, a cyclic adenosine 3′,5′-monophosphate binding protein and studies on intracellular and extracellular phosphodiesterases. Can J Microbiol 20:1567–1576, 1974.
39. Alobaidi ZS, Berry DR: cAMP concentration, morphological differentiation and citric acid production in Aspergillus niger. Biotechnol Lett 2:5–10, 1980.
40. Turian G, Bianchi DE: Conidiation in Neurospora. Bot Rev 38:119–157, 1972.
41. Terenzi HF, Flawia MM, Torres HN: A Neurospora crassa morphological mutant showing reduced adenylate cyclase activity. Biochem Biophys Res Commun 58:990–996, 1974.
42. Dietzler DN, Leckie MP, Magnani JL, Sughrue MJ, Bergstein PE, Sternheim WL: Contribution of cyclic adenosine 3′:5′-monophosphate to the regulation of bacterial glycogen synthesis in vivo. Effect of carbon source and cyclic adnosine 3′:5′-monophosphate on the quantitative relationship between rate of glycogen synthesis and the cellular concentrations of glucose 6-phosphate and fructose 1,6-diphosphate in Escherichia coli. J Biol Chem 254: 8308–8317, 1979.
43. Leckie MP, Ng RH, Porter SE, Compton DR, Dietzler DN: Regulation of bacterial glycogen synthesis. Stimulation of glycogen synthesis by endogenous and exogenous cyclic adenosine 3′:5′-monophosphate in Escherichia coli and the requirement for a functional crp gene. J Biol Chem 258: 3813–3824, 1983.
44. Robison GA, Butcher RW, Sutherland EW: Cyclic AMP. Ann Rev Biochem 37:149–174, 1968.
45. Nimmo HG, Cohen P: Hormonal control of protein phosphorylation. Adv Cyclic Nucleotide Res 8:145–266, 1977.
46. Hofer HW, Allen BL, Kaeni MR, Harris BG: Phosphofructokinase from Ascaris suum. The effect of phosphorylation on activity near physiological conditions. J Biol Chem 257:3807–3810, 1982.
47. Mansour TE, Mansour JM: Effects of sertonin (5-hydroxytryptamine) and adenosine 3′:5′-phosphate on phosphofructokinase from the liver fluke Fasciola hepatica. J Biol Chem 237:629–634, 1962.
48. Dumont JE: The action of thyrotropin on thyroid metabolism. Vitamins Horm 29:287–412, 1971.
49. Hillensjo T, Ekholm C, Ahrem K: Role of cyclic AMP in oocyte maturation and glycolysis in the pre-ovulatory rat follicle. Acta Endocrinol 87:377–388, 1978.
50. Mansour TE: Studies of heart phosphofructokinase purification, inhibition and activation. J Biol Chem 238: 2285–2292, 1963.

Received 26 August 1983.

GPSR Compliance

The European Union's (EU) General Product Safety Regulation (GPSR) is a set of rules that requires consumer products to be safe and our obligations to ensure this.

If you have any concerns about our products, you can contact us on ProductSafety@springernature.com

In case Publisher is established outside the EU, the EU authorized representative is:

Springer Nature Customer Service Center GmbH
Europaplatz 3
69115 Heidelberg, Germany

Batch number: 10755910

Printed by Printforce, the Netherlands